Dominance Relations

Garland Series In Ethology

Series Editor: Gordon M. Burghardt

Primate Ecology and Human Origins: Ecological Influence on Social Organization
Edited by: Irwin S. Bernstein and Euclid O. Smith

The Development of Behavior: Comparative and Evolutionary Aspects
Edited by: Gordon M. Burghardt and Marc Bekoff

The Behavioral Significance of Color
Edited by: Edward H. Burtt, Jr.

The Behavior and Ecology of Wolves
Edited by: Erich Klinghammer

Handbook of Ethological Methods
By: Philip N. Lehner

Cat Behavior: The Predatory and Social Behavior of Domestic and Wild Cats
By: Paul Leyhausen

Dominance Relations: An Ethological View of Human Conflict and Social Interaction
Edited by: Donald R. Omark, F. F. Strayer, and Daniel G. Freedman

Species Identity and Attachment: A Phylogenetic Evaluation
Edited by: M. Aaron Roy

Other Garland Books in Ethology

Primate Bio-Social Development: Biological, Social, and Ecological Determinants
Edited by: Suzanne Chevalier-Skolnikoff and Frank E. Poirier

The Face of Man: Expressions of Universal Emotions in a New Guinea Village
By: Paul Ekman

The Dog: Its Domestication and Behavior
By: Michael W. Fox

Ethological Dictionary: In English, French, and German
By: Armin Heymer

Dominance Relations:

An Ethological View of Human Conflict and Social Interaction

edited by

Donald R. Omark

University of Illinois
Urbana, Illinois

F. F. Strayer

University of Quebec
Montreal, Quebec
Canada

Daniel G. Freedman

University of Chicago
Chicago, Illinois

Garland STPM Press
New York & London

15 14 13 12 11 10 9 8 7 6 5 4 3 2 1

Library of Congress Cataloging in Publication Data

Main entry under title:

Dominance relations:

(Garland series in ethology)
Bibliography: p.
Includes index.
1. Social Darwinism. 2. Human behavior. 3. Social interaction. 4. Primates—Behavior. 5. Aggressive behavior in animals. I. Omark, Donald R. II. Strayer, Fred F. III. Freedman, Daniel G. IV. Title: An ethological view of human conflict and social interaction. V. Series.

HM106.D65 301.11 70-14352
ISBN 0-8240-7048-8

Published by Garland STPM Press
136 Madison Avenue, New York, New York 10016

Printed in the United States of America

To other students in the social sciences
who not only can comprehend
an interdisciplinary perspective,
but who are themselves
interdisciplinary researchers.

Contents

Preface

Dominance relations were among the first readily apparent social structures noted in nonhuman primate groups. They also are apparent, though sometimes in more subtle forms, in human groups. This book attempts to examine such relationships from phylogenetic and ontogenetic perspectives and within various ecological settings. Whether in other primate groups or our own, dominance relations between individuals are obviously only a small part of the total set of interactions in which individuals engage. One of the purposes of this book is to illustrate for the student the variety of techniques that can be brought to bear upon the topic. Each of these techniques produce a slightly different set of results, which in total provide some feeling for what dominance relations may mean at the human level, albeit mainly for younger humans. The techniques can be applied to other social phenomena and in turn, need to be applied to dominance relations and their developmental transformations in a variety of social situations.

A second purpose of this book is to provide research results on dominance relations. As will be seen, this is not a subject about which everything is already known nor is it directly inferable from the study of primates. Dominance relations at the primate and human level have now been studied for close to 40 years. In that period of time one should begin "to know something," but during the same time period there have been major changes in the way individuals and groups have been conceptualized. Therefore, what was "known" may no longer be accepted as fact. Different questions from different perspectives are still needed, especially since the discipline of human ethology has only existed for a decade and is still in the process of conceptualizing what it is and what it should be investigating.

A third purpose is to introduce, to at least some readers, a few ideas about what it means to consider behavior in ethological or evolutionary terms. A major problem in such a consideration will be the prior training of the reader. If the reader is a psychologist, there will exist a strong bias toward seeking causal explanations; if the reader is an anthropologist or sociologist, there will probably be a bias toward seeing discontinuities between man and other animals. If one is a sociobiologist, there will be a preference for dealing with genetic levels of explanation rather than behavioral levels, except when they serve as examples. For those who come from a zoological/animal ethological tradition there will be the problem of what should be considered valid data when examining Homo sapiens as a species.

What follows will be an attempt to shift some conceptualizations, broaden some perspectives, and illustrate a variety of approaches to one aspect of primate and human behavior—dominance relations. Such relations do not occur in a vacuum but generally within social groups, and what this might mean to these groups will be discussed. In Part I a holistic perspective of human ethology is briefly presented, followed by a consideration of the group. An evolutionary approach suggests that phylogenetically related organisms—primates—should be examined. Comments on their patterns of behavior will be presented in Part II of this volume. Developmental findings appear in Part III, with a discussion following on the differences and similarities between observable patterns of behavior and children's conceptualizations of their experiences. Part IV deals with other settings to illustrate the range and limitations upon human behaviors, and Part V relates dominance relations to other aspects of human behavior. We hope this volume will lead to others that examine a wide range of human behavior from an ethological perspective.

We would like to thank each of our contributors for their forebearance during an overly long editorial review. The Spencer Foundation must also be thanked for their patience and support. Linda Gould and Pierrette Précourt made many of the final manuscript corrections and this is greatly appreciated.

List of Contributors

Rona Abramovitch

Wesleyan University
Middletown, Connecticut

Jerome H. Barkow

Dalhousie University
Halifax, Nova Scotia
Canada

Carol Barner-Barry

Lehigh University
Bethlehem, Pennsylvania

Harold R. Bauer

Rockefeller University Field Center for Ecology and Ethology
Millbrook, New York

Irwin S. Bernstein

University of Georgia
Athens, Georgia

Linda Camras

DePaul University
Chicago, Illinois

Thomas R. Chapeskie

Willowdale, Ontario
Canada

Carol L. Cronin

Detroit, Michigan

Mark S. Cummins

McGill University
Montreal, Quebec
Canada

Patricia Draper

University of New Mexico
Albuquerque, New Mexico

Daniel G. Freedman
University of Chicago
Chicago, Illinois

Harvey J. Ginsburg
Southwest Texas State University
San Marcos, Texas

Elizabeth A. Missakian
SYNANON
Badger, California

Donald R. Omark
University of Illinois
Urbana, Illinois

Richard Parker
Watsonville, California

Rich C. Savin-Williams
Cornell University
Ithaca, New York

Andrew M. Sluckin
University of Edinburgh
Edinburgh, Scotland

F. F. Strayer
University of Quebec
Montreal, Quebec
Canada

Janet Strayer
Simon Fraser University
Burnaby, British Columbia, Canada

Brian E. Vaughn
University of Minnesota
Minneapolis, Minnesota

Everett Waters
State University of New York
Stonybrook, New York

Glenn E. Weisfeld
Wayne State University
Detroit, Michigan

Part I

Introduction

Chapter 1

Human Ethology: A Holistic Perspective

Donald R. Omark

Human ethology is a new discipline, with its forerunners—animal ethology and primatology—not being much older. All three disciplines rest upon Darwin's work a century ago, but they are still in the process of defining themselves. This volume attempts to add to the debate of what human ethology should include by emphasizing one aspect of social behavior, dominance relations. We expect that these debates will continue.

Setting the Stage

Human ethology, first of all, has to be ethology. As Tinbergen (1951, 1963), Eibl-Eibesfeldt (1970), and others have stated, ethology is the study of behavior. To understand behavior properly, the researcher has to consider a variety of factors that may affect the behaviors that are being observed, tested, or examined. These factors, as paraphrased from Tinbergen, are (1) the external or internal events which "cause" some "effect" on the individuals' behavior, (2) the phylogenetic, or prehistorical, forces which predisposed the individual to act in a certain way, (3) the ontogenetic, or historical, experiences of the individual, (4) the survival value of the behaviors in terms of how they permit the individual to continue to live and reproduce, and (5) the biological function of these behaviors as they appear in everyday life. The fourth and fifth factors are closely intertwined. They are separated here, however, because some behaviors may permit the individual to continue to live but not necessarily to reproduce. The five factors listed suggest the range of research areas that human ethologists should explore.

A point to be remembered in any research endeavor is that we (Homo sapiens) are a biological species. What we do, create, produce, speak, and so on, is done by biological organisms. At no time during our many years do we stop being biological organisms. We are evolved human organisms at every stage of life, and there is no more reason to focus research upon infants or nursery schoolers than upon any other age. In fact, following Tinbergen's injunction, every other age needs to be studied for a full ethological description of us as a species. Similarly, establishing a "cause" for an individual's apparent behavior provides only a partial explanation for the appearance of that behavior, and human ethologists must continue to explore the other factors listed.

Human behavior exhibits elements which range from the labile to the more stable (i.e., stereotypic of our species), and these need to be explored as products of a biological species. In the past the contrast between labile versus stable behaviors has been disguised as a difference between those behaviors which are learned versus those which may be innate (see Burghardt, 1977). Such a dichotomy is a false one, for the ability to both learn and to produce labile behavior patterns is part of our species' specific characteristics (i.e., both capabilities may be inherited) and typical for many other primates and other species. Learned behaviors and labile behaviors need not, and perhaps should not, be lumped together. As we will illustrate, behavior genetics suggests that there should be some lability in any set of behaviors. What is learned can be conceptualized in different ways and is obviously limited by the inherited capabilities of the organism, a point that will be considered later. Language and many nonverbal patterns of social interaction are examples of behavior which are labile and some of which may be learned in their specific forms, but they still are species' specific characteristics which need to be examined by human ethologists. For example, nursery school children talk as much, if not more, than adults. What they are saying, how, to whom, and particularly why they are saying it are important problems for study.

Many human ethologists, in the past, have limited their studies to observable and nonverbal behaviors. What this book will attempt to do is to illustrate that a variety of techniques can be used to explore human behavior. The rest of this chapter will discuss contrasts between the conceptualizations of research as conducted by other social scientists or animal ethologists and holistic human ethologists.

Data: Observations and Verbalizations

What children or adults are able to say about what they are doing should certainly be of interest to human ethologists. Animal ethologists have for decades observed animal behavior and speculated about the animal's motives, needs, drives, intervening variables, and so forth. With the uniquely linguistic animal—man—that unique ability should be used wherever possible. Language, as an obvious behaviorial fact affecting social interactions, needs to be added to the human ethogram (see Bates, 1976; Omark, n.d.; Searle, 1969). Language not only can be studied in its own right and in relationship to other behaviors, but can also be used to probe cognitive, emotional, memorial, and other processes which can only be inferred when observing behavioral actions (cf. Nisbett and Wilson, 1977).

Early animal ethologists and comparative psychologists focused upon individuals' patterns of behaviors (in order to develop an ethogram), noted some group-level phenomena, but primarily sought explanations at the physiological level. Primatologists (primate ethologists) also examined group-level phenomena, and finally human ethologists have begun to use verbal reports to explore the meanings of behavior at still another level. All of these approaches need to be carefully considered. As human ethologists, we should use every means at our disposal and not limit ourselves to just naturalistic observations or just physiological measurements. While care must be taken in whatever techniques are chosen, research at other levels should not be restricted. Altmann (1974), Omark, Fiedler, and Marvin (1976), and Omark and Marvin (n.d.) suggest problems with even some of the more common observational techniques, and Blurton Jones (1972d) discusses problems with the verbal techniques. The main point is that no scientific discipline is defined by a single methodology (see Weisler and McCall, 1976).

From time to time at ethological meetings one hears in essence the following: "But the children (or adults) may lie to you when you ask them questions." This comment, or the investigator(s) who made it, assumes two things: first, that what the child said may be untrue (from the child's perspective) and, second, that observable behavioral actions are, in some sense, "truer." Let us consider the latter first. Anyone who has observed children's play behaviors at 2 years of age, and even earlier, knows that much of what a child does is not real, "going to the store" when playing is not really going to the store. At any age, much is bluff and posture that need not reflect anything "true" in the sense that it will predict future behaviors. Those who have examined dominance relations in the wild know that other organisms also bluff and posture (see the example of Mike in van Lawick-Goodall, 1971). An act may be a lie (to an outside observer in the sense that the act makes behaviors unpredictable), as well as can any verbal statement.

But what if behavioral actions are predictable while verbal statements contradict them? Is what the observer sees "truer" than what the child verbally reports? Such questions illustrate the difference between "etic" and "emic" approaches to research. In an etic approach (from phon*etics*) the researcher basically decides what actions, symbols, phonetic structures, and the like will be recorded and then analyzes them in terms of some preconceived structure. An emic approach (from phon*emics*) assumes that there are structures which have meaning to the individuals being observed, and the researcher attempts to understand this meaning. The etic approach then just reports observations (as phoneticians sometimes do) or places meaning, from the observers point of view, on the phenomenon being investigated, while the emic approach attempts to gain meaning from the participant's perspective. Observing children versus asking them what they observe, thus, involves an etic / emic shift in perspective which taps very different parts of an ongoing process, with neither part being necessarily truer. Meaning can be sought in many ways. If children present verbal reports which do not match with observable phenomena, then one can begin to explore the development of cognitive structures. When the child "lies," then other factors can be explored.

Such differences as these will be presented in Chapters 8, 13, and 24. The results of the research in these chapters amount to a broadening of the ethogram to include emic as well as etic types of data. They are only a first step; obviously, they are limited by the questions the investigators asked. If cognitive and other processes are to be explored within an ethological framework, then the emic approach needs to be expanded and contrasted and compared with the etic approach of just observing behavioral actions. Such an approach not only should provide firmer ties with the rest of social science, but also should provide suggestions for research at other phylogenetic levels.

Other Theoretical Perspectives

Just as one should not limit oneself to one type of data (particularly when studying humans), one should not limit oneself to one theoretical perspective. While one, as an ethologist, takes an evolutionary perspective on the organism and its interactions with the environment, still one can use other theoretical findings when considering particular aspects of behavior. For instance, ideas about how an organism remembers, stores, and processes information should be considered subtheories of evolutionary theory and these ideas should be reexamined in terms of the survival and evolution of the species. Two examples would be learning theory in its various ramifications and Piaget's theory of cognitive development. Both views assume that some interaction between the organism and the environment occurs such that the organism's behavior is different following such interaction. For the learning theorist the changed (learned) behavior becomes part of the repertoire of the organism after a relatively brief period of time. For the cognitive developmental theorist the organism (usually a young child) proceeds through a series of hierarchically organized stages. The organism gains the necessary organization of experiences within one stage to enable it to proceed to the next stage after some period of time. While these two theories are not entirely compatible, they can provide insights into how organisms relate to their environmental conditions. However, as Blurton Jones (1975) suggests, the human ethologist continually has to question how realistic (in terms of external validity) are the "findings" of such other approaches.

The major problem that the student will encounter when attempting to use other theories are the philosophical perspectives upon which these theories are based. The two major influences on modern social science have been the writings of Descartes and Locke. From these writings follow the varieties of current research efforts directed at examining the relationships between mind and body ("internal causation" of behavioral phenomena) and the role of experience (external causation) on the developing organism. Internal causation, as an explanation of observed behaviors, is a consideration of various factors, such as physiological, neurological, or hormonal, within an individual which may "cause" the observed behaviors. External causation would then be those stimuli emanating from the environment to which the organism responds. In both cases of internal and external causation there is a conceptualization of some physical entity (light, sound pressure, hormone, and so on) causing or evoking the observable behaviors.

For example, the social sciences generally investigate factors external to the organism (e.g., the culture or language) which may be learned, or they create external factors (stimuli) to which the organism is to respond in some more-or-less predictable fashion. The medical model, also followed by some social scientists, generally directs research toward (1) external agents that have entered the body, e.g., bacteria, viruses, (2) malfunctioning, or "nonnormal," organs, tissues, or other bodily structure, or (3), in the case of psychiatry, either of the above or some earlier external factor that may have produced the disturbance being investigated.

All of these approaches assume that a particular factor can be identified as the causal agent of the effect or behavior noted in the client or patient. Frequently experiments will be made to isolate the hypothesized causal agent. While the medical model may work when an external agent has been internalized or when some organ definitely malfunctions, such approaches should not be applied indiscriminately as a way of explaining and investigating all of human actions or those of other organisms. Alternate approaches will be examined in the next two sections.

The Mind/Body Problem: Dualism versus Holism

In the history of Western thought the writings of René Descartes (1596–1650) played a crucial role in directing the course of research that the social sciences would follow for the next 300 years. Descartes formalized the religious distinction between soul and body and moved this distinction toward the secular level, where the mind was considered as an entity separate from the material body (although hovering close by if not hidden within the body—ESP research being one example of the hovering close by). At the same time, and continuing to the present, man was viewed as being definitely distinct and unrelated to other animal species. John Locke (1690) further defined the mind as a "white Paper, void of all Characters" (see Rachlin, 1976, for a more complete historical discussion). These two philosophical perspectives set the stage for the behavioristic approach to research wherein external stimuli are considered as deterministic for what the mind will contain.

Research in the area of psychosomatic medicine suggested that the boundary between the mind and the body was not impenetrable. Depressing and other kinds of thoughts appeared to be able to affect bodily functions, and bodily actions (e.g., responses to drugs) could obviously lead to euphoric or other kinds of thoughts. While still conceptualizing the mind and body as separate entities, many researchers could see a linkage between the two.

As long as the two are viewed as distinct entities, and mind appears only in man, then causal statements can be constructed about stimuli causing behaviors to appear. For example, on the external side experimental psychologists and earlier ethologists could say that a particular stimulus produced, elicited, or released a particular response. On the internal side the corresponding statement would be that this drug, this nerve impulse, this IRM, or this hormone, released this particular set of behaviors. For man, with mind included, this perspective suggests that the thoughts occurring in the mind (and those sets of behaviors more complex than reflexive actions) are always caused by some stimulus originating either within the body or in the outside environment (cf., Griffin, 1977; Marler, 1977). The mind is always reactive rather than proactive.

Such perspectives or philosophical approaches are, literally, statements of faith; after years of schooling from a particular perspective they become easy to accept. If all of the research results that one reads are phrased in terms of a comparison of causes and effects, then it becomes difficult to think about the individual and what individual differences may really mean. Similarly, if one only reads about how the mind can be controlled and manipulated (or how behavior can be controlled and manipulated), it becomes difficult to reverse the process and think about alternative models.

As an alternative model, we would like to suggest that the student consider a holistic perspective. By this we do not mean a materialistic (just body) or spiritualistic (just mind) perspective, but rather a perspective which considers the total organism as a single entity. Both Descartes and Locke preceded Darwin so their attempts to describe human behavior should not be faulted; instead they should be modified in terms of Darwin's evolutionary theory and Mendel's later genetic theory.

From an evolutionary perspective it is the entire organism which is selected for or against by natural or social forces. It is an entire entity called man (or bee, or fish, or monkey) which manages to reproduce or not reproduce the next generation. This selection over time matches the organism to the environment (not perfectly, Jacob, 1977; Lewontin, 1978; Waddington, 1975) such that it is capable of perceiving and responding to the aspects of its environment that are important for its survival. But perhaps most important, in this brief summary, was Darwin's notion of the continuity of nature wherein virtually every animal species is more or less distantly related, and man is a descendant within the primate family.

Evolutionary forces act so as to limit the genetic material provided to the next generation. The genetic material, in turn, interacts with the nutrients and experiences provided to create the new organism. From this perspective the only place from which the *capabilities* of the mind can spring is from the genetic material that the previous generation has passed forward, and since this material is the same one that encodes all of the other possible bodily functions, there is little reason to consider the mind and body as separate entities.

It is the capabilities of the organism which clearly demarcate this holistic evolutionary perspective from a Lockean perspective. The mind, or rather that part of the organism which is prepared to receive, process, store, and engage (along with other parts) in the production of behaviors, received instructions from the genes as to its basic structure. The total organism can be examined at various levels of organization, i.e., chemical, structural, physiological, psychological. All levels will be found to be simultaneously active and hence, correlative rather than causal models would be more appropriate (see Hinde, 1974, concerning precursors). Within this perspective an organism is always proactive.

Innate versus Learned Behavior

Externally produced information from the environment which is received by the organism has obvious value for the survival of the organism, or at least part of this information does. This will be explored in more detail in the chapter on the Umwelt; what will be discussed here is the second kind of dualism that has appeared in the literature which presents certain paradoxes that might be resolved from a holistic evolutionary perspective.

This dualism led to the controversy over what portions of any behavior were innate (built in) versus what portions were learned (acquired from the environment). This view sees the organism as absolutely distinct from its environment. The organism can acquire information from its past through its genes about how to produce certain behaviors (instinctive behaviors, generally viewed as stabile, stereotypic, or labeled as species' specific characteristics). In contrast, some behaviors are totally acquired through interactions with the environment (à la Locke).

Within studies of genetics or sociobiology the contrast between the internally and externally caused behaviors came to be studied with a variety of "heritability" models. Wilson (1975) defines heritability as "the fraction of variation of a trait within a population—more precisely, the fraction of its variance, which is the statistical measure—due to heredity as opposed to environmental influences. A heritability score of one means that all the variation is genetic in basis; a heritability score of zero means that all the variation is due to the environment" (p. 586).[1]

The basic dualistic model would then be

$$\text{Behavior} = f(O + E), \text{ or Behavior} = f[\mathrm{O} + \mathrm{E} + (\mathrm{O} \times \mathrm{E})]^{2}$$

Behavior is a function of the organism's internal biology *plus* environmental factors (O + E) *plus*, perhaps, the interaction of the two (O × E). If one accepts this model, one can ask such questions as Is this behavior inheritable? (i.e., Is most of the observed variance due to genetic—biological—factors?), or Is the behavior subject to environmental manipulations? (i.e., Can an individual, such as the experimenter or teacher, control the production of this behavior while ignoring the biology of the organism?).

One problem with this model is that heritability scores can be obtained on anything, e.g., how often people wash their hair, go to the store, or get up at 6 A.M., IQ scores, and on and on. As Wilson uses such a model, he says, "Notice that its magnitude depends on the character selected for measurement. Different characters in the same population vary drastically in their heritability scores. Notice also that heritability depends on the environment in which the population lives. The same population, with an unchanged genetic constitution, can yield a different heritability score for a given characteristic if placed in a new environment (Wilson, 1975, p. 69). Hirsch (1970), in a detailed examination of the heritability concept, points out that this concept is *not* a measure of traits (characters) and cannot make comparisons between individuals in different environments. Further, it cannot indicate whether genes are involved in the traits observed because if all of the parents of the offspring to be observed were identical genetically, the variance due to heredity would be zero! This result would suggest that heredity played no part in the resultant trait even though it might be directly involved in the appearance of the trait.

What the heritability models will do is supply you with a number as a result. So will any statistical test applied to a set of data. This does not mean that the number need be "sensical." To those ethologists (and ecological psychologists, see Shaw and Bransford, 1977) who took a holistic perspective on the organism, such a model, and the questions *derived* from it, appear "nonsensical." The way out of the controversy is not to accept *a priori* the model but, instead, to question the nature of the organism in relationship to its environment. After all, most organisms have spent eons being subjected to the forces of naturalistic selection such that they should fit into the environments in which we find them. Their Umwelt (that which they can perceive and of which they can conceive) has also been selected to be that which it is (see Chapter 13, this volume).

Questions then about learning versus innateness are nonsensical because the learning aspect is assumed to be only environmentally related. It is, however, only the biological organism which can learn, and then only within the limits of its biological Umwelt. From a holistic perspective that which is biological can never be separated from that which is environmental because anything external only has meaning to the organism in terms of how that organism perceives and responds to it. Thus, a more appropriate model might be one in which the only factor is an interactive one:

$$\text{Behavior} = f(\text{O} \times \text{E})$$

In this model the two factors O and E cannot be separately observed. As Kummer (1971) said:

> It takes a drum and a drummer to produce a sound. Nobody would try to differentiate between sounds produced by the drummer and sounds produced by the drum. But we can very well discuss whether two recorded performances sound different because of a new drummer or a new instrument (p. 12).

This means that a variety of individuals need to be studied in the same and different environments. For groups of children engaged in developing social relationships it is not easy to manipulate experimentally their environment (experiments generally being short term rather than long enough for real social relationships to develop). Contrasting classrooms and long-term projects are suggested.

For the experimentalist who thinks he can observe variations in O by completely controlling E, one can only say that he or she does not. First, it cannot be done because E is usually completely artificial and not that in which the organism has been selected to function or in which it developmentally grew. Results from experimentalists who never view the real world should seriously be questioned as Blurton Jones (1975, p. 74) has suggested. Second, those who control organisms through various environmental manipulations are equally suspect—primarily because they assume that the group of subjects they are observing (frequently college sophomores) are equivalent to any other group, age, sex, etc. Piaget's results should have warned them otherwise, but such research continues.

Overall, what is presented in the chapters that follow are attempts to investigate real organisms interacting in their natural habitats, with experiments designed to investigate particular phenomena only after they have occurred in naturalistic settings. We think this is the essense of ethological research: observe, note, question, postulate—first in the real world, then, if possible, in contrasting real worlds, and finally, and only as needed, in experimental settings. The student should always be both wary and aware of the *a priori* models that subtly guide research. The questions that one asks are derivitives of the model that one assumes.

Typological versus Populational Models

> The phenotype (appearance, structure, physiology, and behavior) of any organism is determined by the interaction of environment with its genotype (the complete genetic endowment). Each genotype is the end product of many mechanisms which promote genotypic diversity in populations (Hirsch, 1963, p. 1436).

Hirsch's statement emphasizes individual differences and is a primary principle within much of the research done on evolution and behavior genetics. It is an emerging principle in ethology as researchers begin to examine similarities and differences between individuals and populations, this book being one example. The investigation of dominance relations (as observed and as perceived by the subjects) addresses itself to interactions between individuals and attempts to move beyond earlier group-level research addressed to questions such as "Are boys or girls more aggressive?"

Unfortunately, Hirsch's principle is antithetical to much of the research within social science. For a variety of historical, religious, and philosophical reasons research on genetic diversities has been ignored or suppressed in social science research and the average (mean) of grouped data has been investigated as the ideal which can be sought (see Hersen and Barlow, 1976; Hirsch, 1963, 1970). Some of these reasons were suggested above; in this section the contrast between typological and populational models will be discussed.

Thinking about clusters of individuals as types appeared as long ago as Aristotle,[3] but it was not until the late 1800s that it was reemphasized in modern social science. Quetelet (see Hersen and Barlow, 1976; Hirsch, 1970) suggested the average man as an ideal toward which nature strove. Galton and Pearson expanded on Quetelet's ideas about the normal curve and the "error variance" that individuals represented. R. A. Fisher, working as a statistician within an agricultural setting, basically completed the process of shifting a large segment of the social sciences toward thinking about averages rather than individuals. Hirsch pointed out that an "average" increase in one plot of land with one genetic strain of plant over another plot treated differently or planted with a different strain may be important economically, but it does not consider what happens to individual plants.

Making comparisons between groups generally involves comparing the mean score of one group with the mean score of another group. In essence, this involves producing a composite caricature of each group and then seeing whether the caricatures differ enough to be considered "significantly different." No individual within the group need resemble the caricature, and no such individual need even exist! The variance within the population may enter into the final formula for making comparisons, but skewness or kurtosis are generally ignored. Populations or groups which deviate from the "normal curve" expected by Quetelet in the last century can often be made to appear "normal" through various statistical manipulations so that they can be compared with other "normal" populations.

The need to find normal distributions of behaviors, traits, and the like and the need to compare means, or caricatures, of groups has become the *sine qua non* within much of social science. Such strivings remind one of the story told by Hans Krummer about when he had taken a student's dissertation to Adolph Shultz (who had done very seminal research on primates). Shultz read through the dissertation, which was filled with a variety of statistical tests, and finally said, "That's all very nice, but where is the animal?"

The assumption behind many statistical tests is that the average of groups should be compared. Research based upon such tests frequently strive to produce some aspect of behavior in one group (the experimental group) and not in another group (the control group). If such studies succeed in producing "significant differences," they are reported in the literature. If not, they are generally not reported or are rejected when submitted for publication. A perusal of any of the social science journals will show that successful experiments outnumber unsuccessful experiments by a 10 to 1 ratio, while personal experience with doctoral dissertations suggests that half of all experiments fail. If half fail, then what is reported is nothing more than chance variation about a variety of behaviors or traits. The animal has been lost, and the literature presents only half the story. Faced with that, what can human ethologists do?

The other approach to research would be to examine behaviors in terms of a populational model. Within such a model the individual and its patterns of behavior become important. Because of independent assortment of chromosomes during meiosis (the process of producing sperm or eggs prior to copulation) and because chromosomes are both fragile and "sticky" (and hence can break apart in combinations different from the parental chromosomes), the final set of genes which an individual is apt to acquire bears less than "1 chance in 70 trillion" (Hirsch, 1963) of resembling a sibling in its total genotype. In this sense every individual is unique, even if, as discussed below, some populations may have characteristics in common. In a sense, then, one need not randomly assort individuals to experimental or control groups because the process of meiosis has already randomly assorted everybody.

This does not mean that populations may not differ. Any group that is reproductively isolated from another group, whether in ghettos or on islands, may produce subsequent generations which over time may "drift" to contain only one or a few of the various alleles (alternate forms) of *any particular* gene. A population may then be identifiable by some characteristics (but not the IQ test, Omark *a*, *b*, in press). This is evident in the various breeds of domestic animals. At the same time every individual is still unique in *total* genetic configuration from every other individual and becomes more or less so as the phenotype interacts with its environment.

The uniqueness of the individual has certainly been glossed over in experimental psychology, where even species differences are frequently ignored, and even in animal ethology, where such commonalities as infant rearing, courting and mating, and defense have been sought rather than differences in infant-rearing patterns, and so forth. A primary reason for the failure to find "general laws" that hold irrevocably across species or individuals certainly results from the uniqueness of each individual's interaction with its particular environment.

Two obvious problems have occurred in separate social science disciplines (psychology and linguistics) because of the failure to consider the genetic uniqueness of the individual. Within psychology, John B. Watson's writing's in the 1920s moved much of psychology toward "behaviorism," or the study of the observable, because Watson found that observers' comments about the feelings of the animals observed differed widely while their descriptions of behavior all agreed (Rachlin, 1976). "Introspectionists," those within psychology who attempted to describe how they felt or how they perceived the world in different settings, or when presented with varying stimuli, soon found their approach to the investigation of the "psyche," or mind, abandoned because every response made by them differed slightly from others' responses. If everyone's responses vary, then "general laws" are difficult or impossible to formulate.

Labov (1971, p. 162) pointed out a similar problem on lack of agreement about the grammaticality of language samples when he said that "there is no evidence that consistent and homogeneous judgments can be obtained from native speakers on such crucial matters [as to the grammatical correctness of language samples] . . . it is now evident that *the search for homogeneity in intuitive judgments is a failure*" (italics in the original). While communication obviously has to follow some rules so that a signal from a transmitter is responded to appropriately by a receiver (Altmann, 1967), still each producer and perceiver of such signals should be expected to perceive both the rules and the signals slightly differently because of their unique phenotypes.

In the same chapter cited above, Labov goes on to argue that more attention be paid to language as people produce it rather than to intuitions (linguists call such language "speech" to differentiate it from the purer form that they are intuiting about). Such an appeal to record the "observable" rather than the "introspected" follows 50 years after psychology's similar appeal to research directions. It parallels the equally recent human ethological appeal to record nonverbal behaviors rather than either verbal statements or what individuals may say about their perceptions and conceptions of their interactions with aspects of their environment.

A populational model derived from evolutionary and behavior genetic theory would predict individual differences at the various levels of individual structural, cognitive, and behavioral organization. Intuitions should be expected to vary, as should preceptions and conceptions of peers—even when all of the peers are together half of the waking day for almost a full year. The same set of experiences do not provide the same input to each of the individuals. This means that "general laws" that will predict everyone's behavior can never be found. What can be sought is a partial picture, the range of reaction of individuals within different environments (Parker and Omark, this volume, Chapter 24, being one attempt). Because developmental processes, as well as genetics and environment, are involved, a complete picture will never be known. Studying a variety of individuals in various settings at different age levels, however, should provide a more accurate description of us as a species. This description will be more like a matrix rather than a single general rule.

Communicative patterns involve interaction between individuals and ultimately interactions with an entire group or culture. For a signal to have some effect it has to be understood by the broader group even if it is not responded to uniformly by all of the group's members. For example, the courtship displays of a male bird may be rejected by five females and accepted by the sixth. This 5 to 1 ratio of no response to acceptance should not be seen as a signal which was "non-sensical" most of the time. Rather, one has to wait and see what is its purpose and function. The broader rules of communication may then be mixed in with details of idiosyncratic production and idiosyncratic responses to such production, but the "generality" of such rules should *per force* have to be broader than a study of individual differences might suggest. It is where interactions are examined that one moves beyond the field of behavior genetics and into the fields of ethology, primatology, sociology, and anthropology—while still keeping in mind the contribution that a populational model and individual differences can make. While communicational rules (or cultural rules) have to transcend individuals (be understood by more than one individual) if communication is to be understood, still it is the individual who has to both produce and respond to such signals (cf. von Bertalanffy, 1964). Hence, any communicational system ultimately depends upon the individuals involved and does not exist as a thing apart from them.

Communicational signals and culture (as a broader term) may appear to change or evolve at rates different than the speed with which biological evolutionary change can proceed. It should be recognized that humans are not alone in their ability to change communicational signals. Two similar groups of rhesus monkeys reared separately in the Madison, Wisconsin, zoo seem to be developing related but distinct "dialects" of communication patterns.[4] This should be expected because different individuals live in each group rather than a bunch of monkeys here and an *identical* bunch there. Any changes, however, have to be within the range of capabilities of the producers and at least some of the receivers. If man can change a system rapidly (e.g., learn another language or learn to use a different tool), this means that *man*—and not culture—has evolved to a certain stage of flexibility. The recent work with a variety of chimpanzees (Lana, Washoe, etc.—Sebeok, 1977) suggests that they too have reached a stage where other communicational systems can be handled. An individual may not produce a new form of culture but still may be able to reproduce the current form. (The length of time spent in each of the various styles of stone tool production suggests that invention is much rarer than production.)

Only a movement away from comparison of means (e.g., are boys more aggressive than girls) toward considerations of individual performances and individuals in interaction with one another will save human ethology from falling into the trap of seeking "general rules of behavior," supposedly applicable to every individual. An understanding of how genetics and evolution operate are an *absolute necessity* for anyone who would claim to be a human ethologist.

Conclusion

A holistic approach to the study of human behavior would, first of all, suggest that the entire organism should be studied at a multitude of levels, from the cytoplasmic to the behavioral. No single level should be taken as "causal" for the phenomena of interest, but examinations should be continued at other levels for correlated phenomena. Scientists tend to study the level they are best equipped to investigate, but further investigations will probably reveal other factors of interest. For example, while lithium may relieve depressive frames of mind, still what the individual is depressed about has to be dealt with if the individual is to continue to function in society.

The way individuals perceive their world will be the result of their own particular constitution and set of unique developmental experiences. That "general rules" fail to explain any particular individual's responses should not be surprising; individuals are not general but particular. Their particular views of the world are, nonetheless, interesting; they permit us to see how the individual is structuring his / her world and what cultural and other factors are being incorporated. Their views should not be ignored simply because they do not fit some preconceived theoretical view of what the world should be like. A behavioral genetic approach to the study of human ethology requires that individual differences be examined for what they may reveal about man.

Interactions between individuals should reveal some general principles because both communication and group formation should necessitate some behavioral regularities. These will be discussed in the next section. Most of the chapters in this book deal with interactions between individuals: dominance relations. Part of what is being examined are the patterns of communication—primarily nonverbal—and part is whether, and when developmentally, group structures appear. A few of the chapters consider individual perceptions in relation to group structure, but more work is obviously needed in this area.

Homo sapiens should be viewed at both the individualistic levels and at the group level. While each individual is genetically unique, that individual generally exists within some group structure. The group can be supportive of a wide variety of individual differences and can shift the selective pressures operating upon the individual. The group, as the setting where dominance relations occur, will be examined in more detail in the next chapter.

Acknowledgments

In my chapters of this book R. Fagan, L. Fourcher, J. Hirsch, J. Locke, R. Marvin, R. Savin-Williams, and G. Weisfeld made many cogent and insightful comments, which were much appreciated. The errors that remain are mine. My thanks to O. Crespo for drawing the figures.

Notes

1. Hirsch (1963, 1970; McGuire and Hirsch, 1977) provides corrective statements for Wilson's misunderstanding of heritability.
2. What is presented here is what I and others (see Baer, Letters, *Science*, April 28, 1978) see as the conceptual model behind the sociobiological approach and much of current educational and psychological research. Despite Barash's comment (Letters, *Science*, June 9, 1978) about Baer's conceptualization perhaps being simply a confusion over "heritability," I think that the model as presented follows directly from Wilson's definition. Barash's reformulation of Baer's statement leads to the same point that my holistic model will arrive at in a few pages. The reader is referred to Note 1 for references which more completely discuss the concept of "heritability."
3. L. Fourcher, personal communication, 1978.
4. R. Goy, personal communication, 1978.

Chapter 2

The Group: A Factor or an Epiphenomenon in Evolution

Donald R. Omark

In this chapter groups of organisms, as they occur in naturalistic settings, will be considered from a variety of theoretical perspectives—the genetic, the sociobiological, and the ethological. Darwin (1859) suggested that it is the individual who is the important unit of selection, Mendel (1866, see Stern and Sherwood, 1966) proposed the basis—the genes that were passed on to the next generation by each individual—and Huxley (1942) posited, probably, the first real synthesis of evolutionary thinking with research on genetics.[1]

If individuals are the ultimate transmitters of genetic material, and also the ones who are stopped from transmitting by the forces of natural selection (i.e., they die before reproducing), then why do groups of individuals appear among so many species? That groups do appear cannot be questioned. There are numerous examples, ranging from social insects through fish, bird, and mammalian species. In many of the groups that have been studied dominance relations between group members have been found. Initial, and even recent, conceptualizations of the function of such relations have been that the results of dominance interactions aid particular individual's survival. Dominance relations would then appear to be antithetical to group maintenance, from at least some perspectives.

Differences between individual and group evolutionary processes will be explored in this chapter. The basic questions are (1) how and whether group structures affect the forces of natural selection such that they may be different for individuals living in groups than they are for individuals living alone (i.e., whether selection can be for group structures as well as for individual structures) and (2) whether particular patterns of behavior, e.g., dominance relations, are supportive of or antithetical to group formation. There are widely differing views on these questions that can only be indicated in the present volume. The student is invited to read the references cited and decide for him- or herself which approach to follow. In any case, some of the comments made about other perspectives will only make real sense when these other perspectives have been examined in detail.

Hypothetical Phylogenetic Changes in Group Development

Starting from the therapsids, mammal-like reptiles, the evolved *basic* mammalian group pattern is the mother–infant relationship. This is the *sine qua non* of being a mammal (or marsupial) since the initial feeding of the young is by the mother. A male is necessary, but only during the brief periods of procreation. Many social mammals still exhibit this primarily male-absent pattern including some carnivores (e.g., lions) and primates (e.g., patas monkeys), albeit in slightly modified form.

As with many passerine and other birds, when aid and caring for the young was also provided by the male, the young could (1) be born in a more immature state with a potentially longer growth period which could provide (2) opportunity for more learning from the adult models. In this progression in group size a distinction could probably be found between those who had to learn something about their environment and those who already had to "know" how to act toward their environment.[2]

The next quantum jump in group size occurred when a number of females and their infants traveled together. The females could cooperatively provide warnings to each other about potential predators (the multiplicative effect of having many eyes, ears and noses attuned to the outside environment). Various deer and other ungulates form such seasonal male-absent female groups, with the males either foraging alone or in their own groups. Similarly, predators have many sense organs probing the environment for prey, while also watching out for other predators; infants traveling with adults share this advantage of increased group size.

For plant-eating mammals the required intake of forage to sustain life is generally so great that food sharing is impossible, except when it has been concentrated as milk. Even normally herbivorous birds shift to being insectivors (predators) when feeding their young.[3] The passage of food other than milk seems to be relatively rare among mammalian herbivores, and the young are soon expected to forage on their own; there is frequently also predator pressure to be mobile very early. (See Blurton Jones, 1972c, for predictions of human behavior based on comparative data from other species.)

The balance between how much food can be converted to milk while still fulfilling the nutritional needs of the mother should mean that, for many species, there will not be a great deal left over. In those species that live in sparse surroundings (e.g., goats) mechanisms have even developed to reject strange offspring; there is simply not enough to provide care for strangers or orphans. Within some primate species there have been reported instances where males have killed the offspring of their predecessors within the group (Hrdy, 1977a; Marler, 1976).[4] This might be viewed as not simply *not supporting* someone else's offspring (who are not related) but as increasing the potential milk supply for their own offspring.

At an early age we should expect herbivorous animals to forage on their own, and we should not expect the young to have food shared with them, except insofar as it is necessary for them to learn what plants are edible. For species who feed on only one or two plants the learning can be very brief or even built-in. For species with more variety, like most primates, a more extended period is necessary. Apparent reversals in the dominance hierarchy should be expected (rather than be a surprise) in encounters between infants and adult males where the infants are "permitted" to take food from the adult. It is obviously of more immediate survival value to learn what to eat than it is to learn whom to avoid or defer to, and both infants and adults seem to know this.

With carnivores a different pattern can potentially emerge. Meat, like milk, is nutritionally closer to the final form in which it will be utilized than are plant products. Hence, food—either as meat or as milk—can be more readily shared. Bears, omniverous but generally solitary species, will provide milk to infants other than their own (Tanner, 1976). Lions similarly do this (Schaller, 1972), and wolves may share prey (Mech, 1966).

The argument for "inclusive fitness," that it may be good to aid kin,[5] has been raised (Clutton-Brock and Harvey, 1976) and that such sharing occurs because the lion "pride" is genetically related. This pride, however, has presumably been reared together (at least the female *hunting* portion of it) and hence the skill of lions at interacting during the hunt and subsequently may be due as much to familiarity as to genetic relationships. Whether one or the other of these factors is more important, or whether both are important, remains to be tested. That is, will unrelated but familiar individuals hunt together and then share their kill with offspring, as bears share their milk? The hypothesis is that they would. If so, then relatedness is a by-product of the way animals of certain species may typically grow up. It may also have been the initial propelling factor toward a new form of social interaction. Related animals obviously have more chance to interact socially, refine their forms of communication, develop mutual hunting techniques, and so forth. Once the underlying bases for these skills have been refined by the forces of natural selection, and incorporated into the genotype, what then?

The "what then" is obviously that it is a species which evolves, not any individual species member (cf., Klopfer, 1976, who prefers "lineages" which evolve). If characteristics of a species are that it forms groups, dominance relationships, affiliative and cooperative relationships, can tolerate differential access to resources, and has means for including or excluding nongroup members, then these species' characteristic phenomena have to be taken into consideration. These phenomena are all extra-individual, as is the concept of inclusive fitness. Relationships between individuals have to be examined not only in a way which conceptually will maximize the continuance of any individual's genes, but also in a way which will maximize the continuance of group structures and interrelationships between individuals. In essence, the group has to be selected for.

While the species is the ultimate gene pool which is subject to the forces of natural selection, and provides the diversity from which the next evolutionary step is taken, the operating unit within many of the vertebrates is the group, herd, gaggle, bevy, or troop. It is these latter, more or less relatively inbred sets of populations, upon which ecological pressures are brought to bear. Such pressures are probably not uniform across any species (given local drought conditions, concentrations of predators, and so on), and so it is to the groups as they exist that one must turn in order to investigate such pressures, and responses to them, in action. It is the groups that survive which then provide the genetic material for the next generation.[6]

For bisexual species a group of two from time to time has to exist for mating to occur. This means that species recognition has to exist and that some rudimentary communicative processes have to appear, at least until such time as copulation has been completed. For many organisms such a bare minimum suffices. But when one studies various social insects, fishes, birds, and particularly carnivores and primates, it becomes obvious that the young are cared for. This automatically increases the size of the groups which natural selection has created. Behaviors produced by the young, as well as those produced by the parent(s), can then be acted upon by natural and social selection so that those behaviors which maintain this minimal group are continued into the next generation.

Trivers (1972), Wilson (1975), and others have discussed parental investment where one or both parents invest more of their total reproductive potential in single rather than multiple fertilizations (although multiple births may occur). They have extended this concept to include "investing" behavioral action in the maintenance of related offspring (an aunt aiding nieces and nephews or grandparents aiding grandchildren). Investment in related offspring can then include the simple genetic contribution resulting from copulation, the actual behaviors exhibited by one or both parents in support of their own offspring—"parental investment"—and the combination of genetic linkage and behavioral actions which individuals contribute—"inclusive fitness"—to related individuals. And so the group grows to include more and more distant relatives.

Genetic Factors in Group Development

Groups can then be seen to have increased in size in some vertebrate lines. The suggestion has been raised by the sociobiologists (Wilson and others) that such increases may have a genetic base (or more correctly, that genetic correlates may be found). This section will consider some of the limiting parameters that genetic theory places upon within-group copulations.

It is within the group that most copulations, obviously, occur. If the group is isolated from other species members, then the Sewall Wright Effect (i.e., random fixations of particular genes at a single locus, or very few loci, in small isolated populations) may appear. These genes are then subject to the forces of natural selection, as is any gene package that a group may be carrying. If a group is not isolated, from time to time gene exchange will occur within the larger deme,[7] but with most gene exchange still occurring within the group. Isolated groups should "evolve" or change faster than groups which have access to the genetic diversity of larger interbreeding populations (see Simpson, 1953).

Within the group, as the major interbreeding entity, a fine line exists between the establishment of a trait which will promote the continuance of the group (whether selection be natural—due to ecological pressures—or social—due to sexual or other preferences) and inbreeding too much. Most mutations, and many genetic combinations, may be deleterious in themselves (as with the spread of hemophilia among the "royal lines" of Europe) or may eliminate natural defenses against, say, disease or reactions against particular kinds of predators. For example, the Indians of North and South America through the Sewall Wright Effect and/or the absence of selection pressure may have lost their immunity to measles and a host of other diseases; hence, they may have been conquered as much, or more, by disease as by the guns that the "white men" brought with them. It may not have been "cultural advance" that shifted the balance of power and control on the North and South American continents but a genetic anomaly that occurred 20,000 or more years ago.

In looking at groups one can see that for a group's continued viability there is a need for both the expression of those behaviors needed for the group's immediate survival within its particular environment and some contact with the larger deme (potential breeding population). While sociobiologists like Wilson and others have concentrated on how particular aggressive or altruistic "genes"[8] may be fixed as the norm within a population, the other side of the evolutionary coin is that it is generally desirable (from the species point of view) to have some genetic variability (heterozygosity) in a population. This heterozygosity is important because it both (1) minimizes the chance for deleterious genes to become the norm and (2) provides the variability necessary for meeting changed environmental conditions.

The American Indians noted were probably a large enough migratory group to prevent the occurrence of deleterious genes becoming the norm. They, however, may have suffered from the lack of those particular kinds of genetic variability necessary when the environment included an invading, genetically different, group. Disease may seem to be the immediate cause of the decimation of vast numbers of Amerindians, but equally important were the genetically programmed responses to various diseases. Caspari (1972) discusses incestuous matings and suggests that there is "a high probabilitiy of actual biological damage to [such] matings." Very small populations with no immigration may produce similar results if deleterious, or inappropriate, genes are present in the initial breeding population.

Being a "species" implies that there is a genetic and chromosomal compatibility between species members that is high enough to insure that matings between any two randomly selected individuals will, in general, produce viable offspring. This means, in the strictest sense, that all species members are related to one another. This degree of relationship is then obviously not zero but some higher figure. The degree of relationship between an aunt and a niece or nephew on the average is .125. The difference between .125 and a relationship larger than zero (i.e., .004, .01, etc.) is probably not much for populations outside a large urban population (a very recent evolutionary phenomenon). Given the ethnic patterns of settling within urban settings, even here the degree of general relatedness is apt to be high.

As an example of how close you might be related, pick a friend and trace back your "roots"; when was the most recent chance that you might have arisen from common ancestors? Given the migratory patterns of humans, and the recent rapid expansion of the human population, as little as eight generations may separate, for example, some whites from blacks in a United States population (the degree of relationship then being, on the average, .004). In many cases the possibilities will be higher, and they would certainly be expected to be higher for primates, carnivores, and even humans occupying contiguous territories. The point is that the range of relatedness between any two individuals who meet, of whatever species, is apt to range between 1 out of 10 to 4 out of 1,000. Such a differential may be enough for the purposes of natural selection, or the differential may fall in that grey area where the difference between killing a conspecific or driving him or her away and permitting him or her to exist, or even aiding him or her, is questionable, and where permitting the latter may prove to be to the group's advantage.

On the other side of the argument is the upper bound on inbreeding. Lethal genes (Ginsburg, 1966; cf. Bischof, 1975) can appear within a population. These genes are lethal to the developing organism because they interfere with one or more of the metabolic processes necessary for life (e.g., sickle cell anemia is due to a gene that prevents normal red blood cell formation). Inbreeding increases the probability for occurrence of homozygous combinations of such genes and incest has to be avoided (.5 relationship between, say, father and daughter and even .25–.125 between uncle and niece). Such genes seem to occur in many, if not all, species. Most of the state legislatures of the United States recognize the deleterious effects of first-cousin marriages (and incestuous conjugations) and place an upper bound on such unions. It should be noted that inbreeding does not create lethal genes; it only provides the opportunity for them to appear as both alleles on corresponding chromosomes, as with sickle cell anemia and hemophilia.

From either direction, then, there appears to be boundaries within which any genetic or "inclusive fitness" argument can operate. In man of virtually any culture these rules are explicitly stated. There are only, generally, certain others with whom one should copulate for x number of reasons; but all exclude incest, and most seem to require some degree of exogamy (these cultural rules are, of course, not all genetically accurate). That most primates and other species seem to behave in terms of similar rules should not be surprising; deleterious effects result from too close matings, and natural selection seems to have produced behavior patterns that tend to preclude incest. For instance, male offspring rhesus monkeys rarely become dominant to their mothers and hence rarely copulate with them (Sade, 1968).

In considering the arguments about genetic or "inclusive fitness," what should be kept in mind is not only the degree of relatedness between individuals (who it is "good" to aid in order to perpetuate one's own genes), but also who it is "good" not to copulate with in terms of the deleterious effects this may have on offspring. This range may be quite narrow and may require the consideration of other parameters than just the "genetic" as parameters within which species operate. The following sections will elaborate upon selection and its consequences in terms of the individual and the group. Other factors or parameters aiding group selection will then be presented, and the role of dominance relations at various levels of group organization will be discussed.

Individual or Group Selection?

Darwin suggested that it is the individual organism which is ultimately selected for or against by the forces of natural and social selection; if the individual does not survive and reproduce, its genetic material will not appear in the next generation. *Survival* is the first key term here, followed by *reproduction*. The individual has physiological structures and elements of behavior which may aid in its survival and also in its ability to reproduce. In essence, the individual has to be strong enough to survive environmental forces and also competition with rivals in order finally to mate or even to aid others so that they can mate. As Emlen and Oring (1977) say:

> What accounts for the differing intensities of sexual selection found in different species and, frequently, between populations of the same species? We hypothesize that one important cause is the *ability of a portion of the population to control the access of others to potential mates* (p. 215). (italics in the original)

Dominance relations between individuals may be a key factor in individual survival and reproduction, but, as suggested below, it may also be a confounding factor in considering individual versus group selective factors.

What Is a Dominance Relationship?

Interactions between organisms of any species can take a variety of forms. They can meet and copulate, fight, ignore each other, flee, and so forth. The early ethological literature stressed the displays which one organism may be producing, to which another organism may respond—and respond in a particular manner. These counterdisplays by the other organism were found to have a certain predictability to them, and, hence, it could be assumed that some form of communication was occurring between the organisms involved (Altmann, 1967). At this point it could be said that an observable predictable relationship existed between these organisms. What the relationship was "about" depended upon the outcome of the final "consummatory act" (Eibl-Eibesfeldt, 1970), be it copulation, nursing, following, fighting and fleeing, or staying.

Within the ethological literature discussions of dominance relationships appear very early (Carpenter, 1937, see 1964; Schjelderup-Ebbe, 1922; Zuckerman, 1932, see 1963). The earlier observations were that some form of hierarchical arrangement, based upon observable displays of behavior, could be discerned in groups as diverse as chickens and monkeys. When two individuals are brought toegether (in experimental studies) or come together (in naturalistic studies), a dominance relationship can be defined as existing *if an attack by one individual is followed by flight or submission of the other individual*. There are, of course, numerous combinations and permutations of what can happen during an interaction; the attacker may, in turn, be attacked and may ultimately flee or resort to other action. Attack per se may not define the relationship but only the final attack / retreat sequence (see Bernstein, this volume, Chapter 3). In this sense the submitter defines the interaction as did the consummatory act define what preceded it. Whether or not submitters maintain the dominance relationship is an empirical question, whose answer may vary from species to species (cf. Rowell, 1972, 1974).

Other definitions of dominance relationships occur throughout the animal and human literature. Most imply that there is a functional reason for the relationship, i.e., the relationship differentially imparts some ecological advantage to one of the members of the interacting pair over the other member. Such interactions can be described as taking the form of one of the animals gaining "prior access" to food, a place to sit or sleep, a mate, and so on. An accumulation of such advantages through consistency in behavioral displays may, in the long run, lead to differential reproductive success, which means that the genes for dominance, aggression, agonistic displays, and so forth, should increase in a population. Since one of the goals of an ethological study is to demonstrate that there is a functional reason for particular behavioral displays, and hence something which the forces of natural selection can act upon, it is understandable why examples of prior access to environmental resources have been focused upon by researchers. However, assuming this functional relationship between dominance relations and prior access to some environmental feature, like a mate, is not the same as demonstrating that this assumption is valid. Dominance relations may have other functions; so the assumption must be tested and other factors explored (see Kalmus, 1969, for the application of this concept to other behaviors).[9]

The other reason why acts involving prior access, rather than dominance / submissive acts, per se, have been so frequently recorded is that they occur more frequently; hence, they provide a larger data base upon which to develop subsequent arguments. In part, this higher frequency of occurrence results from the fact that the various species that have been observed have been naturally occurring dyads or groups. With experimental manipulations where individuals unknown to each other are placed together, the rate of aggressive interactions are very high, but with ongoing natural dyads or groups this type of interaction has occurred in the historic past, before the observer arrived on the scene and, outside of new young individuals entering the adult hierarchy, the rate of fighting might be quite low. Quite simply, by now all know their own and everyone else's position—hence the reason for turning to other, more visible, behaviors.

While a great deal of ingenuity has been shown by various researchers in determining what other behaviors might represent the outcomes of earlier dominance / submissive interactions, questions have arisen about the vailidity of such an approach. As suggested, the first question is whether, in fact, those who act in a dominating fashion at some earlier ontogenetic period have later prior access. Weisfeld, Omark, and Cronin in Chapter 11, this volume, suggest that this may be true (at least perceptually) for human boys. Wilson (1975, p. 257) goes so far as to define dominance as "the assertion of one member of a group over another in acquiring access to a piece of food, a mate, a place to display, a sleeping site or any other requisite *that adds to the genetic fitness of the dominant individual*" (italics added). Wilson claims that this is not a circular definition (i.e., a scientist would attempt to demonstrate that the italicized portion followed from the unitalicized rather than simply claiming it as a fact) because of the many demonstrated instances in the animal literature. One immediately wonders what happens to all the subordinates; Wilson's argument is that if they wait long enough within a group, they may get a chance to mate; anyway, they can always emigrate.

A second question arises from the fact that various primate species seem to display different quantities of these prior-access types of behavior. While access to a mate correlates highly with dominance relations in one species, another species provides high correlations for access to grooming partners and vice versa for which scores low (Bernstein, 1970a). If primates are related phylogenetically, why should such different results occur? Each species, of course, has its own phylogenetic history, and subtle environmental factors may have selected different aspects of their behavioral repertoire. Some of the differences among closely related species living in similar environments are so marked, however, that the whole concept of dominance relationships has been called into question. The balance of this chapter, hopefully, will clarify some of the confusion that has arisen around the concept of dominance relations.

Are Groups with Dominance Relations Possible? A Sociobiological Perspective

Prior access to particular environmental features, especially mates, has been taken by many as the key aspect of dominance relations. Using this aspect within sociobiological theory, researchers have developed a number of hypotheses. What will be examined here is whether such a usage should support group maintenance or whether such usage may suggest other results.

At the individualistic level, or at the genetic level as the sociobiologists would prefer, if an individual is to get its genes represented in the next generation, then the individual had better do something to see that this happens. In essence, the gene or genes had better program the organism properly so that it behaves in a way which will see to it that "they" are passed on. Dominance relations at "lower" phylogenetic levels where two male ducks, alligators, frogs, or deer fight over a potential female partner suggest that for some species, and perhaps in our own much earlier phylogenetic history, such struggles probably occurred. Under such genetic programming is it possible to form groups, and what would be the consequences if they did coalesce?

As numerous *experimental* studies have shown, a dominant individual living in a group may gain access to food, mating partners, and whatever else *he* (since males are generally dominant in most primate groups) desires. This is a basic tenet within the sociobiological model. The occurrence of offspring, however, requires that a parent of each sex survives for some period of time. If environmental conditions are severe enough to push for individual survival over group maintenance, a dominant female may survive over other females, but she would certainly lose out to numerous other males since that is the basic structure of most primate groups. If survival through prior access is the sociobiological model that is suggested, then a species should go extinct as soon as the next highest ranking male usurps access to those parts of the environment necessary for survival.

If the dominant female manages to survive, copulating with her may not increase the dominant male's genetic fitness,[10] i.e., if the male were number one and the female number eight in the total group hierarchy, their male offspring (on a strictly probabalistic basis) would tend to rank fourth or fifth within the surviving group.[11] Since generations are relatively long among primates (approximately five to eight years), the other older males would still be around and hence would outrank this new individual and would have prior access to environmental features over the new individual. Prior access by a single dominant male within a group where other males manage to survive would then appear to act against the survival of females and his own young.

Because the dominant male acted in a dominating fashion, he has also usurped any chance for increasing his inclusive fitness. So, while dominance may gain some momentary survival value under severe conditions—for the dominant individual acting as an individual—this strategy would be expected to be selected rather quickly against for a male living *in a group containing other males*. For males of nongroup species (e.g., orangutans) or groups where other males have been driven off (e.g., patas monkeys and gibbons), complete dominance over other males is a perfectly reasonable strategy, but behavioral adaptations have to be made to confront the environment without the support of other males.

Suppose the dominant male supported, in agonistic encounters, not only himself but also his "family." Quite surprisingly, from a Western-oriented, nuclear family perspective, this evolutionary pathway seems to have occurred rather infrequently in mammals. Any number of bird species follow this strategy (e.g., cardinals, blue jays, robins), and some even do this for life rather than just for a season (e.g., some of the geese). Upon closer scrutiny it can be seen that many of the birds which follow this strategy only do so during the mating season, while those mammals which do so are a bit more constant. Thus, jackals, wolves, patas monkeys (to some degree), gelada, and hamadryas baboons, and some men remain attached to their families for life. Of these, only the latter four are really group-living species. Therefore, some other strategy must be explored for the prevalence of dominance relationships among even ourselves and the hamadryas baboon, and perhaps wolves and geladas.

When environmental conditions are reasonably good (such that the mature females will survive), it may be to one or some of the males' advantage if he or they can be dominant to the other males if, in turn, this dominance provides access to more females. Access to other environmental objects (food, sitting and sleeping places, and so forth) is probably immaterial under "good" conditions, and if such access occurs, it is apt simply to be an offshoot of the dominance relationships or due to entirely different factors. Having access to the females, especially during the proper point in the estrus cycle, would obviously increase the genetic fitness of a male.

If he is the exclusive male to mate, then the next generation will contain one-half of his genes. These genes will not be the same genes in each of the offspring (because of the chance factors involved in chormosomal assortment during the mitosis phase of reproduction, etc.—Hirsch, 1963, 1970). Still, any genes necessary for "dominating" behavior should be overrepresented in the next generation as compared with the parent generation, and increasingly so in later generations. If *a* male is the dominant breeding individual, there is obviously no support of kin or reproduction by kin and, hence, no "inclusive fitness" that can be measured. This point is immaterial since producing and, if necessary, supporting one's offspring is better, genetically, than supporting a brother's offspring (i.e., an average one-half representation versus an average one-quarter representation). However, if one cannot completely dominate the troop, since one's brother also received some of the father's genes which may have led to the display of dominating behaviors, then one probably has to share some of the females with the brother. The next generation will contain one-half of the dominant male's genes (or slightly less if the brother also mates), some proportion of which will be the genes necessary for displaying appropriate dominance behaviors. One does not have to "aid" the brother, one simply needs to do nothing while the brother mates.

From the sociobiologist's perspective, if this process continues for a few generations, the gene(s) producing dominance behaviors will become fixed at high levels. But then there would have evolved a group of all leaders and no followers! Who is to dominate the group—since everyone will want to dominate—and what is to become of it? Obviously, such a group would immediately stop being a group as each individual male or female wanders off in its own direction with the mate(s) it has managed to win. Such may have been the initial case for jackals, coyotes, numerous bird species (see Emlen and Oring, 1977), and a modified form of this behavior may occur in hamadryas (Kummer, 1968).

One could hypothesize that such behaviors evolved in rather benign environmental conditions where the female as well as the male had access to enough food, sleeping places, and so forth. Then, what is there to fight over? One could argue for marginal conditions from time to time, but in any bisexual species where either genetic or inclusive fitness[12] is to be maximized then dominance relationships should either lead to "family" support or should be nonviable in terms of forming groups as illustrated above, when only the individual's genes are considered. That such is not the case is obvious from the current zoological record. Dominant individuals do occur in groups in many species, and they fill a variety of roles within such groups.

To suggest how groups might have begun, the sociobiologists present two alternate pathways—inclusive fitness, and altruism and reciprocal altruism. In inclusive fitness—aiding kin rather than, or besides, the self—the group has to be closely related. Can the group be reduced simply to considerations of the inclusive fitness that x number of individuals within the group may have? This question is further complicated by considerations of when the economic costs, in terms of time and energy expended by continuing particular behavioral patterns, (e.g., aggressive actions) to those x individuals outweigh the continuance of such behaviors (see Clutton-Brock and Harvey, 1976). In essence, this means that for evolutionary success (i.e., spreading one's own genes into the next generation) one should be sure to act aggressively enough toward others (nonrelated) so that one has the best chance of copulating with as many females as possible. But, one should not be so aggressive as to wear oneself out; in any case, the one you are fighting may be your brother or uncle and so your genes will be propagated regardless. This certainly seems to be an example of having your cake and eating it, too. How is one to test such a theory which says it is both good (in evolutionary terms) to be aggressive and good not to be *too* agressive because you (or your genes) are bound to win.

Such theories obviously depend upon the consideration of only one gene and only one factor at a time (e.g., aggression, sexual dimorphism, spiteful behavior) with references drawn from the literature as need be for the purposes of each argument. While the literature on any species is obviously incomplete in many details, still each species is an entity which has managed to reproduce itself over thousands and possibly millions of years and should be considered in holistic terms rather than picked apart to support one's own particular argument. Any hypothesis then becomes testable in terms of whether it explains the behavior of a particular species rather than bits and pieces of various species' behavior. This is the comparative approach basic to biological studies. For example, Clutton-Brock and Harvey (1976) state, "However aggression is defined, it usually falls within the category of disruptive behavior." They then argue that a "sifaka [prosimian] . . . which fights to gain access to females . . . or the mangabey [Old World primate] . . . which successfully disputes a clumped food source by threatening its neighbor . . . is likely to increase its own fitness. Behavioral traits which enable individuals to do this can spread through the population even if they diminish average reproductive success" (Hamilton, 1971). If such a line of reasoning were true, we should expect primate species, as group-dwelling individuals, to have long since disappeared unless some other modifications were made in their behavior patterns.

The second sociobiological pathway is "altruism" and "reciprocal altruism" (Trivers, 1971). Wilson (1975) defines altruism as "self-destructive behavior performed for the benefit of others" (p. 578) and reciprocal altruism is "the trading of altruistic acts by individuals at different times. For example, one person saves a drowning person in exchange for the promise (or at least the expectation) that his altruistic act will be repaid if the circumstances are reversed at some time in the future" (p. 593).

An altruistic gene (for self-sacrifice for others or for cooperation within a group) has to counteract individual selection (basically a dominance gene that makes each individual want to win a mate, nesting site, and the like). Depending on which gene is present (or which gene is dominant, see Wilson, 1975, p. 111), the individual will either cooperate or be aggressive. If the former, then groups can occur; if the latter, then breeding pairs (plus offspring) should be the largest group units. Both of these approaches (altruistic and dominance) are based upon the behavior of an "actor." For bisexuality to work at all, the "receiver" has also to behave. If the receiver also behaves altruistically, one has reciprocal altruism. If one behaves altruistically and the other dominantly, the dominant wins but the group continues to exist. Both kinds of genes may continue for some time if the group is composed of kin, or if very specific and, supposedly, rare evolutionary conditions are met (see Wilson, 1975, Chapter 5). In the presence of "dominance genes," which strive only for prior access to environmental resources, the dominant individuals should generally win out. If both individuals act in a dominant fashion, one will win (because of differential abilities), and in many relatively solitary species the other will flee to a different site (e.g., territorial behavior will appear). If both act altruistically (reciprocal altruism), then it appears that a group can occur, but it should lack any features involving dominance relations.

Hence, when the sociobiological model is examined, it does not appear to be able to account for groups which have dominance relationships, which act cooperatively in a variety of ways, and which maintain themselves as groups. Picking and choosing among behavioral examples from any species is not the way to demonstrate that a model works. What the student must do (and what the sociobiologists should do) is demonstrate that the total model is applicable to a single species. Primate groups or human groups certainly present this opportunity since they manifest many of the behaviors within each species that the sociobiologists only consider in isolation in various species.

Wilson (1975) seems to have a bias to establish his pet hypotheses, and "The Will to Power" is but one example, when he says "the young males are *routinely* excluded by their dominant elders. They leave the group and either wander as solitary nomads or join bachelor herds" (p. 287, italics added). This is in reference to "a wide range of aggressively organized mammal species," including "langurs, macaques, and baboons." Wilson's statement is simply not true for these three primate genera! (For discussions of various primate species, see Dolhinow, 1972; Jay, 1968; Kummer, 1971; Poirier, 1972). Even the logic of perpetuating groups would argue against excluding young males, except for one-male groups like patas monkeys and gibbons.[13]

By now the reader can begin to appreciate the genetic and evolutionary dilemma of trying to explain group formation by considering only individuals and their kin rather than interactions within the broader population. The long-term existence of such orders as the primates, carnivores, ungulates, and others, with their obviously evolutionarily successful group structures, makes it difficult to understand why reductionistic reasoning is still being considered (e.g., Dawkins, 1976). A *single* gene can never act in isolation from the organism in which it appears. In essence, this means that most, if not all, genes have a pleiotropic effect upon their particular organism (i.e., the products of the genes interact within the total organism, and all of this before and while behaviors and structures are being produced). Genes are also not all the same in what they accomplish in that some genes are structural (i.e., direct their effects at producing physiological structures), while other genes are regulatory (i.e., control when the structual genes can operate). What this means is that a structural gene may be permitted by the regulatory gene to appear earlier or later in one species than it does in another. The chimpanzee may have about 90 percent of the same *structural* genes as do humans but the final product can appear quite different because of the action of regulatory genes (cf. Mayr, 1978). There are obvious differences in the ontogenetic appearance of a variety of behaviors (weaning, entrance into peer groups, and so on) that occur between the various primate groups and us. These should be explored in terms of what differences regulatory genes might produce.

Finally, any member of a bisexual species does not exist in isolation but must interact with conspecifics during some portion of its life cycle. For many species this interaction is almost continuous. "Reciprocal altruism" appears to be the most complicated interactional model of genetic behavior proposed by the sociobiologists. Their model, however, builds upon "altruism" and is basically part of a simplistic either / or consideration of either dominance or altruism as the underpinning and behavioral manifestation of primate and other groups. The world may be more complex than simple genetic models would indicate or suggest. As Kalmus (1969) suggested, Occam's Razor (always taking the simplest explanation) may not always be the appropriate rule to follow. Genes do not act in isolation—and neither do individuals. The next section will present some complications that should be considered when exploring interactions of individuals within group settings.

Are Groups with Dominance Relationships Possible? An Ethological Perspective

Flocks of birds, schools of fish, and swarms of insects exist, but since this book is directed to human ethologists, primatologists, and such, the discussion about groups and dominance relations will be limited to mammalian and principally primate groups. A more complete comparative approach should include species at other phyletic levels. As Lorenz (1971), Tinbergen (1951), and Blurton Jones (1972a, 1975) have suggested, an ethological study should start with the observable data. In very gross terms, what has been observed about most primate groups is that, in general, they have members of all age and sex classes present in the troop on a continuing basis, dominance relations almost always exist among the adult males, and less frequently among the females, and both phonological and gestural communication about intra- and extragroup phenomena occurs.

Wilson (1975, pp. 274–276) discusses the shift in behaviors which accompany increasing population densities in a variety of species. The progression proceeds from no aggression / no territoriality at extremely low densities through some aggression in the form of territorial defense at moderate densities to aggressive behaviors resulting in dominance hierarchies within artificially crowded groups. Naturally occurring groups resemble the artificially crowded groups in that many of them exhibit dominance relationships, with the additional possibility of escape from extremely stressful situations (see Crook, 1970b). That such escape does occur has been reported in the primate literature (van Lawick-Goodall, 1971). But the increasing reports of emigration by both males and females in a variety of species suggest that emigration may be as much or more a mechanism for preventing inbreeding as it is the result of stress situations. Bernstein (1974) also suggests that there is no relationship between individual stress and who emigrates from primate troops. In any case, most groups stay relatively intact across the years, and flight out of the group as a result of attack by another group member is not the typical response. It is obviously more important to maintain group membership than it is to get away from some momentarily stressful situation—flight is the exception rather than the rule.

Data from Hausfater and Watson (1976) indicate that, in general, group membership may not be stressful to low-ranking individuals (contrary to the data from basically nongroup dwelling rodents). Hausfater found that the low-ranking group members carried less of a parasite load than did the momentarily dominant individuals. This load, or drain upon the physiological system, suggests that low-ranking individuals may not be under as high a stressful pressure as hitherto assumed (Rowell, 1972). Hence, to be low-ranking and still a group member is not necessarily bad in terms of individual survival. Further, Hausfater found that high-ranking individuals, in part perhaps because of their parasite load, did not maintain their dominance rank for more than three or four years. This means that other individuals could have the chance to enjoy the prerequisites (including possibly producing more offspring) normally assumed to be associated with high dominance status.

It must be stressed that individuals at other status levels *do* have reproductive success. In Hausfater's (1974) data one can see that

> second-ranking males had a higher reproductive success than did fifth-ranking males, but fifth-ranking males had longer durations of rank occupancy, on the average, than did second-ranking males. Thus to achieve any given level of reproductive success, a male may either occupy second rank and reproduce at a high rate for a

> short period of time or occupy fifth rank and reproduce at a low rate for a longer period of time. The end result in terms of total reproductive success, however, would [might] be equal by either strategy (pp. 126–127).

As Hausfater points out, the data are simply not yet available on, even, male primate life histories to permit the development of models which illustrate the "various sequences of rank occupancy" and the "different rank occupancy strategies." Different status ranks do reproduce in many species, and the earlier models limiting reproductive success to only those individuals at the top of the hierarchy are too simplistic to be considered valid any longer.

The group then can be viewed as a dynamic structure where the inclusive fitness of any one individual may be momentarily enhanced but where other individuals will have their chance. This means that reductive models that consider only the potential gene flow from a particular dominant individual are too simplistic when examined from a dynamic perspective. First, that individual is apt to be displaced. Second, the behavior of other group members may not be determined at any moment in time by that individual (Chance and Jolly, 1970; Kummer, 1968), and, third, the group per se can provide selective advantages quite beyond those provided by any individual's particular physiological structures or behavioral patterns.

Simpson (1977) suggests that the group is different from and more than any individual. The group has continuity through time beyond the life of any individual. Environmental information, and responses to environmental pressures, can be transmitted by other than genetic means in many species, and adaptations need not only be under direct genetic control. Groups also constitute one of the major points where prior evolutionary change may have occurred in a species (the Sewall Wright Effect mentioned earlier). No individual, in the absence of others (above haploid species' levels), can produce any change in the genetic composition of its species. It always takes two or more individuals of differing genetic structure to provide the variation upon which natural selection can act. The group, then, should be examined as a necessary part of every primate's life.

First, aggression as a sociological rather than an individualistic phenomenon within intact ongoing primate troops generally appears to be produced not to disrupt the social order, but to maintain it (Bernstein and Gordon 1974). Where aggression is most frequent is along the diagonal (i.e., between individuals hierarchically near each other) of any dominance relationship matrix for primates (Bernstein and Gordon, 1974) and even human children (Savin-Williams, in preparation; Strayer and Strayer, 1976). Such aggression appears to be the result of a necessity to clarify social relationships between almost equal individuals rather than the gaining of ultimate priority of access to environmental resources. If the latter were the case, then most fighting should be between those at the top of the hierarchy and all other individuals as each individual tried to gain access. This simply does not occur, either in primates or children.

Second, such traits as always gaining priority of access do not manifest themselves throughout *any* primate troop as a prime characteristic of primate behavior. Primates are not always striving to get all they can, nor are most carnivores and other species. If any characteristic is common across social species, it is that, first, except for extreme situations or experimental manipulations, the social order is conservative and will be maintained for extended periods of time by the group members. Second, in general, the environment external to the group provides resources sufficient to meet the needs of all or none of the group members,[14] and, third, while initial reports indicated that certain individuals usually gained prior access to, say, females (Hall and DeVore, 1965), subsequent studies have indicated that many individuals may have access (Crook, 1970). Hence, being able to gain prior access may not be the key to the examination of primate, carnivore, or even human group behavior, and reductionistic models that examine only the benefits to individuals, or their immediate relatives, have failed to examine carefully what evolution has selected—the group.

A slight digression is necessary to provide an historical footnote as to why reductionistic (genetic) models have been pursued by the sociobiologists. Williams (1966) is frequently cited by the sociobiologists as part of their rationale for examining the genetic level rather than individual or group levels of selection. The materials presented in this chapter and in the ethological literature suggest that much more is known about group actions than was available in 1966. It is this author's belief that Williams' reasoning might have been different if his book were written today.

Williams says, "A *group* in this discussion should be understood to mean something other than a family and to be composed of individuals that need not be closely related" (p. 93). Following from this, his arguments rest upon "competing" groups and whether there were sufficient numbers of these (>100 groups) to promote selection for particular groups structures. As this chapter will attempt to indicate, one does not start with groups in being and then say that there are not enough of these groups for natural selection to operate to produce group structures! Instead, one has to examine how groups might have arisen (since ethological data shows that they obviously exist) and ask what factors may have been selected for that enhance their continued existence.

A further restriction that Williams (1966, p. 92) adds is that "group-related adaptations must be attributed to the natural selection of alternative *groups* of individuals and that the natural selection of alternative alleles within populations will be opposed to this development." Altruistic models are then proposed as a way around this restriction, which seems to have generated much of the subsequent sociobiological research. Interestingly, much of the sociobiological literature has been directed at "alternative alleles" (like altruism) which do not oppose development. In any case, Williams' statement appears to provide a needless constraint because one has to have groups before selection between different groups can occur.

The sections in this chapter that discuss communication and attachment will suggest how groups may have developed (without "altruism" even being considered). As the model develops, it will be seen that particular group structures can be selected *for* without some groups necessarily being selected against vis-à-vis other groups. One particular "group structure" is not then chosen, by natural selection, over another group structure. Instead, the individuals create the group structure which, in turn, aids in their efficiently dealing with environmental factors. Dominance relationships are an important aspect of many groups' social interactions. Chapter 13 on the Umwelt discusses how cognitive abilities and cognitive distortions may aid the formation of group structures.

For some reason the sociobiologists seem to have accepted Williams' suggestion that the genetic level be examined while ignoring his injunction that group selection may occur within, at least part, of the mammalian line. One wonders why this blind spot when the sociobiologists talk about mammals, primates, and man. Even accepting Williams' arguments about "sufficient numbers of groups," one can see that any species that typically travels in groups of 20 to 40 (numerous primates and, supposedly, early man) will have sufficient numbers of groups very early in their existence as a species to counter Williams' statements. Williams'

(1966) work should be viewed as an historical document that generated some interesting hypotheses (e.g., altruism), but it should not be taken as a firm basis for reductionistic arguments, as suggested by the sociobiologists. This contention will be developed in what follows.

The most striking fact about virtually all primates is that they exist in groups or "troops." Further, all of the individuals in the groups are not "leaders," or overly dominant individuals. The groups also are not, generally, completely closed units; they experience immigration and emigration. Depending on the particular circumstances, the group can then evolve for a while in terms of its particular environmental situation, but it will, from time to time, receive external genetic material to maintain contact with the species as a whole and to provide the genetic variability needed to remain viable. The extremes may have been reached for the human species with the American Indians, who were "cut off" from continued contact with their Asiatic brethren by an ice age, and with some of the Polynesian groups who settled on isolated islands and were "cut off" by their lack of sophisticated navigational instruments (see Finney, 1977). In any case, both such groups have been shown by subsequent copulations to still have remained Homo sapiens.

The point to note is that man and the primates have evolved within group settings that contain both immediate relatives and, potentially, nonrelated individuals. Further, even paternity can frequently be in doubt (see Hausfater, 1975b for additional references). A mother may be sure of her offspring, even if not sure of the father, but outside of the hamadryas baboons (Kummer, 1968), the patas monkeys (Hall, 1968), and the gibbons (Ellefson, 1968), most primates, including man (in 13 to 30 percent of offspring, Hirsch and Vetta, 1978), cannot really be sure that the offspring they are supporting are really theirs. Anyone's efforts then to maximize genetic fitness (plus parental involvement) or inclusive fitness by some set of actions supporting one's own or one's brothers' or sisters' offspring is apt to be fooled. In fact, if a female wants to be sure that she is fertilized, and that her genes will be propogated, her strategy should be to copulate with as many males as possible. Such may well have frequently been the case throughout our own evolutionary history given the mating patterns of primates (e.g., adult females in estrus will even sneak out of sight of the troop to copulate with nondominant males). As long as the mother and her offspring live in a group, the female does not need a single male close by. If maintenance of the group is what has been selected for, a primate female need not worry about bonding with a particular male (cf. Barkow, 1978a). Subsequent human behavior needs to be reexamined in terms of this point.

Since one then cannot be sure of offspring, what are the evolutionary possibilities open for appropriate behaviors that will maximize one's own desires, abilities, potentialities, or what have you? The only recourse seems to be either (1) the perpetuation of the group, of which one is a member; (2) the complete fragmentation of the group as each "leader" fights for his position, his females, and so forth (see Crook, 1970, on responses to attack behavior); or (3) some compromise like that produced by the hamadryas baboons (Kummer, 1968, 1971), where the males closely control their females—to the point where this "harem" group under severe conditions may travel separately from the larger troop—but harem groups can still coexist at sleeping cliffs and during daily marches in normal conditions, within a much larger troop. Gelada baboons (Crook, 1970) approximate this behavior but with looser control by the males over the females.

At some point in the evolutionary past groups became a viable approach for how some species confronted their environment and the pressures that the environment produced for survival. Not every species, or genera, or family responded similarly, nor does every kind of group structure, once selected for, necessarily resemble the group structures of other species, genera, or family. After all, some animals are predators and others prey, some herbivorous, some carnivorous, and some omnivorous, and even those that fall into these broad rubrics live in different environments. Still there are advantages to group life, and these will be returned to later.

When a group follows a genetic pathway where dominant animals are overproduced, recourse (2) above may occur and the group will disintegrate. In the end it will probably consist of a male, one or more females, and a few young. Since the young will also contain "dominant" genes, it will be to the father's advantage to drive off at least male young at some point in time, as occurs in the gibbon. For genetic fitness plus some degree of parental involvement this obviously cannot be too early, but it must occur somewhere near puberty to prevent usurption of the father [see Wilson's (1975) error, p. 287].

Such driving off of subadult males may have been the strategy in numerous species, during their phylogenetic past, including coyotes, foxes, and jackals—as contrasted with wolves and hyenas—and patas monkeys and gibbons—as contrasted with most of the other primates. Some such strategy may have been short-lived among the felines with the lion presenting the only near social phase where hunting is jointly conducted by females; the young can be cared for by any female, but the male interacts only to gain food and a partner. The question remains, then, as to whether a group first appeared and then disintegrated because of the genetic factors which were selected for or whether the group stage was never reached. For the primates, and many of the ungulates, it is obvious that a group stage was important in their evolutionary development. Both families, until presumably recent times for the primates, have been the prey rather than the predator. Predators are at the top of the food chain, which means that it takes many more prey to support a few predators. If an active defense is used by the prey (e.g., some primates) against the predators, more prey than predators should be present to resist the predators. If a passive approach (no defense, no concealment) is used (e.g., many ungulates, Kruuk, 1972), vast numbers of prey should be expected. A passive approach with concealment suggests individuals or very small groups so that they will not be spotted.

Groups of prey can maintain themselves if they employ different strategies when confronting predators. They can actively defend themselves (as will be illustrated in the next section on communication), or passively allow chance and the infirmities or inexperience of group members to take their toll, as each group member actively tries to escape, or run and hide as an individual (e.g., rabbits and the young of deer). Each of these strategies occurs in species that may exhibit dominance relations, some of which may exhibit altruistic behaviors. What each of these factors and other behaviors may mean in group formation will be developed further.

Communication

A prime virtue of a group is that the group provides the opportunity for the development of a system of communication. The simplest group—a male and female of a bisexual species meeting and mating—involves the exchange of a set of signals: species recognition, ready to mate or not, territory available or not, site suitable, and so on. At the lowest levels the exchange of information is about the self. (See Wynne-Edwards, 1962, for a description of signals ranging from the auditory to the visual, and Sebeok, 1977, for more recent findings.) At slightly higher levels of phylogenetic organization the information can include aspects of the environment. At this point in evolutionary development it becomes not just "I" but "we" who have to decide whether conditions are right for reproducing young.

This shift, literally from one mind to two, is probably as great in terms of countering the forces of natural selection as the shift from unisexual reproduction (where if the individual dies, "her" genetic makeup dies) to bisexual reproduction. This shift is also, probably, the major one leading, first, to support of offspring (parental investment), second, to support of related kin (inclusive fitness), and finally, to support of group—including nonrelated kin.

Communication of information is of the essence here. Once the barrier has been broken between two individuals, interested as it were only in terms of themselves and just "their" line of offspring, to broader concerns for environmental conditions, the information code can then be elaborated upon, both in terms of what may be communicated and, even more important, in terms of what may be thought and remembered about the environment. It is this power of communication, as it becomes expanded and elaborated across succeeding generations and species evolution, that can provide the opportunity, and the power, to move the species from the individual selective level to the group selective level. The process goes as follows: Proceeding from the initial species recognition signals, mentioned above, the next round starts with genetic fitness and true parental investment.[15] The parents care for the young until they are viable, can support themselves, and can look for partners. Species can be found at every phylogenetic level which do this as far down as the insects and perhaps lower. The next change is when some of the offspring do not leave "home" for longer and longer periods of time. This provides the opportunity for the parents to impart their knowledge—to communicate about the environment. Bringing foodstuffs back to the nest or the den is one obvious example; the foodstuffs are representatives of environmental objects which are safe to eat. Altricial birds and many carnivores follow this example. Among precocial birds, ungulates, and primates the infants travel with the parents and can observe what they eat.

The next jump is to a sibling-within-a-group level, which is a subset of the above, i.e., brothers and sisters continue to travel together either with or after the demise of the parents. Then they each have offspring (presumably with other partners to prevent deleterious effects of inbreeding); communication then continues. If offspring remain around, however, the information extended is not just about foodstuffs but about the total environment. The elders obviously decide where the groups will travel each day, and in the same sense they have decided what the day's experiences for the young will be (see Kummer, 1968, for a description of how older males determine group progression). "In short, social communication is a process by which the behavior of an individual affects the behavior of others" (Altmann, 1967, p. 326, and rest of chapter). Nonmovement by the elder, as the juveniles scamper off in different directions and then return, signifies that the wrong direction has been taken by them. Movement in a particular direction by the elder probably signifies the day's route and may even signify the destination.

If, in addition, information is imparted about aspects of the environment, e.g., what is safe or what is dangerous, then the whole communicative and experiential process is enriched. For example, in Kortlandt's film (Kortlandt, 1962; Kortlandt *et al.*, n.d.) a "stuffed" leopard was placed where chimpanzee groups would come across it. After the initial encounter the leopard was removed, but the chimps showed excited behavior when nearing the now vacant site, which would certainly serve as a mnemonic device to reinforce the memory of where leopards might be found. It should be noted that such behavioral displays can be seen by the entire group and hence promote both the inclusive fitness of the displayer while also aiding any nonkin member.

For the sociobiologists and those others who are on the fringe between individual and group selection, the next problem area is how to explain behaviors that appear to aid those group members who may not be related to the producers of the behavior. What they have focused upon were "altruistic" genes. Altruistic genes were proposed to explain the rather strange acts (from the sociobiologists' point of view) to nonkin; i.e., if one does something good for someone else, how can this possibly aid the self, or one's own offspring, or one's related kin? The sociobiologists were confronted with the fact that groups of relatively nonrelated kin occur very frequently in nature and something had to be done to explain their existence.

Trivers' (1971) invention, mentioned earlier, was to introduce the idea of "reciprocal altruism." In essence, if I help you now to survive, then you had better help me later with a similar act (and cheaters will be punished by having their genes excluded from the population). The problem with this model is that it rests upon behaviors that are to be reciprocated, but how are those behaviors to be defined and delineated? If a leopard approaches a juvenile baboon and an adult baboon rushes forward frightening away the leopard, is the juvenile baboon supposed to do the same, and when? If soon (in the next year or so), then the juvenile simply will not—it will flee instead. However, the juvenile, especially if it is a male playing on the periphery of the troop (instead of being driven out as Wilson would like), will give an alarm call at the approach of a predator—is that sufficient reciprocation? If it were a juvenile female that was initially saved, reciprocation in kind will probably never occur.

The real problem is what to count as the reciprocated action that pays back the debt. To claim that the alarm call was a form of reciprocation would be a reasonable claim within Wilson's (1975, pp. 120–121) statements about reciprocal altruism.[16] Such a claim, however, would be irrefutable, i.e., not subject to scientific proof. If we assume that individual A has performed some altruistic act, then the actions of individual B have to be examined to see if those actions aid A while harming B. A and B are going down a trail, B. leading. Individual B trips over a log and hurts himself (lowering his fitness). Individual A consequently sees the log and steps over it. A is not harmed, as he would have been otherwise, so his fitness increases. Or another example, B does not pick a fruit off a tree to eat; A picks it. B's fitness is decreased by one unit, while A's fitness increases by one unit. As one can see, any number of actions may fit this model, but they would certainly be meaningless. The giving of one banana to you now in return for one banana from you later seems to be too simple a base upon which to construct the occurrence of groups, especially in groups where, supposedly, one individual continues to grab all of the bananas.

What needs to be recognized within a reciprocal altruism model is, first, that altruistic acts are not one banana now in return for another banana later, but the entire life of the group. Communication is the real exchange product, and it occurs continuously, as in the example of the alarm calls from the juveniles who generally "have no bananas." Second, as Tinbergen (1963) suggested in the factors that must be considered in an ethological approach to the study of behavior, an evolutionary model is also an ontogenetic, or developmental, model—one that continually moves forward to confront the future. Such a model is in direct contrast to the sociobiological model of whether or not I will stay in my group if "he" does not give me a banana. The important outcome of group life is that both related and nonrelated offspring survive into the next generation to support the group that they form.

The final step in our communicational model derives from the fact that communication is constant. It is not simply that an individual aids another individual, but that an individual can aid the group! In turn, the group aids the individual. Communication is always a relational phenomenon—between individuals—and, as we have suggested above, such a phenomenon was initially most important between *non*related bisexual individuals. Working backward through the above scheme, one can ask, first, how the group aids the individual in communicatory terms. Considering prey rather than predatory species, one of the obvious answers is that the group (say, herds of ungulates, schools of fishes, or flocks of birds) possesses many eyes rather than a single pair watching the environment. Coupled with some communicatory signal, danger can then be indicated earlier, and this should certainly prove to be a selective advantage of evolutionary significance. If the group, or some portion of it, then performs a defensive act against the danger (which obviously has to be more frequently successful than unsuccessful), the group aids the survival of all, or most, of the individual group members (see Kruuk, 1972, on prey behavior).

This response to the environment is particularly important when contrasted with the chance of survival for individuals meeting a dangerous predator. For example, despite the size of a large baboon male's canines, a leopard or hyena probably would quickly do him in. Confronted by five adult males, the predator would probably go elsewhere without even bothering to attack. The prey do not always win, even when in groups (see Mech, 1966). Those that are old, sick, young but inexperienced, and the like are lost to predators, as also are some of the active communicators and defenders. This does not nullify what has been said; it simply provides the materials for the continuation of the evolutionary process.

Predators are the other side of the coin. More than one predator working together stand a better chance of spotting prey, and certainly in attacking prey. A willy-nilly attack by a group of predators upon a group of prey, however, is nothing more than one group of individuals chasing another group of individuals and is apt to be unsuccessful. Again, communication is the key, as Kruuk (1972) suggests for hyenas.[17] The group of predators, of whatever species, generally does share the kill, either with other group members or the young, although rank preferences for portions of the kill are obvious in numerous species.

Communication, in contrast to a sociobiological/genetic model just proposing altruism or reciprocal altruism, can then be seen as one important factor in group formation. What the present ethological model so far suggests also contrasts with the typical anthropological approach that considers man as being distinct from the "animals" because man has a verbal language and culture(s). As has been suggested (and is left for the reader to elaborate upon because space is too limited here to do so), communication between conspecifics (members of the same species) is the real turning point between individuals *alone* combating the forces of natural selection and a *group* of individuals with their combined intelligence and perceptual abilities confronting these same forces.[18] Evolution did not wait until the appearance of man to create some counterforce to selection at the individualistic level but produced such a counterforce when the first bisexual species and communication appeared.

How do Dominance Relations Manage to Appear in Groups?

The preceding section discussed communication as an important, and necessary, product of group life, while also being a factor contributing to group development in an evolutionary sense. The interrelationship between communication and dominance relationships will be developed later. It should be noted that both communication and dominance relationships involve interactions between individuals (whether they be individuals who generally are solitary or group members). No single individual ever communicates or establishes a dominance relationship in isolation; hence, both concepts are supraindividual.

The preceding section argued against the sociobiological model. That reductionistic model suggested that groups with dominance relations should self-destruct; the portion of the model leading to group formation—"reciprocal altruism"—seems to be untestable in behavioral terms. The basic question that sociobiologists have been working on, however, is an important one—namely, how can dominance relations occur in groups? Wilson's solution of juveniles waiting or emigrating elsewhere (being driven out) are only partial solutions, particularly the second. The question needs to be examined more carefully at the ethological level (rather than at the genetic level), for it is here that the actual behaviors appear which determine the course of the interaction, and it is this interaction which becomes part of a group's structure against which the forces of natural selection can then act. We will start with the basic question.

We have suggested that "dominance relations" be considered in their simplest terms—an aggressive act by one individual followed by a submissive act by its partner. The behaviors involved in each "act" must be spelled out and will vary from species to species. The conceptualization of these "acts" is as important as the defining of the behaviors to be included within them (the constructing of an ethogram). How important this conceptualization is will be illustrated by several examples of conceptualizations that may have actually interfered with the study of dominance relations rather than clarified its role in dyadic or group encounters.

The first, discussed earlier but so common in the literature that it needs to be emphasized, is the concept of dominance as related to "prior access." As Richards (1974, p. 928) says, "The concept of a social dominance hierarchy, once verified, can be useful in many studies of social behavior, *as a wide variety of social interactions can then be predicted*" (italics added). In a following article Syme (1974, p. 936) states, "However, since the currently accepted definition of dominance is 'priority of access,' it would seem that the prime purpose of a validation is to demonstrate such a priority." The reader is referred to our earlier comments, but it must be pointed out that both of these authors, as do many others, conceptualize dominance relations as providing "priority of access" to something; i.e., if the dominance relations do not meet this criteria, they are valueless, even though they may be an obviously apparent part of social life. The major problem with such a conceptualization is that it is tautological, within an evolutionary perspective. It is *a priori* assumed that dominance relations lead to prior access rather than that dominance relations (or attachment behaviors, courting gestures, and so on) should be continually questioned as to what they may really mean in the life of the species being studied.

The other concept that may interfere with a fuller exploration of dominance relations is that these relations are conceptualized as always leading to flight or escape, Crook (1970) remarks on primate studies,

> Whenever any need-reducing commodity is in short supply (food, water, sleeping place, female, etc.) competition inevitably follows with success going to certain animals and being denied to others. . . . Escape from social positions imposing behavioral constraint appears highly rewarding and means of escape are sought.
> There appears to be three avenues of escape. . . (see p. 152).

Again a particular behavioral outcome is conceptualized as part of the model of dominance relations rather than the question being left open as to what in fact occurs within each species.

Any attempt at an explanation of the structure, cause, and function, of a behavioral phenomenon *per force* limits what is discussed and considered. What we are suggesting is that one not take any "explanation" as being absolutely true and proceed in terms of that model; instead one should continually question what is being said. As you, the reader, will find as you begin doing "Science," you can work within the framework of a model, but you can never "accept it"; you can only "reject it." What follows is another model which we invite you to explore and then reject in favor of a better model.

If we then consider what in fact children, primates, and others do in actual behavioral terms (an inductive ethological approach) rather than proceed deductively within "priority-access," "escape," and other models, a few bits of data that do not fit such models become apparent. The first anomaly can be seen in a wide variety of social species with dominance relations; that is, the individuals who are attacked generally, in fact almost always, do not flee or "escape" from the group.

Emigration and immigration of both males and females does occur in primate troops, and bachelor bands of males can be found wandering independent of mixed sex groups (Kummer, 1971). However, such wanderings do not appear to be the result of dominance interactions over estrus (receptive) females (Hausfater, 1974a, 1975a, 1975b) or to stress-related factors (i.e., fights; see Bernstein and Gordon, 1974). The simplest explanation (which we have suggested should always be suspect) is that such wanderings promote genetic contact with other species members and prevent the appearance of lethal genes in too inbred groups or the tendency for each group or troop to become an incipient species. Groups may be a favorable way to counter the forces of natural selection, but outside of examples like the vast wanderings of the American Indians or the Pacific Islanders, many species probably maintain some linkages to the broader population.

If an individual in a social species does not flee or escape when attacked or punished, in contrast to basically nonsocial species (outside of matings and brief encounters), then some mechanism—structural and / or behavioral—must have existed to initiate this performance of nonflight in the face of adversity. A possible candidate might be the "imprinting" phenomenon investigated by Hess (1973), his students, and others. What Hess found was that precocial birds (e.g., ducks) would go through *a great deal of effort* (or stress) to maintain contact with the mother or "mother substitute" (a variety of experimental objects exposed to the infant very early in life). This imprinting model was extended to human infant and preschool children's behavior patterns by Bowlby (1969), Ainsworth and Bell (1970), Marvin (1977), and others under the rubric of "attachment" behaviors (see Ferguson, 1970, for an overview).

Basically, what these researchers have shown is that the young organism will strive to maintain contact, or reestablish contact, with the "mother" or "caregiver" even when the mother is acting in a fashion which requires the youngster to expend energy that might be better spent elsewhere. For example, in many primate species during and following the weaning process the mother acts inimically to the youngster's desires. The youngster approaches time and again to take the mother's breast or be near her, and time and again the mother pushes the youngster away, ignores it, or gets up and leaves, with the youngster having to expend still more energy to stay near the mother (see Dolhinow, 1972, for one example). The youngster at this point in its life is basically self-sufficient in terms of meeting its own nutritional needs, but it is still attached to its mother as a secure base. Weaning, per se, may not be a key factor since many infants wean themselves. The attachment behaviors seem to stand out at that time in some species. It is the attachment behaviors by the youngster that should be the focus taken. As Kaufman (1974) points out in his study of bonnet and pigtail macques, and as any observer (or parent) of young children knows, "rejection [of the infant by the mother] tends to increase rather than decrease the infant's motivation to establish contact" (Bramblett, 1976b, p. 597).[19]

The infants, or youngsters of all species where parental investment occurs after the young are born, obviously require some form of attachment behavior to the parent(s) if one or a number of conditions are present: (1) the infants are apt to be eaten by adult conspecifics nesting nearby (e.g., as with cichlid fishes); (2) the infants are precocial—able to travel very shortly after birth—but, perhaps, have to learn what to eat and require protection (precocial birds like quail, chickens, and ducks); (3) the infants require nurturence from their parents and may or may not be mobile (for instance, the altricial birds—those which stay in the nest—and all of the mammalian species).

For those mammalian species where the young are almost instantly mobile (e.g., horses, cows, deer, porpoises, whales) the infant has to be attached to the point of following first the mother for milk and then the herd or school for protection, at least until it nears maturity. For less mobile mammalian species (e.g., carnivores and primates) who have to learn some of the complexities of the environment—what to eat, what to hunt, what to avoid—attachment should again exist until the subadult years. For mammals who may be more completely preprogrammed to encounter environmental features (e.g., rabbits and mice), we find that they are frequently raised in "nests" like altricial birds, and weaning and separation from the mother occur almost simultaneously. No attachment behaviors would be needed by these species after the nursing period.

The appearance of a physiological structure or behavior pattern implies that there are structural genes (those genes which control the form of the end product whether it be anatomical, physiological, or neurological) that can be modified by natural selection. At their simplest, "attachment genes" lead the infant to direct attention toward one other conspecific (the mother or caretaker), to emit a limited set of signals indicating needs, and to follow the conspecific if necessary. Regulatory genes determine when the structural genes will operate and stop operating. As Chapple (1976) suggests, regulatory genes are as important as structural genes in determining the final form of an organism.

What we would like to suggest as an hypothesis is that in some species the regulatory genes controlling the structural attachment genes[20] either did not "click off" or clicked off and clicked on again. Attachment would appear to be a simpler basis upon which to develop group structures which contain dominance relations rather than a model which acts contrary to dominance relations, i.e., reciprocal altruism. But let us continue with the model.

If some form of attachment—directing attention toward a conspecific, communicating with it, and following it—is present in a species after the mother–infant period of lactation, then groups of two or more should certainly form. Any attachment genes or their regulators are subject to interaction with the gene complexes carried by the species of interest. Interactions with "dominance" genes in ungulates, carnivores, and others should be different, and we leave it to the student to develop what happens elsewhere while we attempt to develop a model for the primates.

Assuming that attachment in its grossest form continues to operate or reappear at a later date, we can then assume that this provides an affiliative or cohesive factor for group life. In other words, the developing organism will move toward rather than away from other conspecifics at times other than just during infancy or mating. If striving to be dominant is still part of the species repertoire (and certainly an older, phylogenetically, and more "locked-in" part of the repertoire since dominance relationships can be seen in solitary species), how may these two sets of behaviors interact? Specifically, how can an individual manage to approach another who is attacking it when most of the research on nonsocial species would suggest that it should flee?

Such behaviors as approach in threat situations require one more shift in behavior or, possibly, the appearance of a new set of behavior patterns. This new set consists of "subordinate behavior" patterns. Following Ginsburg's (1976) example of cricket communication, "the simplest and surest model to effect synchrony is that of unitary control of signal emission and decoding by the same genetic mechanism" (p. 64). Given that subordinate patterns of behavior are much more recent phylogenetically than the mating signals of crickets (since large groups with dominance relations are probably more recent), then both the emission and reception (decoding) of these signals may still be under the control of some of the same genes. In any case, such subordinating gestures are readily responded to by human children (Ginsburg, this volume, Chapter 20; see also Bernstein and Gordon, 1974), and in part overlap with the attachment behaviors shown by young children to adults when they are punished (e.g., approaching, clinging, following if the parent leaves, and so on). Both orientation toward conspecifics and proper responses in the face of adversity met within the group should be necessary behaviors for the group-living organisms involved in dominance relations with their conspecifics.

If the loser does not flee but instead gives a subordinate gesture and even approaches the dominant, a group can maintain itself and provide whatever advantages a group may confer upon its members. This subordinate gesture is not an altruistic act because it does not immediately aid the other, but instead it aids the immediate survival of the individual producing it. How might such behavior have arisen in the course of evolution? Appeasement (subordination) gestures appear to have three different linkages with other behavioral systems:

1. They can be infantile in form (crying, in many primates and humans) or such as to produce a diminution in size toward that of the infant or juvenile size (Ginsburg, this volume, Chapter 20).
2. They expose a vulnerable portion of the anatomy (the neck, the belly, as in dogs and wolves).
3. They are sexual in form (the "presenting of a rump" in many of the primates, and a reference in much of the scatological, generally denigrating, humor of humans—referenced, however, by the aggressor rather than the aggressee).

Since the "mounter" in (3) is higher than the submissive individual, there may be some connection between (1) and (3), but more research needs to be done on the possible evolutionary pathways of this behavior pattern.

The infantile / diminution response could have evolved from the responses of adults to infants. It is a fact that youngsters are cared for and protected in virtually all mammalian and bird species, some amphibian, reptilian, and fish species, and among the social, and possibly nonsocial, insect species. Thus, a similar response to juveniles or adults performing similar behaviors is probably deeply ingrained at many phylogenetic levels. In most social species one simply does not kill infants or those who momentarily resemble infants, except for rare instances like a nongroup member taking over an entire new group (Hrdy, 1977). Once such a response to a younger conspecific exists, the triggering mechanisms for that response can be selected to act similarly toward older conspecifics who act appropriately.

An explanation for (2) exposing vulnerable parts would rest upon being able to distinguish conspecifics from enemies, perhaps coupled with infantile behaviors, e.g., the puppy which rolls on its back and urinates in your hand when you pick it up (suggested by Marvin, 1977; 5-day-old coyote pups do the same, personal observation). Since distinguishing conspecifics from nonconspecifics is basic to the possibility of animals being bisexual, recognition in the case of conflict between a dominant and subordinate animal should not be a problem. The fact that killing blows may normally be given to prey in the very areas which are exposed by the subordinate suggests that the initial, evolutionarily distant, production of both the killing blow and the subordinate response may have been under the control of the same gene structures. Here the genetic structure necessary for recognizing conspecifics versus enemy or prey controls both the "go/no go" signal for production of either the aggressive or submissive behaviors. Again, the phylogenetic pathways of such behavior patterns need exploration.

The "sexual" behaviors, e.g., presentation, are probably also related to dominance/subordinate actions at a different level. The above two examples (1 and 2) are hypothesized as being direct offshoots of the obvious status differences between adults and infants, and (3) would be hypothesized as a derivitive of sexual encounters; i.e., where the male is dominant to female, the female postures associated with copulation may be expressed by the subordinate male as a way of appeasement during dominance relationships.

A possible approach to group formation has been proposed which permits the existence of dominance relations within the group without introducing behaviors which act in a fashion contrary to dominance relations, e.g., altruistic or reciprocally altruistic in nature. At the same time we have suggested that certain behaviors (subordination gestures) must have appeared, or been modified from earlier behaviors which were present in the repertoire of various organisms, that permitted organisms to approach rather than flee when attacked. Such behaviors permit a group to continue to exist as a unit rather than being continually dispersed, which earlier models of dominance relations submit should happen.

Finally, that a group of any sort should exist would appear to rest upon the occurrence of attachment behaviors from infancy being put to other uses. The important point is that once a structure or behavior appears, the forces of natural selection can then act to modify that structure or behavior. Once "true parental investment" is responded to by the infant through attachment behaviors, those attachment behaviors can be acted upon to serve the purpose of group formation rather than just momentary attachment of the infant to the adult. In essence, Ferguson (1970) asks what attachment behaviors might look like when they occur at later ages. Since she is viewing these behaviors from a development perspective (changes through the life cycle) rather than an evolutionary perspective (changes that might have occurred in the evolutionary past), it is difficult to see that some behaviors might serve quite different but related functions at different ages. One of the functions might be related to the development of "attention structures" (see Chapters 13 and 22, this volume, and below).

One of the primary shifts that occurs developmentally is to whom one is attached. The mother or caregiver is the first object, and in many species this is sufficient to provide the knowledge necessary for the youngster to act independently thereafter. When the infant is big enough, it can simply leave to fend for itself. If attachment to conspecifics continues, however, the infant does not leave the group but redirects attachment to some individual(s) within the group. This new form of attachment appears to proceed through a two- or three-stage developmental process. First, there is attention directed toward and then attachment to the peer group.[21] Second, would be the shift to the similar-sex, adult world. Third, would be the shift in attention and shorter or longer attachments to a mating partner, with the second shift generally continuing in many primate species.

The first hypothesized shift is to similar-age peers because of the limited size of the youngsters; to participate in aggressive interactions with adults would be disastrous. Time is needed to grow up and develop strategies for interacting with others. Further time is needed to learn the group's idiosyncratic set of communicatory signals (its cultural transmissions), and time is also needed to learn about the environment through which the group passes.

The very early stages of this process appear to be asexual (see the chapters in this volume on preschool-age children). In later stages the sexes become more or less separated experientially within the group (Chance and Jolly, 1970), although not necessarily separated perceptually in terms of what is learned about the total social group (Omark and Edelman, 1976). All of this occurs within the protective cover of the group. There is both earlier, and sometimes continuing, parental investment and also group investment in terms of protection and information provided—about both internal social phenomena and external physical phenomena.

The second shift in attachment is to the adult group. The now subadult or adolescent travels with part of its peer group into interaction with adults and ultimately into experiencing aggressive encounters from others. Such experiences are not new since similar encounters occurred in the juvenile male groups with peers or between the young females and mothers (see Chance and Jolly,

1970). Since young humans (ages 2 to 4) can respond in an appropriate fashion such as to minimize physiological damage to themselves (as will be illustrated in numerous chapters in this volume), while maintaining a positive perception of the self (Chapter 13, this volume), this new set of experiences should not prove to be overly traumatic.

What does occur as the adolescent or subadult enters into interaction with adults is a series of dominance encounters—more frequent and overt for the males, less frequent and sometimes more subtle for the female (i.e., the mother takes back her crying infant from a subadult female who had been permitted to hold it, and this might be viewed as a mild form of punishment). Such dominance encounters occur when any set of primates who do not know each other are experimentally placed together, and it even occurs in preschool children, although more delayed in time (McGrew, 1972). Dominance interactions probably occur more separated in time in naturally occurring groups because the group has other tasks to which it must attend if it is to survive. However, dominance encounters are a part of the life of primates, including humans, and need to be explored in terms of what the establishment of such relationships does to preserve the troop or group.

It must be remembered that a primate troop and many human groups (e.g., neighborhood, business, and military organizations) are social structures that continue through time with a low rate of entrance and exit by new or old members. The adult members of the group have an established dominance order into which the new subadult member slowly moves. Initially, the subadult/young adult would be expected to be subordinate to most of the adult population. The adult population, besides being split into sex classes, is probably also split between those who will actively engage in dominance interactions and those who will no longer do so (the "elders"). Given the length of time the subadults have spent in the group, they probably now know who occupies these various categories.

Two of the major functions of dominance relations frequently cited in the literature "are to establish and maintain social structure and to relate it to ecological resources" (Nagel and Kummer, 1974, see their elaboration of this, pp. 175–180). If groups, as the basic breeding unit of their species, are to meet the contingencies of their environment (the forces of natural selection) then it is important that they be relatively stable rather than continually disintegrating per the whims of individuals. Dominance relationships, especially when they produce a dominance hierarchy encompassing most of the group, should provide such stability. Different groups may then "try out" different ways of meeting environmental pressures (e.g., fleeing or attacking in the presence of predators) and these actions, as they are differentially selected for or against, will move the species along its evolutionary path.

So far, we have suggested the following about groups:

1. Groups provide communicatory advantages about environmental features.
2. Groups came into existence as entities with dominance relations present because of attachment behaviors and the appearance of subordination gestures.
3. Groups are relatively stable entities that can be selected for or against as units.

What remains in this broad approach is to consider how these entities decide upon a course of daily action. The basic question within our model, then, is still toward whom one should direct one's attention, form attachments, or become affiliated. In terms of the group surviving, the obvious answer would be that attention should be directed toward that individual or set of individuals who know the most about the vicissitudes of the environment that the group has to encounter.

Given a group containing dominance relations which remains relatively intact, we can then consider how dominance relations and the communicatory abilities of group members may interact to some evolutionary advantage. Chance's idea (Chance and Jolly, 1970; Chance and Larsen, 1976) was that there was a "structure of attention" such that within certain kinds of groups (those with a strong dominance hierarchy rather than more loosely organized groups) attention by most age and sex classes would be directed toward the major "dominant" figure(s).[22] This attention toward a central figure or figures means that more communication can flow between that individual or individuals and other group members than between any other pair within the group taken at random.

This is really a very subtle point that the student must ponder rather than simply underline, especially when examining the ontogeny of the process of developing attention structures. Our examination of numerous classrooms suggests that first graders (age 6 in the United States) rank equally highly "bullies" (those who get what they want) and potential "leaders" (those who engage in numerous fights, but who also help others by breaking up fights, providing suggestions, and the like). By second and third grade the class seems to distinguish between the two (when asked, "Who is the toughest?") by giving the leader a rank of one, while the bully is rated lower down in the group. When lining up to go to, say, recess, lunch, or gym, the leader also seems to occupy a position near the *center* of the line (the best place for communication with the most group members, see Hall and DeVore, 1965, for travel patterns of baboons; and Savin-Williams, Chapter 12, this volume), while the bully could be expected to occupy a frontal position.

When one examines extant groups most fights or dominance / submissive interactions are apt to occur between and among middle- or low-ranking individuals (see Chapters 8 and 10, this volume). The amount of fighting and dominance rank should not be equated (see also Chapter 3, this volume). As we saw in one third-grade Swiss classroom, a single fight (between a newcomer and the former alpha) can establish a new alpha and no further fights are really necessary. It is between near peers that the battles continue, but at a diminished rate. Similarly in primates, once a hierarchy is established the overall fighting diminishes.

Primate and human groups, in general, have existed longer than any researcher's observations have continued. Consequently, while many dominating / submitting acts may be observable by the current researcher, the frequency of occurrence of such acts should not be taken as reflective of a group's structure of attention, or even of its dominance hierarchy. They probably provide the data for assessments about part of the group, but, equally probable, they will not provide data about the upper end of the group, that portion which no longer engages in fighting. Even experimental manipulations (to induce fights) may not provide evidence of where attention is normally focussed or who in fact controls group movements. Longitudinal studies like Kummer's (1968), the Altmanns' (1970), and Hausfater's (1975) continuing investigations can begin to detail shifting patterns of control within troops.

The control of the group, as opposed to immediate individual advantage, is the contrast being suggested here. The phylogenetically earlier advantage provided individuals may always be a factor present—especially under severe conditions. However, for those groups having integrated modes of behavioral interaction, i.e., where the group functions as a unit rather than as an assemblage of individuals, it is toward total group interactions that research must be directed. Even further, it is not just the internal interactions of the group that should be observed but the interactions between the group and the external environment. Of particular importance in the latter are the roles that different individuals perform, e.g., controlling group movements, signaling danger, confronting danger, carrying infants to safety, signaling information about the environment (food and so forth), and on to the more complex roles performed in human groups.

Attention structures, whom attention is directed toward, are obviously multidimensional when only whom one is looking at during some period of time is used as the operational definition. One can look at a mating partner, a grooming partner, an enemy, a friend, an aggressive individual, and so on. There may be differential amounts of looking at various individuals, but given differences in the length of time spent grooming versus time spent watching fights, the differences in attention directed toward others will always be higher for friends than for others. Simply counting frequencies of behaviors may not answer the evolutionarily important questions of how the group establishes and maintains internal stability and also how the group confronts the external environment.

Differential measurements within age / sex classes have to be made to determine the individual or individuals who receive the most attention *when particular kinds of behavior are occurring*. For example, when the group is about to travel somewhere, about to meet another group, about to receive a stranger, and so on. To paraphrase Altmann (1967), the basic question is, Who predictably affects the behavior of others because of his or her own behavior? Such interactions may be very subtle. For instance, while the author watched a third-grade Swiss classroom, a number of the children let the most dominant, active girl know what their grades were on tests that had just been returned, to each of whom she smiled or nodded. What seemed to be occurring was a need for approval from her, but the children's behavior also seemed clearly to be linked to her status. The point is that the investigation of naturally occurring behaviors within group settings cannot rely upon fleeting observations or simple experimental manipulations. One has to observe the group-in-being and then begin to question what is happening.

When one examines dominance relations in ongoing groups, there are both those interactions which are occurring now and those interactions which occurred in the past, before observation began. One should perhaps not be surprised that the upper portion of a "dominance hierarchy" is less aggressive than the middle or lower portions. Nor should one be surprised that some individuals are not even engaged, currently, in such interactions. Within any ongoing group there are probably three "generations"—the newly hired (subadults), those who are striving for recognition from the boss (those who are actively engaged in dominance encounters), and the boss (the one who was dominant in earlier struggles).[23]

The boss, the elder, the experienced individual need no longer engage in dominance interactions once his/her position has been secured and this is evident in third grade and perhaps earlier (see Chapter 11, this volume). Attention *should be* directed toward such individuals if they can aid in coping with environmental exigencies. There is then the continual interaction between establishing oneself as the center of attention within the group while being able to lead the group in its continual confrontation with the external environment. The development of "attention structures" would appear to be at least one way in which the occurrence of dominance relationships have been modified by the forces of natural selection to be different from the phylogenetically earlier use of simply aiding individualistic selection.

Groups are historical products, and this must be kept ever in mind. The older individuals have experienced more environmental changes, know more about where food and water can be obtained, where predators may be found, and so forth. But the older individuals are less capable of fighting and maintaining a position of aggressive dominance. A balance, obviously, has to be struck between experience and aggressiveness. Aggressiveness may determine to whom the next generation will direct its attention for information about group movements, but it cannot be the sole criterion for the present generation's group behaviors. Experience should hold some edge if the maximum amount of environmental information is to be transmitted to the group and to be used for the survival of the group. Where, and when this occurs in the ontogeny of group formation needs to be explored.

If there is something important to be communicated which will ensure group survival, then this should take precedence over internal group management. This is clearly indicated by the response to juvenile monkeys' alarm calls. While the juveniles themselves are obviously subordinate, and rather peripheral group members, they serve to alert the group to predators. Are these simply altruistic behaviors that will be reciprocated at some future date, as suggested by the sociobiologists? Hardly, as has been indicated above.

The upcoming generation, of course, benefits from such alarm calls, but the sociobiologists' models have not yet considered these ontogenetic differences that an ethologist (Tinbergen, 1951) suggested must be a part of any ethological study. Obviously, it is not just the individual (whether in the form of genetic fitness he or she provides or the inclusive fitness of such an individual) that should be the focus of attention, but *also* the group and how it interacts with its environment historically. Research needs to be directed toward how the group establishes and maintains itself in the face of evolutionary pressures rather than toward reductionistic models that focus only upon individual behaviors or genetic factors. We leave it to the student to examine the complexities of group behavior that are only hinted at in the following chapters.

Notes

1. Genetics and evolution were, surprisingly, quite separate disciplines for almost a whole century. While evolutionary thinking needed a "unit" which could be transmitted, the very early Hardy–Weinberg Law (1908, see Stern, 1960) precluded change in a population—hence no evolution—and it was not until later researchers showed how change could occur that genes were seen as the appropriate unit.
2. An innate versus learning dichotomy is not being proposed here; most organisms learn something from their environment, and all organisms come with some predetermined structures and behaviors (cf. Pitcairn and Eibl-Eibesfeldt, 1976). The above statement is also being told the "wrong way around." Variation in individuals' behavior produced situations which could be acted upon by natural selection. Some individuals were selected for and some against. The writings of Darwin (1859, 1872), Mayr (1970), Simpson (1953), and Waddington (1975) can provide the details of how this occurred.
3. The processes described here and in the following sections are always more complex than this brief description can encompass.
4. Infants of almost all primate species are differently colored than the adults. So far as we know, no one has explored whether an experimental change in coat color would produce a change in adult behavior. The reason this point is raised is that there might have been something "strange" in the infants' pelage or in the adult male's perceptual system.
5. Wilson's definition (1975) is "inclusive fitness—the sum of an individual's own fitness plus all its influence on fitness in its relatives other than direct descendants; hence the total effect of K in selection with reference to an individual."

6. Barkow (this volume, Chapter 15, referring to Wilson, 1975) suggests that competition between groups is an important determinant of group survival (following Williams, 1966). Such competition may or may not occur, but is probably immaterial since natural selection is the final arbiter and this latter selector will be the focus of the model presented here.
7. The "deme" is the potential population in which interbreeding occurs. For primates this may be two or more neighboring troops.
8. It is a long metabolic pathway from the genes to final physiological structures and behaviors. Even so simple a factor as white eyes in the fruit fly *Drosophila* was found eventually to involve at least 50 genes. There is obviously not "a gene" for aggression, nor for altruism, nor for attachment (mentioned later in this chapter), nor probably for any other behavior. A "gene" is also not a single entity; the reader is referred to any recent genetic text for a discussion on gene action, mutation, and recombination. It is the organism performing particular patterns of behavior upon which natural and social selection acts, and it is the organism's entire genetic package which is selected for or against. Thus, the focus of attention should be upon whatever behaviors are performed rather than upon any hypothetical gene.
9. As Allee (1977, p. 56) said, "There are fundamental tendencies in the biological universe which are radically different from the disoperative competition based primarily on an individualistic struggle for existence such as Darwin thought resulted from overpopulation. . . . [I]n the phenomena associated with underpopulation we have biological evidence of simple supra-individualistic groupings on which evolution can act to produce more complicated and better knit social systems."
10. "Genetic fitness—the contribution to the next generation of one genotype in a population relative to the contributions of other genotypes. By definition, this process of natural selection leads eventually to the prevalence of the genotypes with the highest fitnesses" (Wilson, 1975, p. 585).
11. Sociobiological models tend to consider only a single or very few genes. If a number of genes are necessary, as would be expected for most structures of behaviors, then genes are more randomly assorted (Hirsch, 1977) such that leaders and followers should continually appear. The genetics model suggests what is evident in most natural groups, while the sociobiological models suggest a progression towards dominance only (or altruism only). If one considers hereditary leaders, what follows from the genetics model is that part of the time the leader will be a weakling, a dominant individual, or someone in between. Reading about, say, the kings of England illustrates how genetics operates and suggests that something is wrong with the sociobiological approach.
12. "Inclusive fitness—the sum of an individual's own fitness plus all

its influence on fitness in its relatives other than direct descendants; hence the total effect of kin selection with reference to an individual" (Wilson, 1975, p. 586). Note that Notes 8 and 10 do not implicitly contain "parental investment."

13. Juvenile males may frequently be seen on the periphery of a troop. They are noisy and active, as are young boys, and they may be driven there (or simply move there to avoid harrassment because of their boisterous activities). In any case, they are still watched over by at least a subadult male (Chance and Jolly, 1970; Kummer, 1968).
14. Hamadryas baboons are one species living under fluctuating marginal conditions. Under extreme conditions the larger group breaks apart into units containing a "family" and, possibly, one nonrelative. Under better conditions the larger group reforms.
15. The "true" is added to imply the positive support of young rather than the more negative sociobiological approach, which implies loss of breeding potential. The purpose, after all, is to get young to survive, not simply to copulate a lot.
16. Wilson (1975, p. 593) defines reciprocal altruism as "the trading of altruistic acts by individuals at different times. For example, one person saves a drowning person in exchange for the promise (or at least the expectation) that his altruistic act will be repaid if the circumstances are reversed at some time in the future." "When a person (or animal) increases the fitness of another at the expense of his own fitness, he can be said to have performed an act of altruism" (Wilson, 1975, p. 117).
17. The number of predators needed to subdue large healthy prey is always surprising (see Kruuk, 1972; Mech, 1966) and suggests that early man may have had a very difficult time of it until he reached the point where communication was sufficiently elaborated to provide the means for coordinated strategems to be used when attacking large prey.
18. Boehm (1978, p. 270) comments that "the overall process appears to consist of decision making and decision implementation at a group level, and is a way of planning vital activities for an entire day or the basis of a great deal of information. . . ."
19. Since species differences, of course, occur, any topic, (attachment, dominance relations, and so forth) needs to be examined within the total life of each species. All that is suggested is that attachment is an important factor to consider.
20. The student needs to again be warned that "structural attachment genes" have not been found to exist as individual entities, nor are they apt to be found. What is being suggested here is the possibility that some metabolic pathway, genetically controlled, may have switched its mode of action such that whatever controls "attachment behaviors" may have acted in a fashion which continues the appearance of attachment behaviors rather than terminating them at some previously selected appropriate phylogenetic level of appear-

ance. What must be focused upon are the behaviors which appear since these are what natural selection acts upon.

21. "Attention toward and then attachment to," in a sense, might be considered as the social counterpart to Piaget's assimilation of and accommodation to physical objects.
22. Hinde's (1974) focus upon attention-"attracting" behaviors, such as van Lawick-Goodall's (1971) description of Mike, the chimpanzee, banging cans together and subsequently rising in the hierarchy is a special case that will have to be dealt with elsewhere. We are trying to describe the more common everyday behaviors of organisms where it is the accumulation of experiences, rather than a bit of noise, that affects group members' behavior patterns.
23. As suggested in the preceding chapter and elsewhere (Omark *a*, *b*, n.d.), every individual is unique—both genetically and ontogenetically. A single rule or simple set of rules will not predict an individual's behavior. To understand any individual's behavior, ethologists must begin to study the individual and move away from group tests that only present caricatures of individuals.

Part II

Dominance in Primates

Chapter 3

Dominance: A Theoretical Perspective for Ethologists

Irwin S. Bernstein

Dominance. Territoriality. Altruism. I have never seen a "dominance" response, a "territorial" response, nor an "altruistic" response—not in the sense that any good scientist would record behavior. What we *can* record is behavior, its structure and form. As such, we can describe the motor patterns and sequences in elemental detail, but we cannot observe causation or motivation directly. We are quite used to making inferences based on data; we are comfortable in the operational definitions we generate to measure conceptual abstractions such as "anxiety," but we sometimes confuse an indicator, a measured hypothetical correlate, with the concept itself. Although we generate operational definitions, few of us restrict the meaning of our terms to our own definitions. As a result, despite the precision of definitions, many words are used as Lewis Carroll's Mad Hatter would, to mean exactly what we want them to mean, nothing more and nothing less.

Kummer (1971), in his discussions of primate societies, clearly differentiates between (1) structure, (2) function, (3) cause, (4) ontogeny and (5) evolution. It would be well if all of us maintained these five distinctions in all of our writings rather than confusing functions, causes, and selective pressures with the behavior itself. How often do we report "threat" or "aggression" instead of what the subject actually did?

The research cited and leading to the view expressed in this chapter was supported by NIMH grants MH 13864 and MH 20863, and in part by NIH Grant RR 00165 to the Yerkes Regional Primate Research Center.

Of course the interpretation of the significance of a response is important. Without this, understanding—or theoretical analysis—would be impossible, and we would have nothing more than collected masses of observations, described and perhaps quantified, but otherwise meaningless. The significance of our observations lies in the theories, hypothetical constructs, and explanations which we develop, but we must never confuse interpretations with observations. Good theories generate hypotheses; good hypotheses may be tested by data collection; but even confirmed hypotheses are not data, and do not "prove" a theory.

Dominance is all too often treated not as a theoretical variable but as an objectively observable response. Much as with aggression, we sometimes confuse the proposed indicator measure with the theoretical variable. Individuals may be said to be motivated by dominance assertion or dominance striving; we may explain the outcome of an encounter as the result of a presumed dominance relationship; but we cannot do more than infer the existence of dominance. A dominance interaction is not defined by its morphological or behavioral characteristics, nor is it measured by any direct observation. Rather it is defined by the hypothesis that a particular interaction in a particular context reveals the existence of a dominance relationship. Indeed, *dominance is a relationship and not an activity or an attribute*.

Definition and Measurement

The definition of dominance and the selection of a suitable measure then becomes crucial. Numerous operational definitions and measures of dominance have been proposed. Some relate to one or another aspect of "peck order" as originally described by Schjelderup-Ebbe's (1913, 1935) accounts of dominance in birds. Others rely heavily on measures relating to "priority to incentives" because of the theoretical functional implications of dominance. Other views relate any ordering of social behavior to dominance, suggesting that any such ranking is a reflection of the universal influence of dominance on social behavior. All of these views, despite their diversity, postulate dominance to be the "cause" of certain interactions which "function" to produce certain outcomes of theoretical significance.

"Priority" models all tacitly assume that dominance relationships strongly influence competitive interactions and that the outcome of such competition must, of necessity, reveal dominance relationships. Naturally, we have to assume that all participants in the competition are equally matched in all respects save dominance rank, that all participants have equal opportunity to compete, and that all participants are equally motivated to strive to obtain some goal we hope is an "incentive." Considering the difficulties in (1) equating motivational states, (2) equating opportunities to perform, and (3) identifying incentives, it is no wonder that our measures are less than perfectly accurate reflections of what we theoretically expect to occur. Repeated trials are often contradictory; subjects do not always seem to respond simultaneously to the trial onset signal, some are badly positioned, some seem to be less well motivated than others, and, worst of all, some seem to be attending to aspects of their lives other than the one we wish to study.

It would be marvelous if we could operationally define and scale "incentive" and "motivation," but since these are internal states or perceptions, we are forced to deal with presumed "causes," indicators, and theoretical interpretations. In our desire to obtain precise objective operational measures, we sometimes forget that we have subjectively defined the drive or the incentive. If only we could equate motivation by starving all subjects a fixed number of hours or down to an objectively measured criterion, such as percentage of body weight! If only we could assure that all subjects will attend to conditions as the experimenters intended! It is much too dangerous to "assume" that this is the case. Consider the food deprived animals in a feeding test. Are they all equally "hungry"? Is the food equally palatable to all? Is a male inspecting female genitalia sufficient for us to postulate "prostitution"? Does the male's tolerance of this female in close proximity, while he is inspecting her genitals, mean that their relative agonistic dominance ranks are now more equal? Does it mean that flight and fight distances have changed radically? Or might it possibly mean that, despite food deprivation, the male is still a complex organism responding to multiple motivational systems which interact to influence his behavior at any one time?

One difficulty with dominance measures is that they often assume the existence of the phenomenon and preclude acceptance of the null hypothesis, i.e., that the concept does not apply to the relationships existing among the subjects. If a measure consists of recording who takes a presumably preferred sitting position, or who feeds first, or who wins a fight, or who obtains an object that the investigator knows that he or she would take if dominant, then any one measure will always produce a winner and a loser or losers. If we use multiple measures, we cannot reasonably expect either that the "dominant" will win on every trial or that wins and losses will be absolutely evenly divided. Neither behavior nor random events are uniform, and variability is a characteristic of both. If we attempt to conduct enough trials to get statistically significant differentiation, we run the risk of changing motivational states and using measures which influence the phenomena being measured. Subjects may "learn" the outcome of a repeated trial, thereby influencing motivation. Intertrial intervals long enough to preclude motivational and learning changes risk changes occurring in the basic relationships as a function of time.

For example, repeated food competition trials are performed with animals whose food motivation must change, if ever so slightly, with each reward received. Food preferences may account for differential striving, and food preferences may also vary from time to time and as a function of recent access to a particular food. If we were to run sufficient trials, we might end up equating gluttony with dominance, and if we change the incentive from grapes to peanuts to lettuce, we may find dramatic differences in the willingness of subjects deprived to the same "motivational criteria" as a function of which preference food was offered first!

If, on the other hand, we try to use "peck-order" criteria, such as the outcomes of fights, we might only succeed in demonstrating that some subjects are better fighters than others, or that outcomes were influenced by the test arenas chosen. Surely neither fighting ability nor gluttony are equal to dominance, although both may relate to the outcome of "dominance testing." Fighting ability, per se, may be instrumental in attaining dominance, and gluttony may relate to "causal" motivations in food competition, but neither can be equated to dominance. Regardless of what measure we use, we may only hope to relate the outcome of our measurements to the relationships which exist among the subjects. The dominance / subordination dimension is only one relationship which may account for the outcome of a particular test series.

Dominance as a Relationship

Since dominance is a relationship, it may be presumed to influence many interactions, but since dominance is not the only relationship possible, it will not be the only factor influencing interactions. The problem, then, is to identify those interactions which are primarily influenced by this one particular relationship and to recognize the contexts in which other relationships may modify the effect of a dominance relationship. Should we fail to find any interactional outcomes which can be attributed to dominance, we must be prepared to accept the null hypothesis, namely, the possibility that dominance does not exist. To argue that we have only failed to produce a measurement instrument sufficiently sensitive to detect dominance is to accept dominance without evidence and to demand that others do the impossible—namely, prove that it does *not* exist. The burden of proof in science is always on the person advocating a theoretical position and proposing an hypothesis, not on critics to prove the person wrong.

If dominance is a relationship, what kind of relationship is it? What are the attributes of "dominant" individuals? Do these attributes express themselves in all relationships, including those with strangers, or are the attributes expressed only when two individuals with a previously established relationship interact? If we claim that an individual is "dominant" even as a solitary, or when interacting with another for the very first time, then we are treating dominance as a relatively fixed attribute and not as a relationship. There are, doubtless, attributes of an individual which correlate with dominance status in a relationship, but we must not confuse the correlates with the theoretical variable.

Dominance Ranks

To demonstrate that dominance is not a permanent attribute of an individual, it is only necessary to consider the dominance relationships in a group of three. In any trio exemplifying dominance relationships, there will be at least one who is dominant to one and subordinate to another. (In nontransitive relationships all three may share this property, for the first will dominate the second, which dominates the third, which dominates the first.) If we recognize a midranking position, then we are saying that this individual is either a dominant or a subordinate depending on with which group member he interacts. Furthermore, if we remove the top-ranking individual, our midranking subject is now the "most dominant" in the group; if we replace the top-ranking individual and remove the lowest ranking individual, our subject becomes the "least dominant" member in the group. The same manipulations with a much larger group would repeatedly transform the same individual into an "alpha" and then into an "omega" with regard to dominance rank. This should be ample demonstration of the fact that dominance rank is the property of an individual only within the context of a specific social organization; it is not a permanent characteristic of the individual.

Rankings of individuals may be generated through a variety of means. Some are subjective, but, nonetheless, they may be highly reliable, producing good agreement among multiple raters. Other systems are based on objective measurements of performance on scales which have good face validity because they relate to functions which are of theoretical importance, or to outcomes which we feel, intuitively, must reflect dominance. No one would deny the biological importance of reproductive success, or the strong sexual motivation of many individuals, or the reinforcement which is associated with sexual performance. There are, nonetheless, some who would dare to suggest that dominance relationships may not absolutely determine sexual behavior; that reproduction can take place without a dominance hierarchy and that since it requires two individuals, both may contribute to success even though there may be a disparity in dominance rank.

In fact, social orders have been generated based on a number of diverse forms of social interactions. Some of these orders are relatively stable over time; others are not. Some of these rank systems correlate significantly with one another, but to varying degrees, whereas others do not. Some social rank systems may reflect dominance relationships; others may reflect different, but no less important, social relationships. Social relationships have been described in a wide variety of animals, and dominance relationships have been described in a host of vertebrate forms. Birds and socially living mammals have received particular attention. Both are complex organisms wherein multiple factors may be demonstrated to influence any particular behavior. Primates have received considerable attention due to their complex social organizations, the plasticity of their behavior, and the inclusion of our own species within this order. An analysis of the stability of selected social hierarchies and a review of the concepts of status which have been applied to primates was published by Bernstein (1970a). The conclusion drawn was that there were multiple social hierachies of unequal stability and independent causation and function. Different measures of dominance also produce different, albeit significantly correlated, rank systems.

Aggression and Dominance

If we return to the original peck-order measures of dominance as the basic and most direct measure of the phenomenon, our attention focuses on aggression, which is itself a hypothetical variable. Aggression must be treated as an independent variable distinguishable from dominance, for if dominance is a relationship, it cannot be measured by aggressivity per se. We must be prepared to find aggressive individuals who are not very successful in their aggression and aggressive individuals who are not necessarily dominant in a relationship. Aggressive expressions will surely be influenced by dominance relationships, and some aspects of their relational expression may be used as measures of dominance. It is, therefore, necessary to hypothesize the particular aspects of aggressive behavior that will reveal dominance relationships and to insure that these measures are compatible with our definition of dominance as a relationship.

First, if dominance is a relationship between individuals, then the mere frequency of an individual's aggressive acts, summed over all possible partners, will not necessarily relate to a particular dominance position in a group. The frequency of aggressive expression may relate to aggressive tendencies and to the fact that there are available targets for such activities, but unless aggression is a uniform response to random encounters with subordinates, the frequency of aggressive acts will not be a direct function of the number of subordinates present.

We may, nonetheless, expect that the top-ranking subject in a group of three, with twice the number of available targets, will have higher scores for aggressive responses than will the second-ranking individual. Naturally, we would not predict twice as many, and just as naturally, we would not expect to differentiate between numbers 33 and 34 in a group of 50 based on their relative frequencies of aggressive expression. In a large social group complex relationships among individuals may inhibit aggressive expressions directed toward some individuals by other particular individuals, and these relationships may produce "scapegoating" by redirecting aggression to targets more acceptable to the group. Altmann (1962) has analyzed possible strategies of response to aggressive signals in rhesus monkey groups, and it is clear that these alternative strategies will also influence the absolute levels of aggressive frequencies directed toward or expressed by, any one individual. Since aggressive expression may be influenced by multiple variables, only one of which is the particular dominance position of an individual, it is not surprising that some individuals "pick on" others and show disproportionate distribution of their aggressive acts among members of the group. The total frequency of aggressive expression will, therefore, not accurately reflect the relative ranks of individuals in a large and complex social organization.

Likewise, the frequency with which an individual shows aggressive behavior toward another need not be a direct reflection of the magnitude of difference in their dominance ranks. Dominance ranks refer to an ordinal scale rather than a ratio scale. Thus number two is not twice as dominant as number four, and the gap between any two rank positions is variable. To assume that this variable separation is directly related to the frequency of expressed aggression is to ignore the relational properties of dominance. To say that a group has a single dominant with all of the others as subordinates is not to deny that the others ever express aggression among themselves. To identify three levels of dominance status, such as employer, straw boss and workers, is not to say that the distance between employer and straw boss is the same as between straw boss and workers. To recognize the existence of fighting among the members of a group is not equivalent to differentation of dominance relationships.

If one individual in a pair is consistently the attacker, that individual still may not be dominant. After all, if the recipient always fights back, and is often the winner in these encounters, might we not ask who, if anyone, was the dominant? Even were the recipient unsuccessful in his own defense, so that we could predict the outcome of future encounters as a consequence of this observed regularity, we might still challenge the application of a dominance label to the winner, as long as the loser continued to resist each attack vigorously. Such resistance might provoke more frequent attack or might inhibit future attack, but the frequency of such attack will not relate to dominance separation in any simplistic fashion.

Although we might assume that the frequent attack of one individual by another reflected the degree of impunity with which the attacks could be launched, we might just as easily argue that frequent attack reflects the degree of challenge that the recipient represents. Both complete submission and vigorous challenge in unsettled dominance relationships may thus both be offered to explain high frequency scores. The reverse arguments would then apply to low frequency observations.

In fact, in my studies of rhesus monkey groups we have seen pairs of animals in adjacent rank positions which rarely exhibited aggressive responses directed toward one another, although the relationships between the pairs was often very different. Bernstein and Draper (1964) reported an almost complete lack of agonistic behavior when a single adult male made undisputed claim to the top-ranking position when introduced to a group of juvenile and adolescent monkeys. On the other hand, in other situations we have seen two animals of apparently very similar abilities occupy adjacent ranks with little overt conflict. In these cases we often inferred an alliance between the two in that they mutually supported each other in agonistic encounters with third parties. Such mutual support and toleration of one another may stem from a variety of causes. A male in sexual consortship with a female may be expected to inhibit aggressive expression directed toward the female, at least during courtship. A mother may be expected to defend her young, and monkeys of adjacent rank may, therefore, be members of a single matriarchy with all of the complexity that such relationships involve revealed in aggressive expressions as well as other social interactions. This is not to say that family members will not fight among themselves, but only that special rules of tolerance apply to family members and group members of certain age, sex, and physical condition categories.

Despite all of this, dominance relationships are dynamic rather than static, and relative positions do change. Changes will require one individual to challenge and bypass another, and these challenges will surely be reflected in agonistic encounters. Successful and unsuccessful challenges and interactions which assert rank or reinforce rank relationships may all be expected to include aggressive components. Inasmuch as individuals of the next adjacent ranks in a hierarchy represent the most likely sources of challenge, then we may expect a disproportionate number of aggressive episodes to occur among group members of similar ranks. If aggressive responses are plotted on a matrix consisting of group members ordered by dominance rank, we may calculate the relative frequencies of aggressive responses directed toward members of N-1 rank position, N-2, N-3, and so on. Such a matrix analysis done with monkeys does reveal a disproportionate clustering of scores along the diagonal (N-1), but due to the factors previously discussed, we cannot expect uniform linearity in the distribution of scores as a function of rank distances. Monkeys of much inferior dominance rank may represent little threat to the dominance status of an individual, but they may also be scapegoated. Individuals just inferior to an individual represent the most likely challenge for an occupied rank position, but they may also be in special relationship to the rank holder as in sibling or alliance relationships.

During periods of social change upwardly mobile individuals may incite repeated attacks by their dominance superiors, and this increase in attacks may be the first visible sign of a challenge. Especially in the study of animal groups, the subtlety of signals alerting superiors to a potential challenge may escape the investigator so that only the response to the challenge is noted. We thus pick up on unstable relationships only because of the noted increased vigor of the dominant's activity in dealing with the subordinate.

Submission and Dominance

Rowell (1973) has suggested that we focus on submissive rather than aggressive responses in our studies of dominance inasmuch as aggression may relate to a dominance assertion, a challenge, unstable or unsettled dominance relationships, or even competition and fighting provoked by other mechanisms which may not be related to, or even influenced by, dominance. Aggressive signals may be elicited in defense of the young, in response to predators, and in other self-defense situations where one would hesitate to invoke the concept of dominance. Yet by talking of dominance rather than subordinance relationships, we have focused our search for measures on active assertive interactions rather than the signals of submission. Without one individual losing or submitting, however, how can we call the other a winner or a dominant? Just as winners and losers exist in complemental relationship, dominants and subordinants can also only be identified by their interactions. As with fighting wherein a winner cannot be recognized until the loser is identified, so also in dominance relationships where it is the signals of submission by the subordinate which most clearly identify the dominant.

A dominance relationship between two individuals is inferred not because one or both "assert" their dominance, but because one readily submits. If, and only if, the subordinate recognizes the relationship, or "predicts" the outcome of an agonistic encounter by immediately showing submission, can we assume that a dominance relationship exists. Relationships may be established only after a prolonged contest or may be immediately recognized such that one individual accepts subordinate position on the very first encounter, sometimes even in anticipation of an aggressive signal. If we can recognize size, age, and bearing characteristics as correlating with rank, might not an animal respond to these same attributes? It is, therefore, the timing and sequencing of submissive signals in an interaction that allows us to infer the existence of a dominance relationship between two individuals.

It is only the submission of subordinates that allows us to argue that dominance may function to partition resources or reduce fighting. Such subordination may be a resultant of a history of unsuccessful agonistic encounters or may result from "prediction" of failure based on the attributes which we too may hope to recognize as correlates of rank. These attributes may be (1) physical, such as size, sex, maturity; (2) social, such as kinship relationships, alliance associations; or (3) individual behavioral, such as the vigor and apparent confidence of the opponent. Dominance may or may not function to partition resources and may either reduce competitive fighting or cause it in dominance contests; but regardless of what functions are served, dominance relationships will be revealed to us only in the patterns and sequences of observed interactions. The utility of the dominance concept will be in our improved ability to predict the outcome and nature of social interactions. If the dominance concept fails to improve our abilitiy to predict future behavior based on a presumed knowledge of established relationships, then it will have little scientific value and will remain an essentially untestable hypothesis. When our predictions fall to chance levels, it is wiser to abandon the concept than to retain it as a description of unmeasureable relationships. A difference, which makes no difference, is no difference.

If dominance is a relationship which is established on the basis of interactions, can we not talk of learned relationships and generalization such that one might learn to be subordinate in all situations? Such behavior would not require recognition of another, nor any special history of interaction with a specific partner. In fact, neither the highest nor lowest ranking members of a group need to identify the partner before choosing appropriate role behavior, so long as they identify the partner as a group member. On the other hand, if the encounter were with a nonmember of the group, both might attack the intruder, who might likewise attack, and we would not be able to describe relative rank vis-à-vis the intruder, *regardless of the outcome of the encounter*. Only if future encounters are modified by the outcome of the initial encounter will we be able to detect the establishment of a relationship. If no changes can be detected in subsequent encounters, we are not justified in claiming that a dominance relationship was established in the first encounter. Likewise, if two omega animals meet and both immediately and simultaneously submit, we will not be able to describe relative ranks nor even to infer that a relationship, vis-à-vis dominance, exists. Neither aggression nor subordination in and of itself is sufficient to define or determine dominance.

On the other hand, when two individuals meet for the first time, one may recognize in the other the attributes usually associated with individuals of superior rank and yield immediately. A juvenile may defer to an adult on the very first encounter. Dominance in these cases may be established with the very first glance, *but it did not exist prior to the encounter*. The investigator may likewise have been able to predict the outcome of an aggressive encounter before it occurred, but in this case the prediction is based on attributes that correlate with relative rank rather than on the inference of an established relationship. The successful prediction of the "winner" is not contingent upon immediate submission, and the eventual loser may not immediately attempt to avoid the encounter. On occasion our predictions of "winners" and "losers" based on attributes which correlate with rank at less than 1.0 will fail, thus serving to remind us that correlational measures are not identical to the items with which they correlate.

Applying the Concept of Dominance

Inasmuch as many social interactions may be demonstrated to be ordered in some fashion and inasmuch as many measures which have been theorized to be related to dominance ranks have correlated with one another poorly, at best, some authors have suggested that dominance and the concept of status hierarchies are not useful descriptions of social organization or relationships. Alternatively, other authors would recognize each proposed measure of dominance as valid and differentiate multiple types of dominance in recognition of the low correlations among measures. Thus, we would talk of "breeding dominance," "feeding dominance," "agonistic dominance," and "water dominance," to name a few. The term "dominance" is thereby reduced to synonymy with rank orders, and we might just as well talk about "weight dominance," "age dominance," or any other sort of dominance when we identified a dimension along which subjects could be scaled.

Differentiating dominance into as many types as there are hypothetical measures relieves us of the necessity to discard some measures as inappropriate to this particular concept and is an excellent compromise to avoid antagonizing friends and colleagues who have strongly identified with one theoretical position or another. We simply agree that we are all correct—that there are many "status" hierarchies, all of which are reflections of dominance rank. We likewise avoid having to explain why some theoretically important functions or causes of dominance fail to operate as predicted when we use certain measures as criteria. Obviously, they *do* function as predicted if we merely search for a measure which will yield the predicted result and then identify *this* measure as the aspect of dominance with which we were concerned.

The term "dominance" is thereby reduced to a meaningless label affixed by each investigator to a particular measure suited to a pet set of theoretical postulates. When we relate the same label to a plethora of concepts and measures, we confuse it until we communicate nothing by its use. *The abuse of the label does not invalidate the concept to which it originally applied.*

Dominance, like territoriality, is a good term which has all too often been misapplied and abused. Like many exciting concepts, it has been inappropriately extended to peripherally related phenomena and to nonrelated but analagous processes. To challenge some usage, however, is not to challenge all usage. Our complaints with misapplication of the label are not a criticism of the concept, and substitution of a new label will only invite new inappropriate use plus confusion with the discarded label. We must not retreat from our confusion by discarding the baby with the bath water.

Dominance is a hypothetical variable to be invoked when we can demonstrate that its usage permits more accurate prediction of the outcome of social interactions than will any other hypothesis invoked in the same situation. As such, dominance relationships can never be assumed; they must be demonstrated in each application. The utility of the concept will be dependent on its consistent and unitary usage.

Correlations among competing theoretical measures are interesting, much as are correlations among independent variables, but they are not crucial tests of the concept. The test of a hypothetical variable lies in the accuracy of predictions. Explaining the failure of a hypothesis to predict outcomes because of "unstable relationships," "competing motivational systems," or similar plausible excuses does not prove the hypothesis. Until we can demonstrate accurate predictions based on the hypothesis, we must accept the null, that the hypothetical variable has no detectable influence on the behavior being measured. Insofar as predictions are concerned, an unpredictably reversable relationship is equivalent to a lack of relationship.

Dominance is not a universal to be assumed as a principle of all social interaction and social organization; it is a hypothetical variable to be demonstrated in each taxonomic group, in each subject population, and in each social organization. Failure to demonstrate a role relationship of subordinate and dominant is not a signal to substitute another form of social order and its measure for a "dominance" measure. Other mechanisms of social ordering undoubtably exist and should be identified. These mechanisms may intercorrelate and reflect the influence of basic principles, but they need not be collapsed into a unitary system.

Many measures that are more readily obtained and objectively measured may correlate highly with dominance relationships. These measures may be extremely useful in developing hypotheses of dominance relationships, especially in quiet and well-established groups, but in every case it will be ultimately necessary to demonstrate the existence of dominance relationships by more direct measurement. No matter how well body weight may correlate with dominance rank in a monkey group (.85 and better reported in some laboratory studies), rankings by weight will not demonstrate a dominance hierarchy in elephants, pigeons, men, or even another monkey group.

Let us, therefore, be extremely cautious in the application of the concept of dominance in our studies of social groups. No matter what measure we use, no matter what group we study, we must first demonstrate the applicability of this hypothetical variable to our particular study group in describing relationships among group members with regard to a dominance dimension.

Chapter 4

Aggressive and Competitive Social Structures in Captive Monkey Groups

F. F. Strayer
Mark S. Cummins

A subadult male strides confidently toward a presenting adult female, he mounts, and the two begin to copulate. At the same moment a young female threatens, chases, and then bites a young male. Finally, a year-old infant pushes and slaps at his mother's arms apparently to gain access to her nipple. The first interaction is a clear example of sexual behavior, the second is certainly aggressive, and the third seems to entail a typical type of mother–infant conflict. But was social dominance involved in any of these episodes?

The concept of dominance has played an historically central but continuously controversial role in the analysis of social behavior among nonhuman primates. The conceptual and methodological problems inherent in many of the early accounts of social dominance became more and more evident with the increase in information concerning social dynamics within stable primate groups (e.g., Altmann, 1967; DeVore, 1965; Jay, 1968; Jolly, 1972; Kummer, 1971). In Chapter 3, Bernstein provides an excellent summary of the major sources of conceptual confusion that have hampered dominance research. His analysis of these problems leads him to the conclusion that dominance should be reserved as a general descriptive term for relationships between members of a stable group. Even with this restricted definition, however, problems remain concerning the appropriate methods for evaluating such social relationships.

Two quite different approaches have been commonly employed to determine dominance relations within monkey groups. The first approach evaluates relative dominance by examining the success of individuals during arranged competitive interaction. Such competitions are usually conducted within the normal group setting and provide a relatively efficient and standardized procedure for determining social rank. The second approach places greater emphasis upon the observational analysis of naturally occurring social conflict. From this perspective, a dominance relationship is defined in terms of characteristic roles that are assumed by two individuals during the resolution of both aggressive and competitive dyadic exchanges.

Although each of these approaches have led to the identification of linear dominance rankings within numerous monkey groups, researchers have only just begun to examine systematically the potential discrepancies between results obtained from the two procedures. Preliminary evidence suggests that a variety of ecological and social factors (i.e., type of habitat, distribution of resources, species differences, group size, group composition, and so forth) may influence the extent to which both approaches yield a similar or identical assessment of social dominance (Christopher, 1972; Richards, 1974; Smith, Harris, and Strayer, 1977). However, since these initial methodological comparisons have dealt exclusively with the final group dominance rankings obtained from each procedure, they have not examined the relative usefulness of each procedure for the assessment of dyadic dominance relationships. The present chapter provides an illustration of both approaches to dominance assessment as well as a summary of analytic techniques that permit a more detailed evaluation of both dyadic relationships and social structure within human and nonhuman primate groups.

Social dominance was evaluated in five captive monkey groups. Groups 1 and 2, each comprised of 16 subadult and juvenile *Macaca nemestrina*, were housed in adjacent colony compounds at the University of Washington Regional Primate Center in Seattle, Washington. Both groups contained an equal number of males and females. Groups 3 and 4 each consisted of 7 *Saimiri sciureus*. Group 3 contained 2 adult males and 5 adult females. Group 4 contained 2 adult males, 3 adult females, 1 juvenile male, and 1 juvenile female. Each group lived in large colony compounds at York University in Toronto, Ontario. Group 5 included 57 *Macaca mulatta* housed in a large outdoor corral maintained by the Wisconsin Regional Primate Center in Madison, Wisconsin. It was comprised of 30 adult animals (5 males and

25 females) and 27 juveniles and infants. All five groups were formed well in advance of the observational period for the assessment of social dominance (at least four months for each of the macaque groups and nearly two years for both saimiri groups).

Analyses of Social Dominance

Dominance among Macaca nemestrina

The evaluation of social dominance for Groups 1 and 2 entailed direct observation of spontaneously occurring social conflict. Two general categories of interaction were distinguished. The first category entailed overt aggressive behaviors (i.e., assault, bite, push–pull, pout–threat, lunge–threat, chase) that led to clear submissive responses by the social target (i.e., rapid flight, fear–grimace, cringe, withdraw). The second category of social conflict included a variety of more diverse forms of activity observed during competitive interactions for the possession of an object or place. This latter type of interaction was coded only when one of the two animals engaged in the competition was clearly able to maintain priority of access. Observations were conducted with each group during a three-week period from inside a visually isolated enclosure.

A total of 632 conflict interactions were recorded for Group 1, and 513 for Group 2. Nearly 40 percent of these interactions were coded in the competitive category for both groups. Approximately 35 percent included forms of aggressive threat that produced submissive gestures but did not lead to physical contact between the two participants. Finally, about 25 percent of the observations in each group entailed more severe forms of physical attack, which characteristically produced the most intense form of submissive reaction and often seemed to inflict severe pain.

These observations are summarized for Group 1 in the two dyadic matrices shown in Figure 1. These matrices were constructed to reveal both dyadic dominance relationships and the group dominance hierarchy. As a first step, all animals were rank ordered according to their total frequency of aggressive or competitive successes and listed in a preliminary matrix as both initiators and targets. Following a systematic analysis of dyadic wins and losses, rows and columns in this preliminary matrix were rearranged to minimize the number of observations that fell below the diagonal. The second rank ordering evident in the final

Aggression

INITIATOR \ TARGET	1	A	2	3	B	C	4	5	D	6	7	E	F	G	8	H	total
1		10	26	15	6	3	3	1	2	2	0	5	5	11	0	2	91
A			12	19	2	9	11	2	4	4	4	2	1	3	3	2	48
2				16	4	2	0	2	1	2	0	1	0	4	0	1	33
3					5	13	4	2	4	3	4	4	1	1	0	1	42
B						2	5	0	1	2	4	3	0	1	1	1	20
C							6	1	5	7	1	0	1	3	2	3	24
4								1	4	0	1=	0	6	14	1	0	27
5									1	0	0	0	0	0	0	2	3
D										2	6	2	0	4	2	7	23
6							1↕				0=	1	0	9	0	3	14
7							1=			0=		2	0	1	↕	2	6
E				1									1	2	0	1	5
F														1	0	13	14
G															0=	1	1
8											2↕			0=		5	7
H										1							1
total	0	10	38	51	17	29	31	9	22	23	22	20	15	54	9	44	394

Competition

INITIATOR \ TARGET	1	A	2	3	B	C	D	6	7	E	4	5	F	G	8	H	total
1		21	30	14	3	0	0	1	1	0	4	2	0	3	0	3	82
A			8	8	4	6	1	1	1	1	4	3	0	4	0	1	42
2				19	2	5	0	1	0	0	3	1	1	1	0	2	37
3					3	2	2	↕	0	1	2	0	0	1	1	0	12
B						5	2	0	1	2	2	0	0	0	1	2	15
C							2	3	0	3	2	1	0	1	0	1	13
D								0=	1	3	0	0	0	3	0	3	10
6				1↕			0=		0=	1	0	0	2	3	1	1	9
7								0=		0=	0	0	0	0	0	0	0
E									0=		1	0	0	1	1	0	3
4												2	0	1	1	1	5
5													0=	1	0	0	1
F												0=		1	0	2	3
G															4	1	5
8																1	1
H																	0
total	0	23	38	42	12	18	7	6	4	11	18	9	3	20	9	18	238

Numbers = ♂ Letters = ♀ = = Tied relation ↕ = Inversion

Figure 1. Naturally occurring dyadic conflict among *Macaca nemestrina*.

matrix yields the set of relationships which provides the best approximation of the general linear model of social dominance. An analysis of the observations which remain below the diagonal in the final matrix provides information concerning both the rigidity (asymmetry) of dominance relationships within the group, and the linearity (i.e., transitivity) of the group dominance hierarchy.

The observations below the diagonals in Figure 1 can be divided into three different classes. The first class consists of occasional instances in which the usually subordinant member of a dyad wins in a particular conflict episode. For the aggression matrix there are only two examples of this type of episodic reversal (E-3 and H-6). In the competitive matrix there are no examples of relationships with episodic reversals (2-A). A second type of observation below the diagonal results from the assignment of an individual to a position in the hierarchy below another individual whom he successfully dominates. Such assignments are necessary only when the dominance hierarchy is not perfectly linear, that is, when a circular relationship exists between three or more individuals. In the aggressive matrix from Figure 1, two instances of such relational reversals are evident (6-4 and 8-7), while in the competitive matrix there is only one reversal (6-3). Finally, a third set of observations entails instances in which both members of a dyad win an equal number of conflict encounters. Such observations reflect cases of *established egality* in dyadic dominance. For Group 1 the only instance of established equality is between 7 and 4 in the aggressive matrix. However, there are a number of other cases (marked with 0) in which one must assume that members of a dyad have an equal dominance relation because they were not observed to engage in conflict and because their relative dominance was not evident in comparisons with other group members. These cases necessarily involve individuals ranked in adjacent positions within the group dominance hierarchy. In both the dyadic matrices for this group cases of assumed equality were observed only between individuals ranked in the lower section of the group hierarchies (7-6 and 8-6 for the aggressive matrix and 6-D, 7-6, E-7, and F-5 for the competitive matrix).

The analysis of aggressive and competitive conflict for Group 2 revealed basically similar dyadic relationships and group status structures. The number of episodic and relational reversals for this second group were less than in Group 1, but again a number of equal dominance relationships were evident in the lower sections of the aggressive and competitive hierarchies. The extent to which dominance relations in Group 2 conformed with general expectations from the linear dominance model was also evident in the similarity of status rankings for the two conflict categories. This correlation was .996 for Group 2 as compared to .923 for Group 1. Thus, in both groups the evaluation of dyadic success in aggressive and competitive conflict revealed a coherent set of social relationships that corresponded well with expectations derived from the linear model of social dominance.

Dominance among Saimiri sciureus

Dominance assessment procedures for Groups 3 and 4 entailed only minor modifications of the descriptive coding system developed in the previous study with macaques. These changes included redefinition of specific activities and responses included in the category of aggressive conflict in order to accommodate species differences in social behavior patterns. Once again the observational procedures were designed to provide 30 hours of information concerning spontaneously occurring conflict in the normal group setting.

A total of only 71 aggressive episodes were recorded for Group 3, whereas 343 instances were observed in Group 4. Similar between-group differences were evident in competitive conflict, where 71 events were recorded for Group 3 and 160 for Group 4. The dyadic analyses of social conflict for Group 3 revealed a pattern of dyadic relationships that closely paralleled the finding from the macaque groups. Only one instance of a relational reversal was evident in the competitive matrix for this group and none in the aggressive matrix. The frequency of episodic wins by a lower ranking animal was very low for both types of conflict. Finally, the correlation between the two conflict hierarchies was .991 for this group of adult animals.

The analysis of social conflict within Group 4 revealed a surprisingly different view of dominance relationships and group structure. The observed conflict for this second saimiri group is summarized in Figure 2. The final aggressive matrix contained six instances of episodic reversals, with most episodes involving one of the two juveniles (3 and D). There were no relational reversals of dyadic dominance in the aggressive matrix and only one case of assumed equality (C-B). In general, the organization of aggressive conflict corresponded well with the analysis in the all-adult group; in addition, position within the Group 4 aggressive hierarchy followed age–sex class differences within the group.

The organization of relationships in Group 4 was quite differ-

Aggression

INITIATOR	TARGET: 1	2	A	B	C	3	D	total
1		3	1	0	3	0	7	14
2			6	6	2	19	43	76
A				2	6	19	56	83
B					0=	13	45	58
C	1			0=		27	72	100
3		1	1	1			4	7
D			4			1		5
total	1	4	12	9	11	79	227	343

Competition

INITIATOR	TARGET: D	3	1	2	A	B	C	total
D		21	3	0	51	0	5=	80
3	10		3	6	21	3	1	44
1				2	0	0	0	2
2	1	3	1		0=	0	0	5
A	1			0=		3	12	16
B	4↕	1					1	6
C	5=	1		1				7
total	21	26	7	9	72	6	19	160

Numbers = ♂ Letters = ♀ = = Tied relation ↕ = Inversion

Figure 2. Naturally occurring dyadic conflict among *Saimiri sciureus.*

ent for competitive conflict. In the second dyadic matrix from Figure 2 both juveniles ranked higher than all adults. This second category of conflict also provided a coherent view of dyadic relationships and a linear group status hierarchy. Eight instances of dyads with episodic reversals, one relational reversal, and one example of established equality were evident in this latter matrix. Finally, it seems important to note that in comparison to the three previous groups the degree of correlation between the rank orderings for Group 4 was extremely low (*rho* = .287).

Dominance among Macaca mulatta

In contrast to the previous assessments dominance relationships with Group 5 were evaluated using standardized competitive testing. With this procedure animals were ranked according to their ability to maintain priority of access to a drinking fountain following 24 hours of water deprivation. Two operational definitions of priority of access were employed in the original analyses. One entailed exclusive access during a 10-second period, while the second involved assessing cumulative drinking time. Since the measures provided virtually identical information, only data resulting from the first criterion will be reported here. Tests were administered once each week during a 6-month period. A single competitive test required a maximum of two hours, but the majority of tests were terminated whenever a 15-minute period had elapsed without further drinking by any animal who had not already reached criterion. This cutoff procedure yielded rankings for only those animals who most actively competed for access to the experimentally manipulated incentive.

The use of a repeated-measures design permitted evaluation of the stability of dominance rankings throughout the assessment period. The 24 testing sessions were divided into six blocks of four tests each. A monthly ranking was obtained for each group member in each of these six blocks by calculating median ranks on each of the four separate tests and reranking all animals in terms of their median score. Examination of relative positions within each of the six monthly rankings showed strong stability in status positions for adult group members (*rhos* > .89), but there was considerable variability among juveniles and infants. Because of the variability in competitive performances of younger animals, more detailed analyses of dyadic relationships and group structure were restricted to the adult animals in Group 5.

A second advantage of a repeated-measures design in standardized competition assessments is that it provides information that can be used for a more precise evaluation of consistent patterns in dyadic priority of access. Such patterns can be represented in a dyadic format and analyzed in a fashion parallel to that usually employed in research on naturally occurring social conflict. Thus, the use of this design for standardized competitive assessments overcomes the limitation of the more simple rank-order procedure that precludes an analysis of the dyadic relationships constituting the group dominance structure.

Dyadic relationships were examined by noting the consistency in monthly competitive episodes for all pairs of adult animals in Group 5. Figure 3 shows the frequency of competitive success for each member of all adult dyads. As with the previous dyadic matrices, the array represents the arrangement of animals which provides the best approximation of the linear dominance model. This matrix was derived from a preliminary array organized with animals ranked according to their median score for all 24 tests. Systematic analysis of observations below the diagonal resulted in the reordering of individuals to minimize episodic reversals of dyadic relationships. Although this matrix contains substantially more observations below the diagonal, it also contains a greater number of total observations. Thus, the overall rigidity of dyadic relationships is quite comparable to that reported for the other four groups. Figure 3 shows 19 dyads with equal dominance relationships. Since 10 of these dyads occupy adjacent positions in the dominance matrix, 9 reflect instances in

INITIATOR \ TARGET	1	A	B	C	D	E	F	G	H	I	J	K	L	M	N	O	P	Q	2	R	S	T	3	4	U	5	V	W	X	Y	total
1		5	5	5	5	5	5	5	5	5	5	5	5	5	5	5	5	5	5	5	5	5	5	5	5	4	5	3	4	3	139
A			4	4	4	3	5	5	4	5	5	6	6	4	5	4	6	6	5	6	6	6	6	6	6	5	6	4	5	4	141
B		2		4	5	4	5	5	6	6	5	6	6	6	6	6	5	5	6	6	6	6	6	6	6	5	6	4	5	4	148
C		1	1		3	3	3=	5	4	4	4	5	5	5	5	5	6	6	5	6	6	6	6	6	6	5	6	4	5	4	130
D		2		2		3	3=	5	5	4	5	5	5	5	6	4	5	5	6	5	6	6	6	6	5	5	6	4	5	4	128
E			1	2	2		3	3	2	3	5	5	5	3	5	3	5	5	5	4	5	5	5	5	5	4	5	3	4	3	105
F		1		3=	3=	2		4	5	4	5	5	5	5	6	3=	5	5	6	5	6	6	6	6	5	5	6	4	5	4	125
G		1	1	1	1	2	1		4	4	5	6	6	5	6	4	5	6	6	6	6	6	6	6	6	5	6	4	5	4	124
H		1		2	1	2	1	1		3	3=	5	5	4	5	3	5	5	4	4	5	5	5	5	5	4	5	3	4	3	98
I		1		2	1	2	2	2	2		3	3	3	3	3	4	4	4	4	4	5	5	5	5	5	4	5	4	5	4	94
J		1	1	2	1		1	1	3=	2		4	4	3=	4	4	5	6	4	5	6	5	6	6	6	5	6	4	5	4	104
K					1		1		1	2			2=	3=	3	4	5	5	4	5	6	5	5	6	6	5	6	4	5	4	88
L					1		1			2	1	2=		3=	3	4	4	5	4	4	6	5	5	5	6	5	6	4	5	4	85
M		2		1		1	1	1	1	2	3=	3=	3=		3=	4	5	5	6	4	6	6	6	6	5	5	6	4	5	4	98
N		1		1						2	1	1	2	3=		3=	5	4	5	4	5	4	5	5	5	5	6	4	5	4	80
O		1		1	2	1	3=	2	1	2	2	2	2	2	3=		4	4	5	4	5	6	6	5	5	5	5	4	5	4	91
P			1		1		1	1	1	1	1	1	2	1	1	2		3=	4	3	5	4	3=	3=	6	4	5	4	5	4	67
Q			1		1		1		1	1		1	1	1	2	2	3=		4	3	5	5	5	4	6	5	6	4	5	4	71
2		1		1				1		2	2	2	2		1	1	2	2		3=	4	4	3	3	4	4	6	3	5	4	60
R					1	1	1		1	1	1	1	1	2	2	2	2	2	3=		4	5	4	3=	6	3	5	4	5	4	64
S															1		1	1	2	2		4	4	3=	5	5	6	4	5	4	47
T									1	1	1	1	1		2	1	1	1	2	1	2		3	3=	4	4	5	4	5	4	47
3												1	1		1		3=		2	2	2	2		4	4	4	6	4	5	4	45
4													1		1	1	3=	2	2	3=	3=	3=	2		3	5	6	3	5	4	47
U					1		1							1	1	1			2		1	2	2	2		3	4	4	4	4	33
5																				2		1			2		4	2	3	3	18
V																				1		1			2	1		2=	3	2	13
W																			1					1		1	2=		3	3	11
X																									1	1	2	1		2=	17
Y																											1	1	2=		4
total	0	20	15	31	34	29	39	41	47	56	57	70	73	64	80	70	96	92	102	97	116	118	115	115	130	116	149	101	132	107	2312

Numbers = ♂ Letters = ♀ = = Tied relation

Figure 3. Dyadic representation of priority of access during standardized competititons among *Macaca mulatta*.

which an animal lower in the hierarchy maintains an equal dominance relationship with a higher ranking group member. Finally, since no relational reversals are evident in Figure 3, all observed relationships conform to the linear model.

Conclusions

The present application of dominance assessment procedures provided directly comparable information for the evaluation of dyadic dominance relationships and group status structures. Table 1 presents descriptive statistics which summarize these two levels of social analysis for each of our five groups. The observed levels of rigidity and linearity for the social dominance assess-

Table 1. Comparison of Dyadic Relationships and Dominance Structures in Five Captive Groups of Monkeys

		N	*Established Dyadic Relationships (%)*	*Rigidity of Dyadic Relationships (%)*	*Linearity of Group Structure (%)*
Macaca nemistrina					
Group 1	Aggression	16	80	99	98
	Competition	16	63	99	97
Group 2	Aggression	16	83	99	100
	Competition	16	74	97	99
Saimiri sciureus					
Group 3	Aggression	7	86	92	100
	Competition	7	62	97	100
Group 4	Aggression	7	86	97	100
	Competition	7	67	85	97
Macaca mulatta					
Group 5	Dyadic analysis	30	100	87	100
	Median analysis	30	—	—	100

ments in each of the groups provide an excellent demonstration of cross-species similarity in the organization of aggressive and competitive activity within stable primate groups. The similarity of dominance structures is most clearly summarized in Figure 4, which portrays hierarchical networks for each group as well as established equality of dominance and reversals of linear dominance relationships. These networks provide a graphic representation that hierarchical dominance structures are more complicated than simple rank orderings of individual group members.

This final figure also demonstrates the extent of correspondence between social structures based upon qualitatively different forms of social conflict. The two networks constructed for each group provide essentially similar information on the structural organization of dominance relations. The single major exception to this general pattern occurred with Group 4, where competitive and aggressive social structures were not significantly correlated. A closer examination of differences between the two conflict networks shows, however, that among adult group members dominance relationships are virtually identical. Furthermore, both examples of dyadic relationships which fail to accord with a linear competitive ranking involved relationships between the

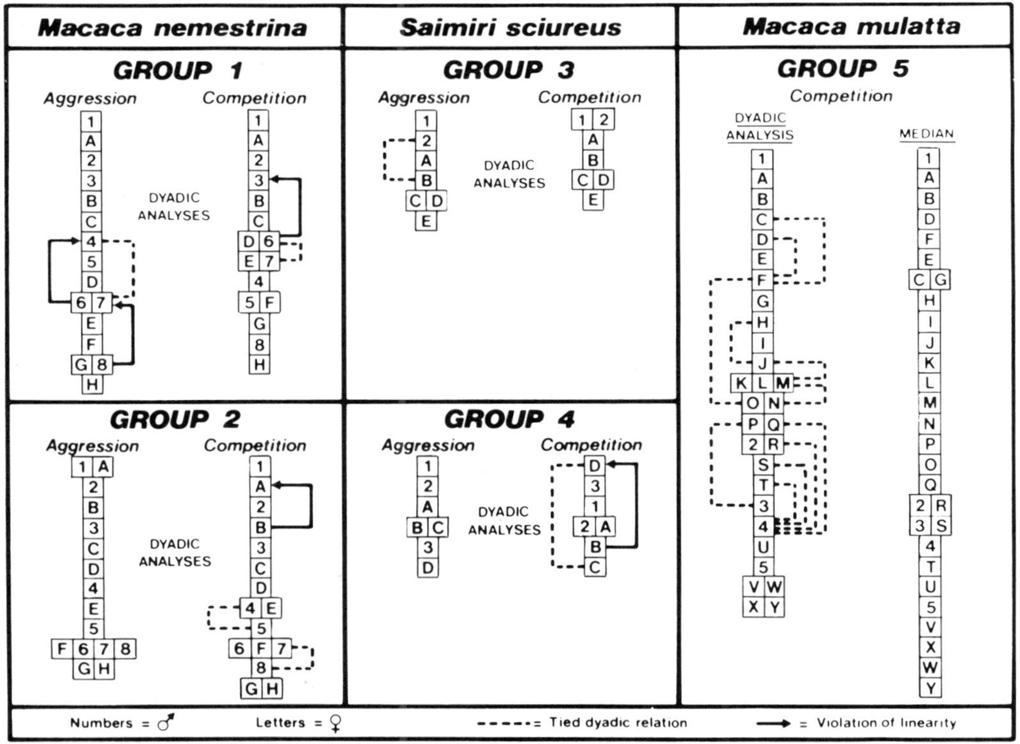

Figure 4. Hierarchical representation of status structures in five captive monkey groups.

higher ranking juvenile and the two lowest ranking adult females. Thus, it seems clear that an apparently unrelated demographic feature of the group may play a major determining role in the coordination of relationships for the resolution of social conflict and influence the emergence of the group dominance structure.

Such apparent anomalies in the integration of particular individuals into a stable group structure may reflect important aspects of social coordination that are necessarily unrelated to social dominance. If parents always successfully competed with their offspring for scarce resources, the ultimate consequences could be disadvantageous for the reproductive fitness of both individuals. Therefore, from an evolutionary perspective we should expect adults to tolerate certain age-related privileges which permit young animals to display the apparently unorthodox disregard for normal dominance relations necessary for success in competitive contexts. Thus, returning to our earlier example, the young infant that successfully gains access to his mother's nipple, in spite of her efforts to wean him, probably provides an example of temporal coordiantion between individual needs rather than an example of social dominance interaction.

Chapter 5

Chimpanzee Society and Social Dominance in Evolutionary Perspective

Harold R. Bauer

This chapter has two objectives. The first is to familiarize the reader with recently discovered aspects of the common chimpanzee's social behavior and social organization that add to the chimpanzee's comparative importance for the understanding of human origins in evolutionary processes. For example, the evolutionary import of kinship ties for the understanding of the primate societies is emphasized by the knowledge that Gombe National Park (Tanzania) chimpanzees are regularly in temporary aggregations, or parties, that often consist of related adult males or mothers and their dependent offspring. These party members regularly split up and come together in new parties within a group of community members, but long-term relationships associated with kin are a common denominator, along with occasionally sexually attractive females, for understanding the cohesion of community members. Human hunter-gatherer societies also regularly split up to forage and occasionally collect to hunt or eat meat in much the same way common chimpanzees do. This life-style of humans has persisted and predominated most of our evolution (Lee, 1968; Lee & DeVore, 1976). The second objective of the discussion is to study party size as it relates to social dominance in adult males of the main (Kasakela) chimpanzee community in the Gombe and then consider the result from a comparative and evolutionary perspective. Both chapter objectives leave human behavioral evolution for the consideration of the reader.

The data and methods in this report grew out of a 27-month field study between July 1970 and November 1972 in the Gombe National Park, as well as observations extended to May 1975 of many Tanzanian and visiting colleagues (Bauer, 1977a).

To make adequate comparisons between ape and human behavior and ecology, the psychological and biological variability within and between wild populations must be known. However, this knowledge generally does not exist. For instance, the social behavior of the common chimpanzee *(Pan troglodytes)* and the pygmy chimpanzee *(Pan paniscus)* differs remarkably. Since social behavior is most often correlated with social organization and ecology, it is likely that these two species of chimpanzee differ along these dimensions. But as yet this has not been verified. If we are to collect data on psychological and biological variability for solid comparative studies, the conservation and preservation of Great Apes and their habitats is essential throughout their distributions. The expansion of human populations and destruction of critical habitats currently taking place should make species conservation and habitat preservation the concern of everyone who has an interest in the subject of human evolution.

Chimpanzee Society

The discoveries by Goodall (1965) of chimpanzee tool making, hunting, and family structure changed many established views on human evolution. The prime one was that humankind had to share the distinction of being a tool-making hunter with our closest relatives. The report of the early results by van Lawick-Goodall (1968) was remarkably thorough and accurate, although the data were primarily collected in or near an often overcrowded provisioning area.

The recent studies are based on extensive field observations during "follows" of generally habituated chimpanzees through their natural habitat. These subsequent observations in the savanna woodland habitat of Gombe confirmed both the earlier discovery of the basic social group of chimpanzees by Itani and Suzuki (1967) and the later modification by Nishida and Kawanaka (1972) as being a unit group, later termed "community" by Goodall (1975). This unit group, or community, is made up at any one time of individuals or parties of individuals that on an hourly or daily basis usually undergo fusion and fission, over a community range of sometimes several kilometers in diameter.

The entire community rarely, if ever, assembles at one time in the same place, but rather it can be found in small clusters feeding on patchily distributed food sources or involved in other routine activities such as sleeping or grooming.

Important sex differences in community identity exist. The adolescent or adult chimpanzee females may temporarily or permanently transfer their home range and community relations from their natal community to that of a neighboring community. In contrast, males never leave their natal community. From this general observation we can conclude that adult males in a community are closer kin than adult females and that the society is patrilineal.

The home range of one chimpanzee community overlaps with that of adjacent chimpanzee communities, although neighboring communities avoid each other. Scouting behavior and violent attacks with cannibalism have been observed on the periphery of a community's range between members of neighboring communities, but the latter does not occur on a regular basis nor at a boundary over the defense of a specific area (Bauer, 1977a). Since community members do not defend what appears to be a specific area but have a fear of strangers and a familiarity with an overlapping community range, chimpanzees are not territorial in the most common sense of the concept (Carpenter, 1958). These intercommunity attacks are based on a variety of different observations (Bygott, 1972; Goodall, 1977).

Social greetings after regular separations of members of the chimpanzee community are called reunions since many different kinds of social behavior occur in them. A most obvious kind of social reunion is one involving agonistic behavior, which refers to an interrelated set of behavior patterns preceding or substituting for attack. Adult males are typically the aggressors threatening subordinate males and adult females (Bygott, 1979). Sexual receptivity of a female, however, appears to moderate these meetings into opportunities to socialize. Attacks are rare and generally very brief between members of the same chimpanzee community. Impressive displays involving charging and branch waving usually suffice to threaten submissive chimpanzees. As noted, the severe and semilethal attacks occur when strange male chimpanzees, on rare occasions, meet a member of an adjacent community.

Reunions and parties occurred for social rather than ecological reasons. Reunions of adult males were times when significant ($p \leq 0.05$) and sudden peaks in agonistic behavior occurred, but these behaviors were also followed by significant increases in social grooming (Bauer, 1976). Interestingly, individuals initiating the reunion were found to decrease their rate of feeding before the reunion, and this rate remained lower for over an hour after the reunion. These two patterns in rates of social and feeding behavior indicate that the formation of parties has a basis in sociality rather than an attraction to a common food source. Like chimpanzees, many human hunter-gatherer societies also disperse to find food and may group to hunt and socialize with kin and familiars (Lee, 1968; Lee & DeVore, 1976).

Although adult males gain attention by performing agonistic charging displays after a reunion, as suggested by Chance's (1967) attention structure theory, the weight of the evidence on social grooming contradicts his ideas. As attention theory would predict, reunions in which adult males did agonistic charging displays upon arrival were followed by these males more often receiving social grooming than nondisplaying males at these reunions, but most social grooming occurred without prior threats. In the forest, most displayers were not groomed afterwards compared to the provisioning area where loitering occurred leading to more grooming. Postdisplay grooming sessions were shorter than those occurring without prior agonistic behavior and most social grooming occurred without prior threatening, charging displays or milder threats (Bauer, 1979). In most instances and for the greatest durations, social grooming in five males occurred without prior threatening or "attention-getting" displays and often between close kin.

Social Learning

Experience during development is important for species with complex learning abilities because this experience tunes the organisms to a particular environment. What distinguishes humankind and the Great Apes from other primates is the long period of adjustment in infancy and juvenility. Wild chimpanzees achieve puberty so late that they are normally young teenagers before experiencing motherhood. The various forms of social experience can be described in terms of learning processes.

Habituation is the most elementary and universal learning process. It is a stimulus-specific response decrement over time and occurs when an initially strong response is not rewarded and thus wanes. In apes and humans habituation to innocuous stimuli provides one basis for understanding the development of complex attentional processes and mature perceptual abilities. Developmentally, it is remarkable that young infant chimpanzees in the natural habitat give species typical responses (e.g. orientation, hair erection, foot stamping, rocking, or branch waving) to a seemingly indiscriminate array of environmental stimuli ranging from their mother farting to the passing of a grasshopper, whereas older infants respond with social behavior patterns more specifically to chimpanzees. Adults are very discriminating in selecting stimuli to respond to. In this sense they are more predictable than young chimpanzees to experienced human observers and most likely adult chimpanzees. Observing a mixture of chimpanzee adults and juveniles is often like watching parents at a cocktail party with their children present. Their conversation is continually interrupted by their curious, exploratory, demanding, and playful offspring. The parents are earnestly engaged in one activity, while in the same time span their young juveniles may have engaged in a series of different activities. In this social context, recognition by the adults of what is unimportant based on previous experience is a most important basis for selective and active perceptual processes. The juvenile's habituation to objects engaging his activity paves a path of experience in play until the more earnest and awkward social processes of adolescence overshadow juvenility.

Many opportunities exist for classical conditioning processes to take place in young chimpanzees with their mothers in the natural habitat. Potent food stimuli first selected by the mother are in regular contiguity with associated stimuli that elicit likely ancilliary responses themselves. Food that an infant manages to grab from the mother might be associated with the routine context of the tree species, its seasonality, food abundance, food preference, food location, number of feeding sites, and subtle social signaling between adults over feeding sites. Such early experience is significant because feeding is by far the most common daytime activity. Classical conditioning may also be important to dominant males, who are often potent stimuli to mothers and their young, ventrally riding infants. These infants typically have a good learning environment from this safe ventral riding position. The mother's submissive vocalizations occur with pulsing, thoracic movements as the infant clings to her chest hair with hands and feet. The orientation of the mother's face to the dominant male is normally imitated by the young infant, which rides just under and behind the mother's head, as the mother submissively approaches higher ranking animals. The infant is thus normally in a great position to socially learn what is important.

Two learning processes which compliment each other in wild

chimpanzees are operant conditioning, requiring instrumental responses for rewards, and latent learning, gaining experience which may facilitate future rewards. Tool use and tool making are clear cases of chimpanzees operating on the environment. Specifically, the finding and refining of appropriate grass stems to use as termiting tools and the selection of termiting mounds and termiting holes indicate large individual and age differences, although generally adult females are more avid termite eaters than adult males (Bauer, 1977a). Many unsuccessful termite mound holes and tools are left by termiters after brief exploration, in contrast to certain successful termiting holes and tools which yield termites for some period of time.

Insight learning, involving the apprehension of relationships leading to problem solutions, is best known from the chimpanzee research of the Gestalt psychologist Wolfgang Kohler (1927). He experimentally found that Great Apes could find absorbing occupation with insightful solutions to problems, some of which involved the manufacture and implementation of tools. Purposeful solutions occurred without repetition of trials and errors. Chimpanzees in the Gombe also select and make tools well in advance of arriving at termite mounds, while infants will unsuccessfully attempt to imitate their mothers once at the termite mound.

Arousal and Development

The intensity of chimpanzee social behavior can be combined developmentally with the concept of arousal (Mason, 1967, 1973). Arousal can be considered as an integrative concept for understanding motivational processes involving chimpanzee attachment systems. Although related to attention processes in general, arousal does not rest on a single central nervous system process, but is related to limbic system and pituitary-adrenal axis functions. Measurements of adult human social dominance in small groups have been found to be correlated with cortisol levels, which may be related to the differing expectations of group members (Barchas and Barchas, 1975). The aim here is to use arousal to better understand the motivational dynamics and intensity of chimpanzee social behavior during development.

Developmentally, the importance of reciprocally calming interactions in the mother–infant relationship and the offspring's dependence on the mother for the regulation of socialization to familiar community members can be interpreted as an attempt by the offspring to maintain optimal levels of arousal. Several behavior patterns are important in the regulation of arousal and the maintenance of the infant's physical and social dependence on the mother in a typically relaxed setting. Infant clinging and suckling while the mother grooms her offspring are the most common and relaxing behaviors in young infants, normally kept in a ventral position on the mother. Infants in their second and third years often leave their mothers for bouts of exploratory or social play, which increases their arousal (Mason, 1973). This valance between arousing play and calming clinging by the infant provides a secure and mobile "classroom" for learning, and permits the perceptual processes to take their course in the rich physical and social milieu of the natural habitat and of the chimpanzee community.

In sharp contrast to this wealth of opportunities for physical development and socialization in the wild, in a limited cage environment socially isolated chimpanzees showed both short and long-term abnormalities (Davenport, 1978; Davenport and Rogers, 1970). These chimpanzees, deprived as infants, showed species typical facial grimacing, prolonged rocking, grooming movements out of social contexts, and gave screams and grunts while alone in an enclosure (called *in vacuo* by Lorenz, 1965; personal observation).

Less atypical are the common responses of captive chimpanzees to human caretakers with such species' typical behaviors as sexual presentation, bipedal rocking, charging, and, most often, food begging. Normally humans do not respond in kind; however, the behaviors are maintained after the socialization of chimpanzees to humans. In contrast to captive chimpanzees often behaving toward humans as fellow social beings, Gombe chimpanzees remarkably ignored human observers following or standing by them quietly in the forest environment. Social behaviors were not directed toward the human observers, who followed in a disciplined manner not imitating any chimpanzee social behaviors that would be intimidating, such as gesticulating, running, rock throwing, or branch waving. The usually safe rule, "do as the others do," for a visitor to a strange human community does not apply to a human observer's social behavior while watching chimpanzees at Gombe.

Protohominid Behavior

The origin of the genus *Homo* is thought to have been facilitated by paleoecological changes. These changes contributed to desiccation, or temperature decrease, and are considered responsible for the division of widespread forested blocks by woodland and savanna. These basic environmental changes are believed to have led to hominization of protohominids on the forest fringes and savanna. These new open country niches probably required some combination of better visual abilities for the grassland, the freeing of the arms to carry food or offspring to safety, the use of free hands to counteract predators, to hunt game more effectively, and to facilitate the already existent tool-using and -making habits (Hockett, 1973). Protohominid life history can only be roughly reconstructed after a comparative study of fossils, paleofaunal-floral relationships, primate behavior, primate physiological ecology, and the paleoenvironmental relationships that can be ascertained from geological records.

Bipedal locomotion probably aided in allowing these protohominid characteristics to occur in new niches that necessitated walking long distances. The comparative functional anatomy of bipedal locomotion has been studied (Zihlman and Hunter, 1972). Although more existent common chimpanzees rarely walk bipedal and then do so inefficiently, certain individuals can stride comparatively well (Bauer, 1977b). This ability of some contemporary, wild chimpanzees to stride bipedally is remarkable, since the *Pan* genus has been evolving over the many millions of years since it was contemporary with protohominids. Specialized chimpanzee adaptations have evolved since this time (as have human adaptations), one of which is chimpanzee knuckle walking (Tuttle, 1970). It is likely that protohominids were superior to modern chimpanzees in their ability to walk bipedally and exhibited a range of traits common to the genera *Homo* and *Pan*. The two chimpanzee species, *Pan troglodytes* and *Pan paniscus,* are quite different, and comparative studies of them would be important in understanding generic differences between *Pan* and *Homo*.

Social Dominance in Relationships between Adults

Social dominance is a regular deference and submission of a lower ranking individual to a higher ranking individual in a social relationship, which is the nature, quality, and patterning of interactions (Hinde, 1975). Chimpanzee relationships are always complex, involving many different qualities, and social dominance is one aspect of these. Overtones of social dominance appear in sexual and mother–infant relationships, for example.

Social dominance in chimpanzees can be understood as a compromising state of psychological conflict between the rare extremes of arousal associated with attack on or flight from another familiar. A schematic, two-dimensional model of these processes is shown in Figure 1, in which Actor I, when in a high state of

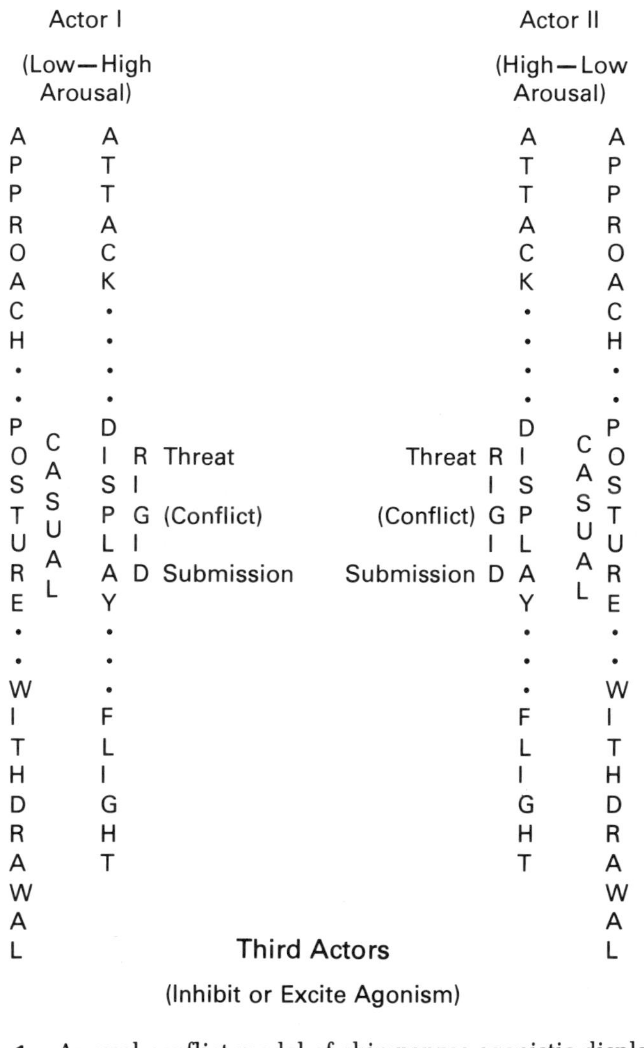

Figure 1. Arousal-conflict model of chimpanzee agonistic displays.

arousal, will give rigidly postured, threatening, or submissive displays as an expression of a conflict between the extremes of attack or flight. In a low state of arousal approaching one another occurs with casual postures, perhaps followed by grooming, sitting, or walking together, and climaxed by withdrawal from each other. Arousal is a psychophysiological, motivational state; as such, many intermediate levels regularly occur. Often complimentary threat and submission displays appear perfunctory and are rapidly followed by postural relaxation, group feeding, or social grooming.

The arousal-conflict model of agonistic displays in social dominance relationships is useful in interpreting greetings of chimpanzees after reunions. Captive juvenile chimpanzees after increasing periods of separation showed a regular ordinal relationship between different social behaviors common to juveniles. The social behaviors of clinging, grooming, and playing were the most to the least common, respectively, for separations varying from long periods to short periods of time (Mason, 1967). When aroused after a longer separation, clinging relaxed the juvenile. Adult Gombe chimpanzees will occasionally cling, touch, hold, or embrace and kiss one another as part of submissive or reassurance displays (van Lawick-Goodall, 1968). These behaviors can occur in greetings, upon seeing preferred food, when hearing strange community members at a distance, after long periods of separation, or in other ambiguous situations that observers cannot relate to an object or event. In both the experimental observation and the naturalistic Gombe observations contact with another familiar reduced arousal. Social grooming is far more common in Gombe chimpanzees and has a similar effect of reducing arousal and maintaining attachments between familiars.

Although arousal reduction is sometimes preferred, in calmer conditions higher arousal is sometimes an attractive state, suggesting that there is an optimal level of arousal that a chimpanzee can regulate by choices between different possible responses. Young chimpanzees especially choose between excitatory and inhibitory responses and, thus, actively regulate arousal. In Mason's (1967) experimental study of the effects of varying periods of separation on juvenile chimpanzee social behavior, social play only occurred more frequently after initial clinging and social grooming waned. In a common form of Gombe chimpanzee reunion, higher arousal followed by relaxation is an attractive and safe event between familiar and predictable community members (Bauer, 1976).

The social relationships of adult male chimpanzees are lifelong since they do not move from their natal communities. The effects of male age on social organization will be evaluated below in more detail, but of recurring importance is the fact that social dominance in these lifelong relationships is age-dependent (Bygott, 1979). Older male juveniles invite play bouts with adolescent or adult males, which are then regularly interspersed with submission and occasional reliance on the mother's social support. Frequently a mother will lead her troublesome juvenile away from such a party. With adolescence, males rarely become playful with older males and are typically, extremely submissive to high-ranking, prime adult males. This general dominance of older males over younger ones continues up until they are postprime, or old males, when younger males about to become prime challenge the dominance of these aging males, who eventually submit but remain important allies. Old males will generally end up being submissive to younger adult males, whom they had dominated in some cases for 15 years.

The strength of a dominant male is not critically important for him to maintain his position. For example, in 1970 Mike was a waning, ill, dominant male weighing only 78 pounds when both his eventual successor and a highly submissive young male weighed 98 pounds.

Not all prime adult males either possess or appear to contend to be the dominant male of the community. These prime adult males are high ranking and appear instrumental in regulating dominance relationships through coalitions in a series of interactions over a period of time. There are two common types of examples of this. A dominant male being challenged enlists other prime adult males in coalitions against the challenger. Also a higher ranking brother may enlist his brother in dominance interactions with challenging rival males of similar dominance. Two near-age but individualistic rivals, Figan and Evered, have for over eight years of adulthood regularly challenged each other, while for much of that time the lower ranking brother of Figan regularly supported him against Evered. Among adult males of a similar age there appeared to be no single factor from which the direction of dominance could be predicted (Bygott, 1979). In spite of these exceptions, age is an important factor in predicting the direction of agonistic behavior in male social dominance.

The aggressive nature of adult male chimpanzees makes their social dominance more obvious than that of adult females. These sex differences become most apparent with the growth spurt and physiological changes during chimpanzee puberty. As is found in other primates, female chimpanzees show cyclic menstration and have accompanying sexual swellings, which, when full, correlate with their being attractive, receptive, and proceptive in social relationships with males (Beach, 1976). Proceptivity is the female solicitation of a male to have sex, and in chimpanzee females it may be related to the laboratory finding that swollen females increase in food dominance over nonswollen females (Crawford, 1940).

Adult female social dominance exists but is not obvious. Some more experienced and older females could be distinguished as higher ranking than others, but since females were much less social than males and generally did not initiate threats in the presence of adult males, female social dominance was infrequently expressed in agonistic interactions (Bauer, 1977a). The inhibition of an adult female to threaten in the presence of an adult male has exceptions, as when a number of high-ranking females form a coalition against a thwarting, young male after he had intimidated them with charging displays. Collectively, females can threaten a younger adult male alone. In such rare cases females will threaten a male with hair erections, "waah" calls, slapping, and chasing. On one occasion the displaced male left the area in which he was threatened by the females and began screaming when he reached the opposite side of the valley that we were in. On another occasion a different young adult male was chased to the top of a palm tree by females, from which he leapt some distance to the ground and fled the area screaming all the while. At other times high-ranking females would vigorously chase adult males after being attacked by them and attempt to grasp or slap them. Most frequently adult females and males would give "protesting," "waah" calls after an adult male did a vigorous charging display or attacked a chimpanzee. This latter observation and the importance of coalitions in male dominance support the impression that dominant males remain so in part by group concensus.

Middle- and high-ranking adult males predictably dominate females at any age. Anestrus females would normally respond immediately to the arrival of middle- or high-ranking, prime adult males in a party with some sort of submissive gesture, posture, or vocalization. Exceptions to this submissiveness could be observed if the party were engaged in dispersed feeding in dense vegetation or included higher ranking males, who might inhibit the arrivee's aggressiveness. Estrus females had more social interactions with adult males than anestrus females, and they would range from mild forms of agonistic behavior related to courtship (McGinnis, 1979) to brief physical attacks followed by social grooming. In the rare case that two adult females threatened each other in the presence of a male, the adult male would threaten the two females and inhibit further female aggression.

Although adult males usually dominated all adult females, adult females sometimes clearly manipulated the male social situation to their own advantage. This manipulation probably provides a mechanism for females to choose male mates, quite important in evolutionary perspective since, after being eight and one-half months pregnant, the mother can usually not have another infant for over three years and will continue to travel and support her juvenile socially for another five to seven years. Compared to the father of her offspring, who does not get directly involved in rearing, she has to invest years of her time and energy into their offspring. Of course, if the pregnancy is aborted or the young infant dies, the mother will not have to invest the full amount of time and can have another offspring sooner (Bauer, 1977a).

Adult females have been observed to manipulate a situation in which a male is attempting to threaten or coax a particular female to follow him away from the community in a special consort relationship. Many conceptions are likely to have occurred in consort relationships (Bauer, 1977a). The solicited female can either follow the male of her choosing, ignore him, or protest his threats. All three choices have been observed. The choice of protesting has been observed to manipulate the situation on at least three occasions by attracting the presence of higher ranking males who inhibited the first male's attempts at consortship.

The Role of the Alpha Male in the Community

With a little familiarization, the most outstanding characteristic of social dominance in a chimpanzee community can be seen to be the alpha male. All other chimpanzees direct submission toward this most dominant male. Coalitions, however, remain crucial in regulating the degree of domination by the alpha male of his subordinate community members, so a successful alpha male strategy appears to be one of a shrewd, "benevolent despot" who does not ignore his fellow male allies. In sociological terms, a system of social exchange appears to exist between the alpha male and his long-time allies.

In the Kasakela community there is a unique history of alpha males in their respective positions between January 1971 and December 1973, with the exception of an ambiguous period from September 1972 to June 1973. This ambiguous period was a transitory period between a former alpha male and his successor. In general, the alpha male stands out from other prime adult, high-ranking males for his lack of inhibition in carrying out threatening charging displays after a reunion, no matter who is in the new party he has joined.

Rules that characterize the role of the alpha male are empirically based and can be usefully discussed in cognitive terms, as long as the limitations of the discussion are regularly grounded in observational or experimental confirmation or disconfirmations. This hypothesis testing approach to nonhuman primate cognition has been recognized for some time (Mason, 1976). Associated with studies of animal cognition are such investigators as Yerkes, Heinroth, Kohler, Lorenz, Tolman, and Lashley. Human social psychologists concerned with attribution theory argue that human actors consistently attribute the cause of their actions to situational requirements, whereas observers usually attribute the identical actions to regular individual dispositions of the actors (Jones and Nisbett, 1972). Students of chimpanzee behavior observe regular individual dispositions, rules, or individual differences among peers. However, the varying composition of chimpanzee parties in the field combined with social relationships which endure for decades make the situational requirements of a temporary association of chimpanzees in a party of considerable importance in determining what may socially precipitate an adult male's behavior.

The alpha male role in a chimpanzee community can be characterized in theoretical terms by rule-governed behaviors found in the alpha male's relationships with other males. An observer can theorize that a set of social expectancies develop in a new candidate for the alpha male position, which combine the new candidate's observational experience of previous alpha males with his own temperament, life history, and social experience with his peerage. Observations during the mentioned state of ambiguity during a transitory period strongly suggest that the alpha role is sometimes a difficult matter to define and clarify when powerful coalitions of males vary in allegiance to rival candidates for the alpha male position. In such cases the matter of who will become the alpha male is to a large extent a matter of the consensus of other prime adult males. It is not a matter of the candidate winning any particular contest, although results of a particular fight between powerful males can help experienced observers find patterns in their relationships over a period of years. The alpha male candidate is most often in the running and the position holder can be removed after little more than a year and one-half, or supported over six years.

The performance of "arrival-charging" displays is quite common for alpha males joining a party and it is most likely due to this that adult males are often "glancing around the periphery" of their visual field when on the ground in open areas. They do this so that they can anticipate and avoid difficult situations before they arise. Evidence for this can be found in the fact that rival males would sometimes remain silent and travel in the opposite direction from the source of a series of "pant hoot" long-distance calls of a rival. Chimpanzees individually recognize each other by these long-distance calls and other characteristics (Bauer and Philip, in preparation). A particularly strong reaction of avoidance or vocal reply and approach often occurred if an adult male heard several prime adult males vocalize in the distance. This varied and complex life of a chimpanzee in the forest is a far cry from the monotony endured by chimpanzees in caged captivity. It is not surprising that captive chimpanzees have expectations of their human caretakers and can complete the varied sequential tasks normally carried out by them.

The Evaluation of Social Dominance in Relation to Community Structure

The aggregative tendencies of adult male chimpanzees in a community have been reported to be a positive function of social dominance. To be most dominant, by this hypothesis, is to have the most populated parties (Nishida, 1979). By studying the alpha male chimpanzee, an evaluation of the importance of social dominance to the internal structure of the community can be made. The hypothesis tested below is that alpha males are the most gregarious of the Kasakela community adult males. Developmental bases for future alpha males raised in dominance-related situations could not be considered, unfortunately, in the data analysis. This section will evaluate the hypothesis that male social dominance and party size increase together, taking into account age changes in aggregative tendencies.

Before we begin the evaluation of the hypothesis, a methodological point should be raised in regard to the sampling techniques in field studies of chimpanzee party sizes. Nishida

(1979) based his hypothesis on the average aggregation size that each male was "found" in during three periods of observation. How does one find chimpanzees in a forest? Normally by their vocalizations. It is not clear if the observer found the males in these aggregations at any one point in time or if specific target animals were being followed over a period of time in the day when party sizes were counted. This is an important point, since if it were the earlier case and observers were attracted to parties by their vocalizations in the forest, it would be likely that Nishida's (1979) data overestimated real party sizes. A wide range of sampling questions are worth considering in such hypothesis testing, but the data presented here are based on long continuous samples of adult party size taken on half-hour intervals of specific target individuals traveling throughout the communities' home range.

The social dominance of the chimpanzee male follows an inverted U-shaped function throughout his life, since young males are the lowest ranking, prime adult males high ranking, and old males become less than high ranking. This change in relative rank as a function of age requires that two techniques be used to evaluate the relationship between social dominance and party size of independently traveling adult individuals. To separate the social dominance variable from the age variable, a systematic comparison was made of both (1) the same two males in an alpha male position and in a prime adult male position and (2) different males of the same age in two distinct statuses, which were the alpha male position and the prime adult male position. The first comparison was intrasubject in two different time periods; the second comparison was between subjects in the same time period. In each case the subject sampled was followed and party size data collected on half-hour intervals for hundreds of hours while the adult male independently traveled about the forest through his home range. By using these two procedures involving the alpha male position and prime adult male position, it is possible to isolate more clearly social dominance as the key variable separate from any confounding influence of age changes in aggregative tendencies.

First to be considered is the comparison of alpha males while in and out of the alpha male position over a period of three years, including data from 1971 through 1973, except for the mentioned unusually ambiguous period in late 1972 and early 1973. The importance of using the alpha male for comparison is that his rank is less affected by the particular composition of the party he is in than any other male. The two alpha males during 1971 through

1973 in the Kasakela community were first HM and second FG with a sample size of 1,744 hours of actual observation on the two males by several observers with simple, standardized field checklists. These observations were divided into two conditions: one was alpha male dominance and another was "beta" male submission. The latter was when first FG then HM were not alpha males but just prime adult males. The dominant condition corresponded to the period from January 1971 to September 1972 for HM and July to December 1973 for FG. The "beta male" submissive condition corresponded to the period from September 1972 to December 1973 for HM and January 1971 to June 1973 for FG. Procedures and results are detailed in Bauer (1977a).

Nisida's general hypothesis would predict that alpha males are in larger parties than beta males. The alpha and beta male distributions of scores were plotted in terms of cumulative percent of all observations in the two categories. The results showed that each distribution of aggregation sizes for alpha and beta males closely tracked and crossed each other. A statistical evaluation of the relationship between the two distributions using a nonparametric statistic, U, with a procedure for large samples showed no significant difference ($p > .05$; Goldstein, 1964). Alpha males were not in larger parties than beta males. Separate comparisons of each of the two males supported this result, in that no consistent differences were found in the direction predicted.

The second comparison was designed to consider the influence of social dominance on gregariousness between alpha males and other like-age males. The two conditions, alpha male and prime adult male, were contrasted for aggregation size to see if the results would support Nishida's popularity hypothesis, which in this case would predict that alpha males would be more gregarious than like-age prime adult males. HM and FG were the alpha males which were compared with two prime adult males, EV and FB. A total of 1,246 hours of observations were used to compare the two pairs, each belonging to separate conditions. The results showed that the alpha male and prime adult male distributions of aggregation sizes closely track and cross each other. Although the results indicated statistically significant differences, there was no consistent direction to these differences ($p < .05$; Goldstein, 1964). The result of this comparison does not support the hypothesis that alpha males are more sociable than like-aged, prime adult males.

Age differences are apparent in several characteristics of wild male chimpanzees. Morphologically, several changes occur.

Canines, premolars, and molars of young adult male chimpanzees are well developed. Old males may be missing canines or molars, while remaining teeth may have apparent black or brown cavities. The blacker hair of the young male often becomes brown tipped in prime adult males with occasional grey hairs apparent in old males, who also bald about the scalp and generally have thinner coats. Weights are shown in Table 1 for most of the independently traveling males in the Kasakela community over a three-year period. Young SH stands out as the least heavy male with the most change in weight, probably related to the adolescent growth he was undergoing. All adult males in Table 1 are ranked according to estimated age. SH's brother JJ was the heaviest male, although he was among the lowest ranking. This data should be interpreted realizing that it represents between 16 and 22 observations by a team of observers averaged together over a period of three years and different seasons of those years (Bauer, 1977a).

Given that the influence of social dominance on party sizes could not be confirmed in the above comparisons and that age was a confounding variable in the original hypothesis, the influence of age differences on aggregation sizes was evaluated for young (SH, JJ, and ST), prime (EV and FB), and old (MK and HG) males. While following and observing younger males, one did get the impression that they spent more time alone or in small parties than adult males in the prime of their life. Their extreme submission to higher ranking males was an aspect of their social timidity. The results of the comparison between young, prime, and old adult males showed orderly differences in party sizes. The prime adult

Table 1. Average Body Weight in Pounds for Adult Males Ranked According to Estimated Age

Individual	*Mean Body Weight*	*SD Body Weight*	*Number of Observations*
HG	82.6	4.38	16
MK	81.6	4.44	23
HM	98.4	4.60	18
FB	—	—	0
EV	89.4	7.40	18
FG	82.1	2.90	16
JJ	102.1	5.60	20
SH	64.1	10.90	22

Data collected intermittantly from January 1971 to December 1973 in Gombe Stream Research Center provisioning area. HG is estimated to be the oldest and SH the youngest male chimpanzee.

males were the most sociable compared to the young and old males, accounting for the similarity between them and similar aged alpha males in sociability. The mean aggregation size for prime adult males as they and others foraged and traveled was 3.92, based on 918.5 hours of observation. Old males were below prime adult males in sociability with a mean of 3.63 chimpanzees in the parties they were in over 208 hours of observation. Least sociable were the young males with a mean party size of 3.53 in 766.6 hours of observation. The statistical distribution of the prime adult males clearly differentiated from and did not cross over with that for young or old males. This was not true for the distributions of the young and old males. Overall statistical differences between the three age groups were significant ($p < .05$; Goldstein, 1964). Thus it was male age that was the best predictor of adult male aggregative tendencies and not social dominance.

Social Dominance Related to Adaptation Processes

What evolutionary interpretations might be given to social dominance in adult male common chimpanzees? It is remarkable how low keyed chimpanzee social dominance is in contrast to many other old world primates, who will often fight over food. In this section selected topics will be discussed, but no single answer to the question of the relationship between social dominance and adaptation will be offered.

The distribution and small size of separate food patches give a particular understanding to the fluid fission-fusion social organization of common chimpanzees. This type of social organization is most uncharacteristic of nonhuman primates. The relationship between food resources and social dominance is a complicated one to the extent that party splitting regularly occurs as individuals search for new, preferred feeding sites, thus avoiding real or expected food dominance interactions. Displacements at feeding sites are seen, but for naturally occurring foods these subtle interactions are normally quiet, and low-keyed communications are easily missed by some observers. Since human observers are not experimentally familiar with chimpanzee natural food preferences, no systematic data exist on this characteristic of feeding, the most common daytime activity (Bauer, 1977a).

Early captive studies of chimpanzee social dominance which used food dominance as one measure found that most often there were no overt signals of competition between paired females. The right of access to a preferred food seemed to be quietly understood prior to the food dominance test of adults, and there was rarely an instance for aggression to be expressed. The food-dominant chimpanzee adult was often by, or initiated going to, the food hopper and maintained the position until satiated (Crawford, 1940).

In the field displacements from feeding sites occur. Individuals would control access to food trees or departures from them through position, glances, looks, slight hair erection, mild stiffening of posture, or hunching of shoulders. Although food preference and dominance studies would be of considerable theoretical significance for the understanding of chimpanzee social organization, systematic study of natural food preferences, social dominance, and spacing has not been done. The option of lower ranking animals to find preferred food in the forest by themselves is a good interpretation of the fact that both social dominance and party size as a function of age have inverted U-shaped functions when related to age in adult males. Young, lower ranking males tend to be the least social, while high-ranking prime adult males are the most social. In smaller parties young males would likely have fewer interrupted feeding bouts on more preferred foods, whereas prime adult males intrinsically act more as peers in such situations. Older adult males are intermediate in both rank and sociality. Chimpanzee party size varies with the food species being fed upon (Bauer, 1977a). This, together with the notion of individuals avoiding competition for preferred foods, provides observational insights into chimpanzee dispersion and dominance that require experimental evaluation in naturalistic settings.

Reproductive competition between members of the same sex might be one of the first places to look for a relationship between social dominance and adaptation in vertebrates. The predominance of sexual competition between members of the same sex in primates was emphasized by the early observations on hamadryas baboons in a zoo, which were made not realizing that the introduction of new females to a group of adult male hamadryas baboons with herding tendencies would predispose excessive aggression not typical of other baboon species or primates (Kummer, 1971).

The chimpanzee pattern is somewhat paradoxical compared to other old world primates. No permanent or exclusive heterosexual bonds maintain themselves between estrus females and fellow male members of a mixed party, where females tend toward promiscuity and every male able to leave his mother will mount and thrust their penises into females when they are sexually swollen. Incest between mother and son or sister and brother does not occur and is an exception to promiscuity. Gombe females, however, also form and maintain consortships with selected males away from the community, as discussed earlier. Since these exclusive consortships can last several weeks, during which the female is sexually receptive for a portion of the time, many pairs would conceive a potential future offspring during these times of exclusive sex in secluded areas. Higher ranking males may have a better chance to initiate female following at the beginning of the consort without being inhibited by the presence of other high-ranking males. Lower ranking males would more likely be inhibited in their attempts to consort with a female, if she attracted higher ranking males. More study of the complex communication and effects of social dominance in these important contexts is necessary.

Physiological correlates of chimpanzee social dominance deserve to be investigated systematically, as was done on both rhesus monkeys and Japanese monkeys. Investigating adult male rhesus monkeys, Rose, Holiday, and Bernstein (1971) found that social dominance and blood testosterone levels were positively correlated, but this finding did not generalize to Japanese monkeys (Eaton and Resko, 1974). The chimpanzee testis is quite large relative to body size compared to other primates. The hypothesis that typical physiological levels of testosterone output are positively related to social dominance or meiotic division in the male has not been evaluated. Indeed, a recent review of chimpanzee reproductive physiology states that data on the male are virtually nonexistent (Graham, 1970). In mice all stages of meiotic division decrease with lower levels of testosterone (Gartner, Reznik-Schuller, and Reznik, 1973). In summary, the behavioral endocrinological questions surrounding chimpanzee social dominance are many sided and unanswered.

Both the fraternity and strong kinship of male community members become of great interest in relationship to viewing attacks between communities as a factor in mortality analogous to predation, but occurring within a species (Alexander, 1974). Fear of strange males is probably effective as a mechanism in maintaining intercommunity separation. The very uncertain outcome of a meeting of strange males and the relatively low genetic relationship between males of adjacent communities would both suggest a good chance of a violent encounter in these unpredictable meetings. The recent observations of violent attacks between males of adjacent communities (Bauer, 1976, 1977a; Goodall, 1977), which are more severe than attacks within communities, suggest that members of male kinship lineages are aggressing toward nonkin, either offspring of strange males or strange males themselves.

Nishida and Kawanaka (1972) astutely pointed out that intercommunity separation was probably a sociological phenomenon rather than a result of ecological or geographical factors. Applying the sociological effects of intercommunity separation, the present author investigated the possibility that the structure of chimpanzee parties was related to the presence of adjacent communities. A study was done of the range and party size of adult males and adult females in the Kasakela community by dividing the range into central, or core, areas and outer, or peripheral, areas. The analysis included both adult males and adult females in the Kasakela community (Bauer, 1977a). The more peripheral areas were associated with significantly larger party sizes ($p < .05$), although this finding does not test for the independent effects of food distribution. Both adult males and adult females had smaller party sizes in the core area of the community home range, that is, while they were more distant from home ranges of adjacent, strange communities. The cohesiveness of adult males mediated by their arousal-reducing system of social dominance may, in part, be an adaptation to confronting strange male chimpanzees, who can exploit their own kin group's opportunities to reproduce sexually and have access to food resources. Social dominance, then, is an adaptation to increase the inclusive fitness of individual males and their male community member relatives threatened by males of adjacent communities. The adult females' inclusive fitness, by this hypothesis, would be maintained by giving a female adult chimpanzee potential options of choosing the community and the male in that chosen community with whom she will conceive her offspring while living in a predictable and relatively secure system of social dominance. This hypothesis would also help explain the obvious sexual dimorphism in the intensity of aggression, which has been emphasized (Hamburg, 1971).

In summary, the integration of physiological, developmental, and evolutionary interpretations of chimpanzee social dominance indicates that dominance is a complex, largely male, age-dependent, and possibly male kinship phenomena. To search for the most important function of chimpanzee social dominance is not likely to be a productive approach in such a complex system. Trying to rank the order of importance of social dominance in relationship to food, sex, or male kinship would be like scaling the importance of various essential organs of the body. All of the functions are essential for the system to operate, and, in the case of social dominance, experimental studies will be necessary to understand the ethological factors controlling them. The results support the view that middle-age males in the prime of their lives are central and most gregarious in a chimpanzee community, and these males include the then alpha male. This finding was predicted indirectly over 30 years ago in a laboratory study. Nowlis (1941) found that young chimpanzees choose dominant or subordinate chimpanzees equally in tests of social preference, emphasizing the theoretical and methodological importance of coordinated laboratory and field research on common chimpanzees.

Acknowledgments

The author acknowledges the support of the Grant Foundation, the Yerkes Regional Primate Research Center, the Gombe Stream Research Center, the National Geographic Society, the L. B. S. Leakey Foundation and, in the final preparation, the Rockefeller University Field Research Center and the Alexander von Humboldt Foundation. I acknowledge the Tanzanian Government for permission to conduct research in the Gombe National Park and the cooperation and support of Drs. David A. Hamburg and Jane Goodall.

Part III

Studies of Dominance in Children

Chapter 6

Animal Threat Displays and Children's Facial Expressions: A Comparison

Linda Camras

Studies of animal behavior report that many social species have repertories of threat displays which they may use in the process of conflict resolution (Tinbergen, 1952, 1953). Threat displays are thought to replace physical fighting between individuals during some hostile encounters. In an idealized dyadic interaction a threat display by one individual will produce submission or withdrawal by the other. The deference shown to a threatening animal presumably defuses the hostile interaction and reduces the probability that a physical struggle will occur. The use of threat displays thus helps prevent bodily injuries which might be sustained by individuals engaging in a physical altercation.

Although no study to date has examined the relationship between children's facial expressions and animal threat displays, several investigators have reported preliminary data suggesting that facial threat displays may be used by children. Blurton Jones (1967, 1972b) found an association between particular components of facial expression and aggressive behavior in nursery school children. Specifically, the expression components *frown* (i.e., lowered eyebrows) and *fixate* (i.e., an abrupt unsmiling look of about 2 seconds) were associated with the actions *hit* and *push* in his study. In addition, Grant (1969) published a descriptive list of "aggressive" expression components which he reported to predict an increasing probability of subsequent attack by nursery school children. Grant's list included 10 aggressive elements of facial expression (e.g., *aggressive frown, lips forward,* and *intention bite*).

The work of Grant and Blurton Jones suggests that some facial expressions[1] used by children may be analogous to threat displays used by nonhuman species. "Aggressive" expressions sometimes might warn others that physical attack will follow. Children observing these expressions could withdraw from a potentially dangerous interaction and thus avoid being attracked by the expresser. Avoiding attack would have adaptive advantage for human children just as it has advantages for members of other animal species.

This chapter will discuss some issues regarding the "aggressive" expressions described by Grant and Blurton Jones and their possible status as children's threat displays. In the ethological literature the term "threat display" often has been applied to all acts which appear to intimidate another individual without involving actual physical assault. However, some investigators have specified a set of well-defined features that they feel characterize acts for which the term "threat display" is appropriate. Three features described by Moynihan (1955b) will be used here as a starting point for discussion. Since Moynihan's description is exceptionally articulate, it offers a useful conceptual framework for analyzing the "aggressive" expressions of Grant and Blurton Jones. In addition, Moynihan's description is of interest because it represents the original concept of a threat display formulated by the Tinbergen school of ethologists. Where necessary, however, the limitations of the original formulation will be pointed out and alternative approaches will be discussed.

According to Moynihan (1955b), the following three features are characteristic of threat displays:

1. Threat displays are acts which have been ritualized in the course of evolution, i.e., they have been specialized in "form or frequency" (Moynihan, 1956, 1960) to serve a signal function.
2. Threat displays are signals "designed" to intimidate the individual to whom they are directed; they increase the recipient's tendency to escape or withdraw from the displaying individual.
3. Threat displays reflect a combination of attack and escape tendencies in the displayer; they are the result of a conflict between these tendencies.

Ritualization

According to Moynihan's description, threat displays are acts which have been ritualized to serve a signal function. Though not all signals are ritualized displays, the process is important in the development of signal repertoires in many species (Wilson, 1975). Ideally, demonstrating the ritualized status of a signal act involves (1) identifying a related action pattern that served as its evolutionary origin and (2) specifying the changes in frequency or form that occurred as this action pattern evolved into a display. As Hinde (1974) points out, neither of these is easy to do. Tracing the evolutionary history of a signal involves an inevitable degree of uncertainty since signal movements leave no fossil records. Two methods are commonly used to generate hypotheses about the evolutionary origin of a signal act. Each involves comparing the act in question to movements in the species' repertoire that are not thought to be signals and/or comparing the act in question to behaviors of other related species. During the process of comparison one looks for actions that are similar in form to the target act and that allow one to make hypotheses about the target action's evolutionary origin. The evolutionary origin of a ritualized act usually is presumed to be an act which served a direct noncommunicative function for the performer (e.g., a locomotor movement, a cleaning movement, or a protective movement).

One approach to establishing the ritualized status of children's facial expressions is to compare them to expressions used by other primate species. Some comparisons of human and primate expressions already have been reported by investigators interested in the phylogenetic basis of emotional expression (Chevalier-Skolnikov, 1973; van Hooff, 1972). For example, Chevalier-Skolnikov (1973) has examined human facial expressions of emotion and "contextually equivalent" expressions of several primate species.

Table 1 lists the "aggressive" expressions described by Grant and Blurton Jones and uses them as a starting point for interspecies comparison. Next to each children's expression are listed some primate expressions that are similar in appearance and that are used in hostile situations. While some of the primate expressions may not have all features of a threat display described by Moynihan, all expressions are associated with aggression or threat.

The morphological similarity between the children's expres-

Table 1. Type 1 Expressions Described for Children and Similar Primate Expressions Which Occur in Health Situations

Type 1 Expression	*Author Describing for Children*	*Similar[a] Primate Expression*	*Species Reported*	*Behavior or State Associated With Primate Expression*
Lowered brow	"Aggressive frown" (Grant, 1969)	Tense-mouth face (van Hooff, 1967)	Wide variety including gorilla, chimpanzee, and rhesus macaque	Expresser usually attacks. Recipient usually flees.
	"Low frown" (Blurton Jones, 1972)	Threat (Hinde & Rowell, 1962)	Rhesus macaque (*Macaca mulatta*)	Expresser may attack or flee.
		Possibly stare (Schaller,1963)	Gorilla (*Gorilla gorilla beringei*)	Expresser uses as a threat; may sit, stand, or approach expresser. Recipient responds with submissive gestures.
		Staring open-mouth face (van Hooff, 1967)	Chimpanzee (*Pan troglodytes*)	Expresser often attacks, occasionally flees or avoids. Recipient frequently flees.
Both lips thrust forward (Lips apart and thrust forward)	"Lips forward" (Grant, 1969)	*Possibly* staring open-mouth face (van Hooff, 1967)	Chimpanzee (*Pan troglodytes*)	Expresser often attacks, occasionally flees or avoids. Recipient frequently flees.
Small mouth (Corners of mouth drawn in to center so mouth appears small)	"Small mouth" (Grant, 1969)	Tense-mouth face (van Hooff, 1967)	Wide variety including gorilla, chimpanzee, and rhesus macaque	Expresser usually attacks. Recipient usually flees.

Table 1 (Continued).

Type 1 Expression	*Author Describing for Children*	*Similar[a] Primate Expression*	*Species Reported*	*Behavior or State Associated With Primate Expression*
Lips pressed together (Lips pressed together, may be slightly rolled under)	"Tight lips" (Grant, 1969)	Glare (van Lawick-Goodall, 1968)	Chimpanzee *(Pan troglodytes)*	Expresser may chase or attack; uses prior to copulation. Recipient may show submissive behavior or approach and present for copulation.
Lower lip drop (Jaw thrust forward and lower lip dropped exposing teeth)	"Intention bite" (Grant, 1969)	*Possibly* threat (Hinde & Rowell, 1962)	Rhesus macaque *(Macaca mulatta)*	Expresser may attack or flee.
Nose wrinkle	Two forms: "Sneer" (Nose wrinkled, teeth exposed) "Wrinkle" (Nose wrinkled, teeth not exposed) (Grant, 1969)	*Possibly* snarl (Bernstein, 1970)	Celebes macaque *(Macaca maura)*	Expresser is aggressive.

[a]Primate expression is equivalent to children's expression or includes children's expression as a component.

Table 1 (Continued).

Type 1 Expression	*Author Describing for Children*	*Similar*[a] *Primate* Expression	*Species Reported*	*Behavior or State Associated With Primate Expression*
Face thrust	Three forms: "Head forward" (Head brought forward slowly)	Head bob (Rowell, 1972)	Rhesus macaque (*Macaca mulatta)*	Expresser may attack or flee.
	"Threat" (Head brought forward quickly)	Jerking head down and forward (Devore & Hall, 1965)	Baboon	Expresser uses as threat.
	"Chin out" (Chin pushed forward) (Grant, 1969)	Jerking head in direct of opponent (Schaller, 1963)	Gorilla (*Gorilla gorilla beringei)*	Expresser uses as a threatening gesture.
Stare	"Fixate" (Blurton Jones, 1972)	Tense-mouth face (van Hooff, 1967)	Wide variety including gorilla, chimpanzee, and rhesus macaque	Expresser usually attacks. Recipient usually flees.
		Staring open-mouth face (van Hooff, 1967)	Wide variety including gorilla, chimpanzee, and rhesus macaque	Expresser often attacks, occasionally flees or avoids. Recipient frequently flees.
		Threat (Hinde & Rowell, 1962)	Rhesus macaque (*Macaca mulatta)*	Expresser may attack or flee.

Table 1 (Continued).

Type 1 Expression	*Author Describing for Children*	*Similar[a] Primate Expression*	*Species Reported*	*Behavior or State Associated With Primate Expression*
Stare *(continued)*		Glare (van Lawick-Goodall, 1968)	Chimpanzee (Pan troglodytes)	Expresser may chase or attack; uses prior to copulation. Recipient may show submissive behavior or approach and present for copulation.
		Staring bared-teeth scream face (van Hooff, 1967)	Wide variety including gorilla, chimpanzee, and rhesus macaque	Expresser may attack or flee; expression also seen in some circumstances where expresser is thwarted but has no tendency to escape. Recipient may flee, avoid, or attack.
		Staring (Struhsaker, 1967)	Vervet monkey *(Cercopitheous aethiops)*	Expresser uses as aggressive or defensive threat. Recipient response is variable; expression terminates or leads to further agonism.
		Stare (Schaller, 1963)	Gorilla *(Gorilla gorilla beringei)*	Expresser uses as threat; may sit, stand, or approach expresser. Recipient responds with submissive gestures.

[a]Primate expression is equivalent to children's expression or includes children's expression as a component.

sions in Table 1 and the aggressive or threatening expressions of primates suggests that they may have a common evolutionary basis. These similarities, however, are not sufficient proof that the expressions of children are ritualized displays. Despite their resemblance to primate expressions, the children's expressions in Table 1 may have been specialized for communication by only nonevolutionary processes or may never have been specialized for communication at all.

Regarding the first possibility, consider the expressions *face thrust* and *lower lip drop*. Face thrust may originate as a component of lunging attack (Rowell, 1972). Lower lip drop (Grant's "intention bite") may have its origin in real biting movements. During development, children may observe these abbreviated forms of biting and lunging and learn that they frequently are effective in obtaining desired results. Thus biting and lunging may develop into the expressions lower lip drop and face thrust through a process involving ontogenetic rather than evolutionary modification. Expressed in general terms, some expressions of children and primates listed in Table 1 may be similar in appearance because they are derived from similar precursor acts. However, the processes leading to specialization for signaling may differ. Cross-cultural and developmental studies are needed to explore the influence of biology and culture on children's facial expressions. These studies would provide important evidence supporting or disconfirming the hypothesis that the expressions listed in Table 1 are ritualized signals.

Regarding the second possibility indicated above, proposals of noncommunicative functions have been made for some of the children's expressions listed in Table 1 (Andrew, 1963; Darwin, 1872; Ekman, in press; Redican, 1975). Consider the expression *lips pressed together*. Redican (1975) has suggested that pressing the lips together is a protective response used by an animal which "intends to engage in an imminent physical altercation" (p. 128). Darwin (1872) reported that pressing the lips together seemed to aid muscular exertion. In either case, children and primates may habitually press their lips together in all situations involving exertion and/or protection of the teeth. These would include hostile situations. Lips pressed together thus might convey information to the recipient, indicating that the expresser will engage in activity requiring exertion or protection. However, the movement may never have been specialized in form or frequency to convey this information.

If lips pressed together can have a direct function in hostile situations, then to support the ritualization hypothesis one should show that the lip-pressing movement has been modified in some way to enhance its communicative value. Evidence for modification might be obtained by comparing lip-pressing movements occurring in hostile social situations with lip-pressing movements occurring in nonsocial situations that require exertion or protection but not aggressive signaling. In making this comparison, one would look for distinctions in "frequency or form" between pressing the lips as a signal and pressing the lips only as a protective movement or a movement aiding muscular exertion. As yet, there has been no attempt to make this type of comparison. However, recent advancements in our techniques for coding facial expression may facilitate future attempts to distinguish differing forms of lip-pressing movements (as well as different forms of some other expressions listed in Table 1). The Facial Action Coding system (FACs) developed by Ekman and Friesen (1978) is an anatomically based coding system that enables an observer to describe facial movements more accurately than previous systems have allowed. FACs will enable observers to distinguish among different forms of lip-pressing movements that may have been considered the "same expression" in previous studies. If appropriate distinctions in form can be demonstrated, one reasonably could retain the hypothesis that lips pressed together is a ritualized signal. However, studies would be needed to explore possible cultural influences on the development of signal forms of this facial expression.

To summarize, an evolutionary basis for at least some of the children's expressions listed in Table 1 is suggested by their resemblance to hostile expressions described for nonhuman primate species. This resemblance suggests, but does not demonstrate, that the children's expressions are ritualized signals. While further research is necessary to resolve this issue more conclusively, some children's expressions may have the first feature of a threat display described by Moynihan. However, even if unritualized, these expressions still may play an important role as threat signals influencing the behavior of their recipients.

Intimidating Effect

The second feature of a threat display described by Moynihan is its intimidating effect on the individual to whom it is directed. Moynihan notes that threat displays do not always frighten the recipient enough to produce overt flight. Recipient behavior following observation of a threat display also will depend on other factors. However Moynihan and most other investigators use the term "threat" for actions which frequently are followed by the flight of the recipient.

To explore the possible intimidating effect of the children's expressions in Table 1, Camras (1975, 1977) studied conflicts that arose between pairs of children in a controlled-setting study. In this study 72 pairs of unacquainted kindergarten children each were seated at a table and allowed to play for a brief period. The children were given a box containing a pair of live gerbils that could be used by only one child at a time. Conflicts between the children occurred when the child who was playing with the gerbil box ("the player") resisted an attempt by his partner ("the nonplayer") to get the box. For each conflict episode the player was scored as either using or not using one or more of the facial expressions listed in Table 1.

Data analysis showed a relationship between the facial expressions used by players during conflict and the postconflict behavior of nonplayers. Players tended to win conflict episodes (i.e., keep the gerbil box) no matter what type of facial expressions they used. After conflict the nonplayers usually waited for some time before making a new attempt to get the gerbil box. Nonplayers tended to wait longer after observing the player use one of the expressions in Table 1 than after observing only other types of facial expressions. One expression not listed in Table 1 (*oblique brows*) also was associated with nonplayer hesitancy in renewing the attempt to take the gerbil box. Oblique brows (involving contraction of the medial portion of the frontalis muscle—*m. frontalis*—with or without accompanying contraction of *m. depressor glabellae, m. depressor supercilli,* and *m. corrugator supercilli*) usually is considered to characterize an expression of sadness (Darwin, 1872; Ekman and Friesen, 1975; Izard, 1971; Tomkins, 1963).

The data suggest that the expressions in Table 1 may have intimidated their recipients, increasing nonplayers' hesitancy to take the gerbil box. It is possible, though, that nonplayer hesitancy did not always reflect intimidation since oblique brows (the sad expression) was followed by the same nonplayer behavior. It is also possible that nonplayers hesitated in response to some aspect of the player's behavior other than his facial expression (see Camras, 1976, for further discussion of this issue). Yet, despite these reservations, this study's findings indicate that the expressions in Table 1 may have functioned as threats (i.e., produced intimidation) in the children's conflict encounters. Thus, these expressions may have the second feature described by Moynihan as characterizing animal threat displays.

Underlying Behavioral Tendencies

The last of Moynihan's three features concerns the motivational basis of a threat display—a conflict between attack and escape tendencies. Recent experimental studies (Blurton Jones, 1968) have shown that an escape tendency does not always underlie all threat displays. Nonetheless, many threat expressions described for primates do seem to involve both tendencies (see last column in Table 1).

The studies of Grant and Blurton Jones provided evidence that the children's expressions in Table 1 are related to an attack tendency. Both investigators reported that these expressions are temporally associated with aggressive behavior (e.g., hitting, kicking, etc.). Temporal association frequently is used as evidence for the motivational basis of a display in the ethological literature (Hinde, 1970, 1974; Moynihan, 1955a). In the Camras study, however, the expressions listed in Table 1 were not associated with physical attack. Indeed, no physical attacks occurred. Instead, these expressions were related to persistence in the expresser's attempt to retain the disputed object. When players used these expressions during a conflict, they tended to resist the nonplayer's subsequent attempt to get the gerbil box. When players did not use these expressions, they might either resist the nonplayer's next attempt or allow the nonplayer to take the box.

Although the children's expressions in Table 1 were never followed by attack in Camras' study, possibly they still reflected an increase in this tendency. Perhaps attack never actually occurred because the expressions usually were "successful": they intimidated recipients and thus removed the stimulus that had produced the attack tendency. Alternatively, perhaps overt attack was precluded by situational constraints present in the controlled setting. During the play sessions the two children were seated at opposite ends of a small table and were asked to remain in their seats. Although this instruction frequently was disobeyed, it may have lessened the likelihood of any vigorous activity such as fighting.

Ethologists generally recognize that signals reflecting an attack tendency are not always followed by the overt expression of this tendency. A display is presumed to reflect an attack tendency if it sometimes accompanies attack and if one reasonably could assume the presence of an attack tendency whenever the display is seen (Hinde, 1974). In the studies of Grant and Blurton Jones the expressions listed in Table 1 were associated with attack. A tendency to attack might be expected in the conflict encounters of Camras' study. Can the presence of an attack tendency reasonably be assumed in all situations involving the expressions in Table 1?

Although no study has documented the complete range of occurrence of these expressions, there are preliminary observations which are relevant to this question. For example, the author recently observed *lowered brows* to occur quite frequently during cooperative social interactions between mothers and children. During one cooperative play session a 2-year-old boy lowered his brows as he tried to balance one object on top of another—a task which required concentration. A 5-year-old girl lowered her brows as she pointed to an unfamiliar toy and asked her mother, "What's this?" The author also observed the expression *nose wrinkle* in situations that did not involve conflict. For example, a 5-year-old boy wrinkled his nose when told about "a girl who ate a rotten apple." Nose wrinkle generally is thought to characterize the emotion of disgust by investigators of facial expressions of emotion (Darwin, 1872; Ekman and Friesen, 1975; Izard, 1971). Lowered brows has been hypothesized sometimes to reflect perplexity of thought, concentration, or anger (Darwin, 1872; Ekman and Friesen, 1975).

These observations suggest that at least some expressions in Table 1 are not restricted to situations where one would expect an aggressive tendency. Thus there appears to be no simple direct relationship between these expressions and a disposition to attack. However, there are several possibilities regarding more complex relationships which may exist. For example, it is possible that the individual expressions listed in Table 1 are not invariably associated with attack but configurations involving simultaneous use of several expressions are. Thus lowered brows and lips pressed together each may be used in nonaggressive situations. However, when used in combination, the expressions may invariably reflect a tendency to attack in the expresser. Alternatively, perhaps the expressions listed in Table 1 differ in intensity, duration and/or timing depending upon whether or not they reflect an attack disposition. For example, the lowered brow of anger may be more intense than the lowered brow of perplexity and may have a more rapid and abrupt onset. Thus, distinctive forms of the expressions listed in Table 1 may be uniquely associated with an attack tendency in the expresser. Further studies are needed to explore these several alternatives. At this time it seems possible—perhaps even likely—that there is no invariant association between an aggressive tendency and at least some of the individual expressions listed in Table 1.

Regarding an escape tendency, there are virtually no data that indicate the presence of this tendency in a child producing one of the facial expressions being considered. For this reason, regardless of their relationship to an attack tendency, these expressions do not have an important feature of threat displays described by Moynihan. However, as indicated above, an escape tendency does not always underlie threat displays used by nonhuman primate species. Thus, if the expressions listed in Table 1 do reflect a tendency to attack—though not a tendency to escape—they still could be considered threat displays.

To summarize, the children's expressions listed in Table 1 do not appear to result from the conflict between attack and escape tendencies in the expresser. Thus, they do not have the third feature which Moynihan described as characterizing animal threat displays. However, some of these expressions may have a distinctive temporal form or intensity which they take when they reflect the tendency to attack. If so, these distinctive forms of the expressions in Table 1 might be considered threat displays.

A Message/Meaning Analysis of Facial Expressions

The preceding analysis indicated that the expressions in Table 1 may not be intrinsically tied to either an attack or escape tendency. If so, could these expressions still function as threats (i.e., produce intimidation) in certain circumstances? Camras' (1975, 1977) study suggested that they indeed may intimidate recipients in some conflict encounters. Nonplayers observing these expressions showed increased hesitancy in taking the disputed object. If we assume that nonplayer hesitancy reflected intimidation, why did the facial expressions produce this reaction? Is there a model of display communication which could explain this phenomenon?

W. J. Smith (1965, 1977) has proposed a general model for display communication which may prove useful for this purpose. According to Smith's model, a display is a nonverbal act that encodes a set of messages. Messages consist of information about the displayer's identity, his location, the kinds of activities he may perform, the probability of his performing one or another of these activities, and/or the probable intensity at which he will perform the activity selected. Messages may designate specific or general activities. For example, some messages designate specific activities such as "attack" or "escape" while others designate general activities such as "attending to a stimulus." As these examples indicate, messages do not necessarily refer to the motivational state of the displayer. Indeed, the displayer's motivational state may vary across different circumstances in which he uses a particular display. Thus, when an individual uses a display indicating attention to a stimulus, he sometimes may feel aggressive, sometimes fearful, sometimes sexually motivated, and so on.

When a recipient perceives a display, he analyzes its message content in conjunction with information from other sources. For example, when analyzing a display carrying messages about general activities, an individual may use contextual information to make a specific prediction regarding the expresser's subsequent behavior. According to Smith, the meaning of the display is the response evoked in the recipient as a result of his analysis of the display. The meaning of a display for its recipient cannot be observed directly, but it is indexed by the recipient's response to the display.

Regarding Camras' study, Smith's model might be used to explain why nonplayers hesitated to take the gerbil box after observing the expressions listed in Table 1. As a hypothetical example, consider the expression lowered brows. Following Darwin (1872), we might hypothesize that lowered brows carries a general message indicating that the expresser has encountered "something difficult or displeasing . . . in a train of thought or action" (p. 222). Encountering something difficult or displeasing frequently may elicit expresser aggression but sometimes it may be accompanied by such feelings as determination or uncertainty. A recipient observing lowered brows would have to use contextual information to determine whether the expresser was aggressive, determined, uncertain, or some combination of the three. In Camras' study, lowered brows was embedded in the context of a conflict situation. Such a situation can contain many "clues" (including nonverbal clues) suggesting a particular interpretation of lowered brows. In this study, a nonplayer might have interpreted lowered brows as indicating that the player was angered by the nonplayer's attempt to take the box, and would respond aggressively if the nonplayer persisted in his attempt. This interpretation could have produced the nonplayer hesitancy which was observed.

The primate literature provides support for the proposition that facial expressions may sometimes function as threats yet carry more general messages than attack and escape. Several investigators (e.g., van Lawick-Goodall, 1968; Bernstein cited in Rowell, 1972) have found that expressions which may occur in hostile situations also are associated with activities such as infant retrieval and/or copulation. In terms of Smith's model, these expressions may encode a very general message such as "seeking interaction." However, they may serve as threats in those situations where fighting appears to be the type of interaction sought.

Conclusions

Considered as a group, the children's expressions in Table 1 may have some features of animal threat displays, but they probably do not have all three characteristics described by Moynihan. To summarize, these expressions may be ritualized acts; they sometimes may intimidate their recipients; but they do not seem to reflect a conflict between attack and escape tendencies.

In their original studies Grant and Blurton Jones called the expressions in Table 1 "aggressive" expressions. This label may be inaccurate. Both the motivational state of the expressers and the expressions' meaning to recipients may vary depending on the situational context. Future studies should examine these expressions both individually and as a group to gain further information regarding their causal basis and role as social signals. Investigators should not begin with the assumption that these expressions necessarily are aggressive or necessarily are perceived as threats.

Though the expressions in Table 1 may not always function as threats, they still may play an important role in the dominance relations of children. Recall that these expressions did seem to intimidate recipients in the conflicts observed in Camras' study. In natural conflicts between children perhaps they also are influential. Consider the case in which a low-ranking child finds herself in conflict with a high-ranking peer. Her decision to defer or not defer to the other child might depend on the type of facial expression that she sees the high-ranking child use. If the subordinate child sees one of the expressions listed in Table 1, she almost always may defer to the other child. However, if the subordinate child sees none of these expressions, she sometimes may defy her opponent. The facial expressions of a low-ranking child also may influence a child of higher status. High-ranking children may seldom defer to low-ranking peers. They sometimes may do so, however, if the low-ranking child uses one of the expressions listed in Table 1. In general, facial expressions may be related to atypical outcomes in conflict encounters between children. If so, their study will prove useful for understanding episodic reversals of dominance relations reported for children (Strayer and Strayer, 1976).

In conclusion, future studies of facial expressions may make valuable contributions to our understanding of children's dominance relations. However, the original model for a threat display probably does not describe the nature and role of facial expressions used in dominance-related encounters. Further study along several lines is necessary to elucidate the nature of children's facial expressions used in conflict and the role that such expressions play in the dominance relations of peers.

Acknowledgments

The author wishes to thank S. Ancoli, P. Ekman, H. Feldman, D. Omark, H. Oster, and W. J. Smith who commented on an earlier version of this paper.

Note

1. For expositional convenience, the units of analysis described by Grant and Blurton Jones will be called facial expressions. The reader should note that these units actually are only *components* of complete expressions; i.e., they are not descriptions of the appearance of the entire face.

Chapter 7

Preschool Conflict and the Assessment of Social Dominance

F. F. Strayer
Janet Strayer

Dyadic dominance relations and group dominance hierarchies are traditional concepts in primate social ethology (Crook, 1970b; Hinde, 1974; Jolly, 1972; Kummer, 1971). Dyadic dominance describes the relative balance of social power between specific members in a social group, while dominance hierarchies summarize the organization of such power relations among all possible group members. Social dominance has been viewed as a basic dimension of primate social organization which relates to a number of other social processes—e.g., group defense (Jolly, 1972), reaction to strangers (Ripley, 1967), social learning (Hall, 1968; Strayer, 1976), social innovation (Frisch, 1968; Tsumori, 1967), social play (Dolhinow and Bishop, 1972), and general reorganization of the social unit (Furuya, 1960; Sugiyama, 1965a.)

Theoretically, the existence of dyadic dominance and the maintenance of a stable dominance hierarchy function to minimize intragroup aggression by establishing a semipermanent sequence of individual prerogatives (Etkin, 1964; Hinde, 1974). Having learned the appropriate power sequence for its social unit, each group member is able to anticipate, and thus to avoid, immediate adverse consequences of severe social aggression. In spite of general theoretical agreement concerning the adaptive signifi-

---

Portions of this research were presented at the biennial meeting of the Society for Research of Child Development, Denver, 1975, and later published in *Child Development*, 1976.

cance of both dominance relations and dominance hierarchies, the empirical application of these concepts in comparative research has been problematic (Bernstein, 1970a; Hinde, 1974; Richards, 1974; Strayer and Cummins, this volume, Chapter 4).

Early researchers often attempted to assess primate dominance in terms of a single type of social interaction. Usually such interactions consisted of artificial competitions in a controlled setting. Unfortunately, such unidimensional indices of dominance proved inadequate because results failed to generalize to other more natural social settings (Bernstein, 1970). In more recent primate research dominance has been reserved as a *general descriptive* concept that summarizes stable social relations evident in the resolution of various forms of naturally occurring conflict (Alexander and Bowers, 1969; Bernstein, 1970a; Hinde, 1974; Strayer, Bovenkerk, and Koopman, 1975). This use of social dominance necessitates the identification of specific behavioral patterns which characterize conflict episodes between members of the species studied. Finally, this descriptive approach stresses analysis of *responses* to initiated conflict as a primary index of dyadic dominance relations (Rowell, 1966, 1974). This emphasis upon response to conflict provides a conceptual distinction between an agonistic episode and an aggressive act. *Aggression* is a class of individual behaviors, while *agonism* is a class of dyadic interactions.

Given the importance of both dyadic dominance relations and group dominance hierarchies for understanding naturally occurring primate behavior, it is surprising that these ethological concepts have not been extended directly to the analysis of human social relations. Although a few early naturalistic researchers (Anderson, 1940; Chittenden, 1942; Jack, 1934) examined dominance in young children, their definition of this term often varied, and their primary interest focused on personality traits rather than on dyadic social relations. Those researchers who dealt with dyadic dominance interactions (Dawes, 1934; Gellert, 1961) failed to extend their analysis to the more general question of social organization in the peer group. In one of the earliest attempts to relate primate and human dominance Edelman and Omark (1973) developed a social hierarchy questionnaire to examine verbal reports of dyadic social power among young children. These investigators questioned children of different ages about the relative "toughness" of selected pairs of classmates. In each group examined the children's reports revealed sufficient communality in the perception of dyadic social power to permit the identification of a single rank ordering on the dimension of "toughness." More recently, cross-cultural extensions of this research have led these authors to conclude that a typical primate linear hierarchy provides an adequate summary of young children's power relations within the peer group.

In a more traditional ethological study of characteristic behavioral patterns among preschool children McGrew (1972) provided a preliminary analysis of human dominance behavior. His assessments of dyadic dominance were derived from naturalistic observation of wins and losses during conflict over the possession of preferred objects. McGrew reported a nearly linear dominance structure among the boys in his preschool samples. In a more recent behavioral analysis Strayer and Strayer (1975) used three general categories of naturally occurring social conflict—*attacks, threats,* and *object/position struggles*—as converging indices of children's dominance relations. Naturalistic observation of such conflict among groups of preschool and elementary school children during free-play periods revealed stable dyadic dominance relations and linear group status structures at both age levels. Dominance appeared to be a more unitary social phenomenon among the older group of children, however, since only with this group did all three forms of conflict converge to yield a single linear dominance structure. In contrast, with the preschool group, analysis of attack and threat interactions produced a single linear status structure, but this structure correlated only .76 with rankings derived from analysis of object/position struggles.

These preliminary studies suggest that there is a structural similarity in the organization of power relations within primate societies and childhood peer groups. However, more definitive evidence for such evolutionary continuity requires more extensive examination of children's conflict interactions. The present research was designed to provide such an analysis through the development of a more fine-grained social agonism inventory. Since our preliminary work suggested the lack of a unitary dominance structure among preschool children, we decided to focus at this age level.

Method

Subjects

Over a three-month period 18 children between the ages of 3 and 5 years were observed daily at their preschool. Systematic observations and video records were obtained for the last six weeks of this period on 17 children, since 1 girl had terminated her enrollment at the center.

Procedure

Repeated observation of the videotaped records of social conflict during the initial six weeks of our study resulted in the elaboration of a final social agonism inventory. This final descriptive framework contained two general classes of behavior observed during social conflict. The first class, which was subdivided into three smaller categories, described aggressive behaviors which often appeared as the first element in a sequence of social conflict. For example, the category of physical attacks included six specific patterns of initiated aggression: *bite, chase, hit, push–pull,* and *wrestle.* The *threat gestures* category contained four forms of activity: *intention hit, intention kick, intention bite,* and *face/body postures.* Finally two forms of competitive conflict were distinguished: *object struggles* and *supplants.* In addition to these forms of initial behaviors, the final agonism inventory included five categories of responses to aggression: *submission, help-seeking, counterattacks, object/position loss,* and *no response.* Submission consisted of seven specific appeasement gestures: *cry, scream, rapid flight, cringe, hand cover, flinch, withdraw,* and *request cessation. Help-seeking* was distinguished in terms of the target—i.e., another child or an adult. *Counterattacks* included any response pattern which was already included in the list of initial aggressive acts. Finally, the categories of *object/position loss* and *no response* were not subdivided.

Although this final agonistic inventory contained a variety of relatively simple behavioral patterns, the conflict which was actually observed often involved more complex organizations of both co-occurring and sequential behavioral acts. For example, the series of drawings in Figure 1 shows an agonistic sequence in which a smaller child pushes a larger boy twice and then threatens him with a closed-fist intention hit. The larger boy fails to respond to each of these aggressive acts. This sequence was actually coded as SA push → BR no response/SA push → BR no response/SA intention hit → BR no response. Figure 2 shows a simple dyadic exchange which contains two co-occurring outcomes. One girl directs a threat to another, who reacts by withdrawing. Although the target of this aggression lost her chair to the aggressor, only the withdrawal was coded in our description of this simple agonistic exchange (KA face threat → TY with-

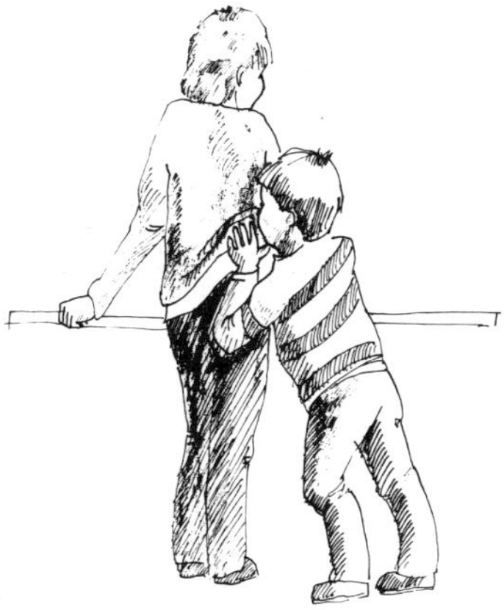
A

B

C

Figure 1. Repeated aggression (SA push—BR no response/SA push—BR no response/SA intention hit—BR no response).

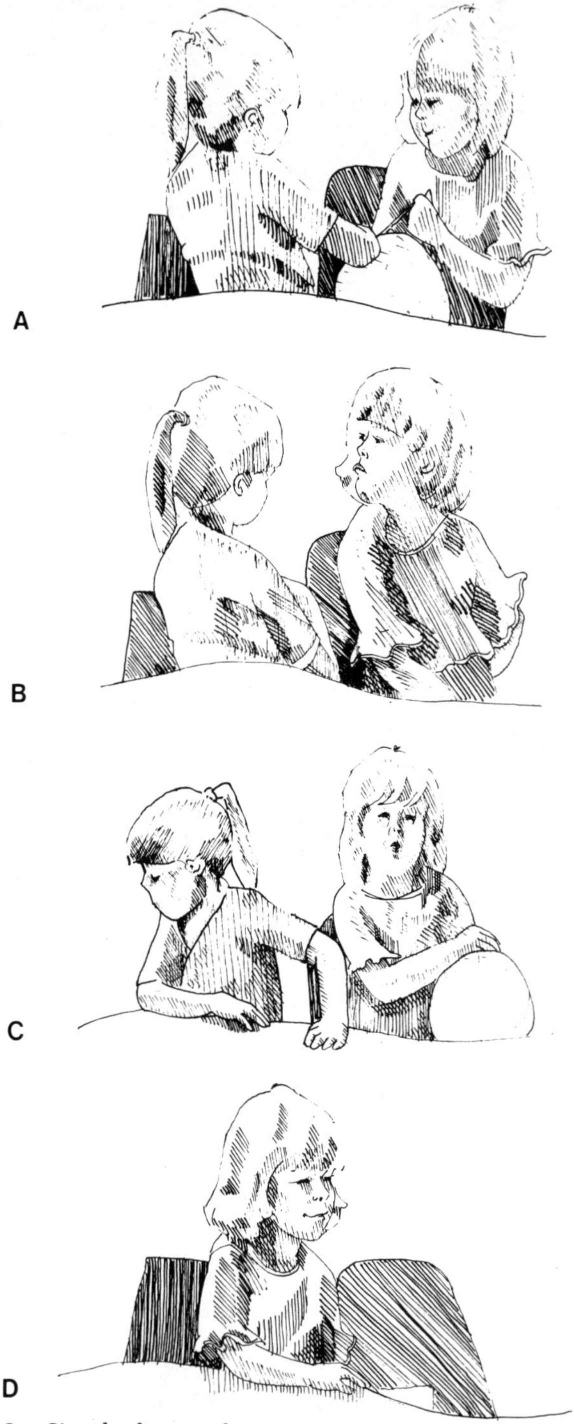

Figure 2. Simple threat-submission (KA face threat—TY withdraw).

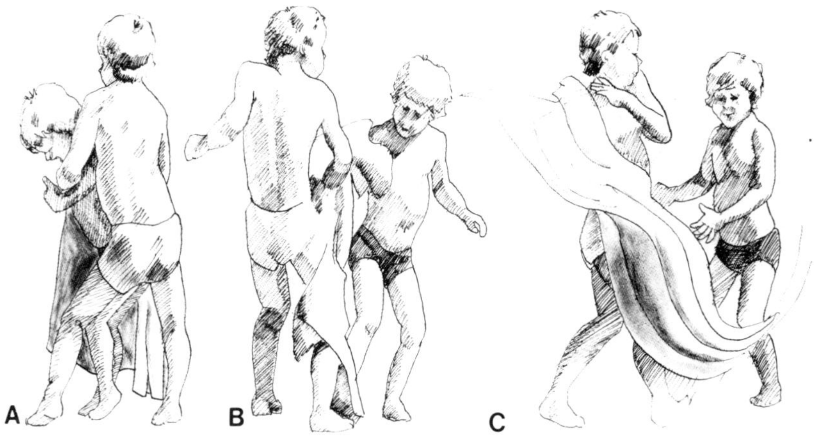

Figure 3. Simple object struggle (IR struggle—GL object loss).

draw). Similarly, the drawings in Figure 3 illustrate a relatively complicated interaction which ultimately was coded as a simple dyadic exchange. Although this conflict bout included body contact that could have been scored as wrestling, the sequence was included as a simple example of competition (IR object struggle → GL object loss). A second less aggressive form of competitive conflict is shown in Figure 4, where one boy gently moves another aside. This second episode also was coded as a simple dyadic exchange (IF position struggle → IR position loss). Finally, Figure 5 shows a somewhat different coding problem. Here a hit–cringe simple exchange leads to the intervention of a third child who hits

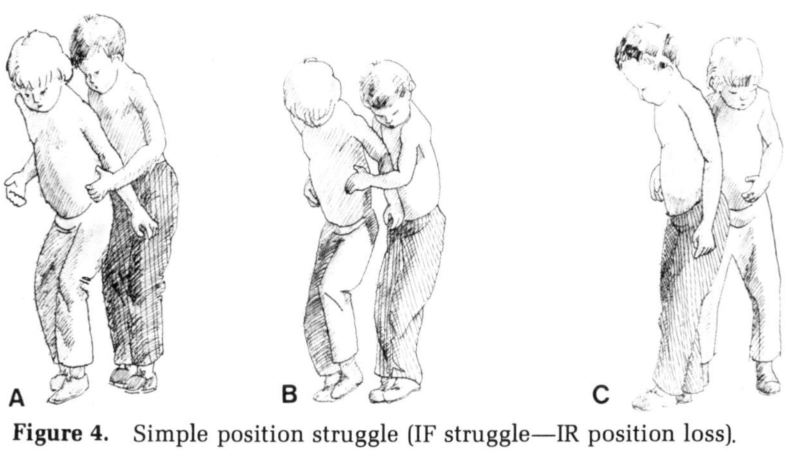

Figure 4. Simple position struggle (IF struggle—IR position loss).

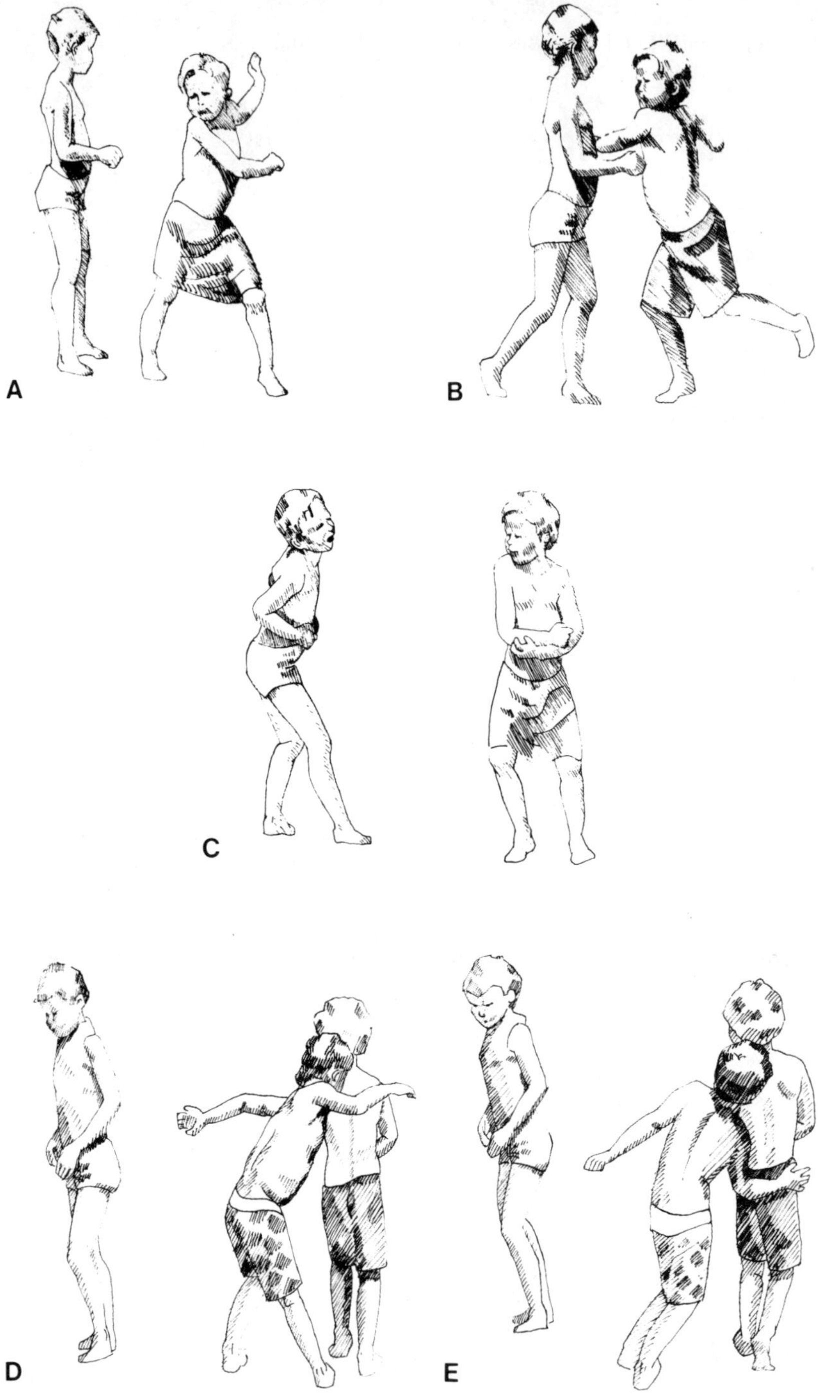

Figure 5. Tripartite interaction (GL hit—SH cringe and IF hit—GL no response).

the original aggressor. Our coding rules broke this complex tripartite interaction into two simple dyadic exchanges (GL hit → SH cringe and IF hit → GL no response).

Dominance relations were assessed through analysis of video records obtained during the second six weeks of observation. These data were collected using a matrix-completion method which allowed simultaneous monitoring of subgroups of individuals in order to maximize the amount of agonistic information available for all dyads within the group. When two agonistic episodes occurred at the same time, the observer selected the episode which provided new information on the dyadic distribution of social conflict. Such sampling decisions favored obtaining representative information on the dyadic asymmetry of social conflict within the social group rather than trying to provide data for estimating behavioral rates for individual subjects (Altmann, 1974). Our records of conflict always included the initiator, the target, the pattern of initiation, and the response of the target in each social exchange. If an episode contained an extended sequence of interaction, each simple exchange was coded separately but the sequence was kept intact by noting the number of exchanges it entailed. Interobserver reliabilities in the description of observed conflict using these procedures ranged from .92 to .96 on four separate tests using half-hour video records.

Results

Social Agonism

During the last six weeks of systematic observation, 443 agonistic exchanges were recorded. Table 1 shows both the total and relative frequency of occurrence for each category and pattern of initiation included in the present agonism inventory. Nearly 40 percent of the agonistic episodes entailed some form of physical attack. Two patterns—hit and push–pull—accounted for over three-quarters of attack interactions. Threat gestures were the second most frequent type of initiated agonism. Approximately a third of the observations were scored in this category. Once again, two patterns—intention hit and face/body posture—accounted for over three-quarters of the observations. Slightly more than a quarter of all observations entailed object/position struggles. However, with this final type of conflict there were virtually no differences between the category subdivisions.

Table 1. Summary of Initiated Agonistic Actions

Behaviors	*Observed Frequency*	*% of Total Agonism*	*% of Total within Category*
Physical attack			
Chase	10	2	6
Push–pull	63	14	36
Hit	80	15	46
Kick	12	3	7
Wrestle	6	1	3
Bite	2	1	1
Total	173	36	
Threat gestures			
Face and body posture	49	11	34
Intention hit	76	17	52
Intention kick	19	4	13
Intention bite	2	1	1
Total	146	33	
Object/position struggles			
Struggle without contact	60	14	48
Struggle with contact	64	14	52
Total	124	28	
Total agonism	443		

Responses to initiated agonism are summarized in Table 2. Seeking help occurred at a surprisingly low frequency—only on four occasions did the victim of an agonistic episode seek external support. It is interesting that on each of these occasions help was sought from one of the preschool teachers and not from another member of the peer group. Three of the remaining response categories (submission, object/position loss, and no response) each comprised about a quarter of the observed responses. Flinch was the most frequent submissive gesture; most of the remaining submissive patterns, with the exception of request cessation, each accounted for about 10 percent to 15 percent of the acts scored in this category. Slightly more than one in five agonistic episodes involved a counterattack by the target. The majority of such counterattacks were scored in the threat-gesture category. Often if an

Table 2. Summary of Responses to Initiated Agonistic Actions

Behaviors	*Observed Frequency*	*% of Total Responses*	*% of Total within Category*
Helping seeking			
Seeks adult help	4	1	100
Seeks child help	0	0	0
Total	4	1	
Submission			
Cry–scream	12	3	10
Rapid flight	18	4	16
Cringe	16	4	15
Hand cover	11	2	9
Flinch	35	8	31
Withdrawal	16	4	15
Request cessation	5	1	4
Total	113	26	
Object/position loss	113	26	
No response	118	26	
Counterattack	95	21	
Total responses	443		

initiated act led to a counterattack, the agonistic episode would run into a more extended sequence with each member of the dyad countering attacks by the other. Such lengthy sequences almost always ended with one individual engaging in a submissive act. In contrast, initiated acts which led to no response by the target were usually short in duration, often consisting of the single dyadic exchange.

Dominance Relations

Analyses of dyadic agonistic interactions were conducted to assess the degree to which our observations corresponded to a linear dominance model. The appropriateness of this model as a summary of group patterns for conflict resolution was evaluated in terms of the percentage of observed dyadic dominance relations which correspond to the linear transitivity rule. This rule requires that if individual A dominates B, and individual B dominates C, then A should also dominate C. Characteristically, within primate groups, nearly all observed dominance relations correspond to this rule (Alexander and Bowers, 1967; Richards, 1974; Strayer, 1976; Strayer, Bovenkerk, and Koopman, 1975). A second question concerned the rigidity of children's dominance relations. Rigidity of dominance was assessed in terms of the number of agonistic episodes in which the usually subordinate member of a dyad was observed to dominate. Thus in a group where one member of each possible dyad wins all agonistic encounters, the dominance relations would be completely rigid. In a second group where determination of relative dominance for each dyad was based upon one member winning only 60 percent of total interactions, dominance relations would be considerably less rigid. Although highly unlikely given the comparative empirical data, it is theoretically possible to find a perfectly linear status hierarchy with relatively flexible dyadic dominance relations.

Examination of relative linearity and rigidity of status rank-

TARGETS

INITIATORS

Ss	Ro	Ss	Br	If	Td	Sd	Pe	Ir	Cs	Ka	Ch	Ty	Gl	Sa	Me	Ju	Sh	Tot.
Ro		1	1		1·								1	1				5
Ss			1·	3	1			4	1			1	3				1	15
Br		1·		1	1				1		6		3	2			1	16
If					2	1	8	2	2	1			4	2		2	1	25
Td	1·					·	3		7	2						1		14
Sd					·		1	1	4	2	1					1		10
Pe					1			1					1·	3				6
Ir									3	2	1							6
Cs						2		1		·	·		1					4
Ka									·		·	9		1		1	3	14
Ch				1∸					·	·		·	5	3		1	1	11
Ty											·		1·	·	2			3
Gl							1·				1	1·		·	·	3	5	11
Sa						1∸					1	·	·		·			2
Me													·	·		·	·	0
Ju															·		·	0
Sh															·	·		0
Tot.	1	2	2	5	6	4	13	9	18	7	10	11	19	12	2	9	12	142

∸ = DYADIC REVERSAL

· = DYADIC TIE

Figure 6. Dyadic dominance matrix.

ings derived from the present conflict categories led to the selection of agonistic episodes which entailed submission following physical attacks and/or threat gestures as the primary behavioral index of social dominance. Systematic comparison of interactions within this class of agonism led to the derivation of the status ranking shown in Figure 6. This ordering of children minimizes the number of interactions which appear in the area below the diagonal of the matrix. Of the 142 observed attack–threat interactions leading to submission, there are only 12 exchanges which lie below the diagonal. Thus, 92 percent of the observed exchanges conform to expectations derived from a linear dominance model. Of the 136 possible dyads in the group, 61 engaged in attacks and threats leading to submission. Since there are two relational reversals of dominance in Figure 6, the linearity of the attack–threat hierarchy was 98 percent. It is interesting to note that both of these reversals entailed a lower ranking child threatening, rather than attacking, a higher status group member. Of the remaining 10 episodic reversals, 4 indicated tied dyadic dominance relations

TARGETS

INITIATORS \ TARGETS	Ro	Ss	Br	Lf	Td	Sd	Pe	Ir	Cs	Ka	Ch	Ty	Gl	Sa	Me	Ju	Sh	Tot.
Ro		1	3	4	1			1	1		1		7		1			20
Ss	1		7	8	2	1	1	12	3		1	1	4	1			2	44
Br	1	4		7	3	2	2		1		8	1	5	5			1	40
lf	3	3	2		3	1	13	3	5	1			8	3		2	1	48
Td	1			3		4	6		8	5	1		1	3		2	1	35
Sd							2	8	11		4		4	3		1		35
Pe	1		1	9	3	4		2			1		7	9	1	1		39
Ir				1	1	1	2		7	5	1	1	1					20
Cs	1	1	1	2	5	11		3					1					25
Ka							1					11		1		1	4	18
Ch	4		4	3	3	2	1					3	11	5		2	2	40
Ty										2			2		8			12
Gl		1	9	3		3	6				11	2		1		7	5	48
Sa		1	4		1	2	1				1		1					11
Me												3						3
Ju						1						1	1					3
Sh									1	1								2
Tot.	12	11	31	40	22	32	36	30	37	14	29	23	53	32	9	16	16	443

Figure 7. Dyadic matrix of total initiated agonism.

with 3 of these ties involving a lower ranking child threatening, but not attacking a higher status child. The remaining tie resulted from an attack–submission interaction initiated by each member of a dyad (SS-BR). Figure 6 also shows that dominance relations were more equivocal among children ranked lower in the hierarchy. The vast majority of sequential ties in dominance are found among the last 9 children. Finally, analysis of rigidity of dominance relations indicated that the outcome of only 10 conflict–submission episodes was contrary to that expected given the observed dyadic dominance relations. Thus, the rigidity index for observed dominance relations was 93 percent. This indicates that submission to attack and threats was almost completely unidirectional at the dyadic level.

Such unidirectionality was not characteristic of total initiated agonism. Figure 7 illustrates the total frequency of initiated conflict for each dyad in the preschool group. Children low in the dominance hierarchy also initiated conflict with individuals of higher status. For example, although RO and SS each initiate the

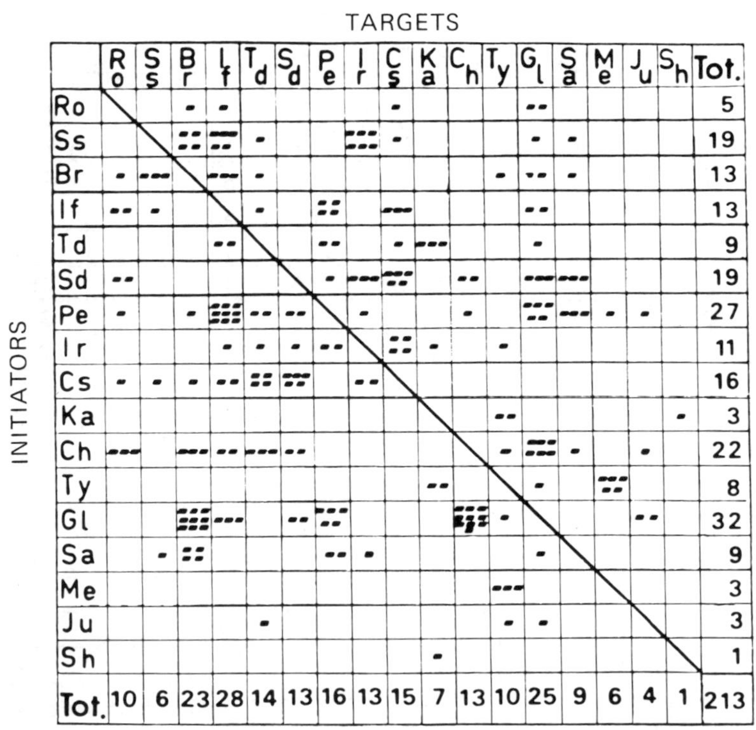

Figure 8. Dyadic matrix of interaction resulting in counterattack or no response.

majority of conflicts with IF, IF also directs agonistic acts toward RO and SS. In the total initiation matrix nearly 25 percent of observations fall below the diagonal. This finding indicated that knowledge of dyadic dominance relations would have correctly predicted the direction of only about 75 percent of the observed aggressive acts. The dominance hierarchy for this group does not suggest a simple pecking order with the direction of initiated aggression always moving downward in the social order. The assessment of relative dominance necessitates noting not only the social direction of aggressive activity, but also the behavioral effect of such activity on the recipient. The difference between the dyadic matrices in Figures 6 and 7 clearly shows the importance of attending to both initiation and response in the analysis of agonistic interaction and social dominance.

Figure 8 shows all initiated dyadic acts which resulted in counterattacks or no responses. These interactions are distributed about equally on both sides of the dyadic matrix. Again, knowledge of dominance relations provides a relatively poor prediction

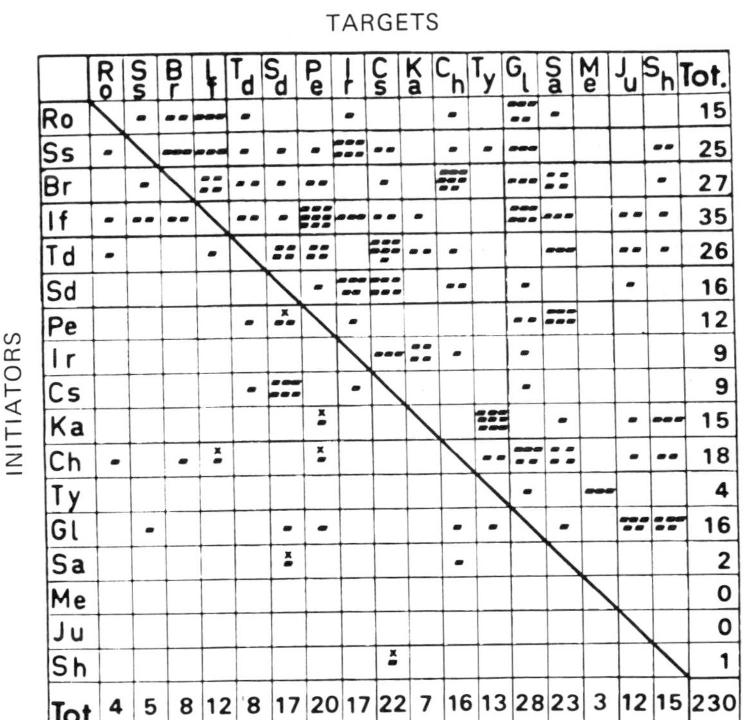

Figure 9. Total initiated conflict leading to either submission or object/position loss. (X indicates dyadic reversal of linear dominance.)

of this class of dyadic exchange. The elimination of no-response and counterattack episodes from the total initiation matrix leaves a dyadic array comprised of only those initiated acts leading to submission and object/position loss. The latter matrix, shown in Figure 9, more closely resembles a typical primate dominance hierarchy. Nearly 60 percent of the possible dyads were observed to engage in this type of conflict. Of these 75 dyads, only 6 show reversals of attack–threat dominance rankings. Dominance relations in Figure 9 are 92 percent linear. Only 31 episodes indicate clear violations of observed dominance relations. Rigidity of dominance defined in terms of submission or loss during social conflict was approximately 86 percent. This rigidity index is only marginally different from that derived from the smaller subset of data in Figure 6.

A more detailed analysis of dyadic interactions was conducted separately for each category of conflict. Approximately 40 percent of the possible dyads engaged in object/position struggles leading to object/position loss. Figure 10 shows that there were 8

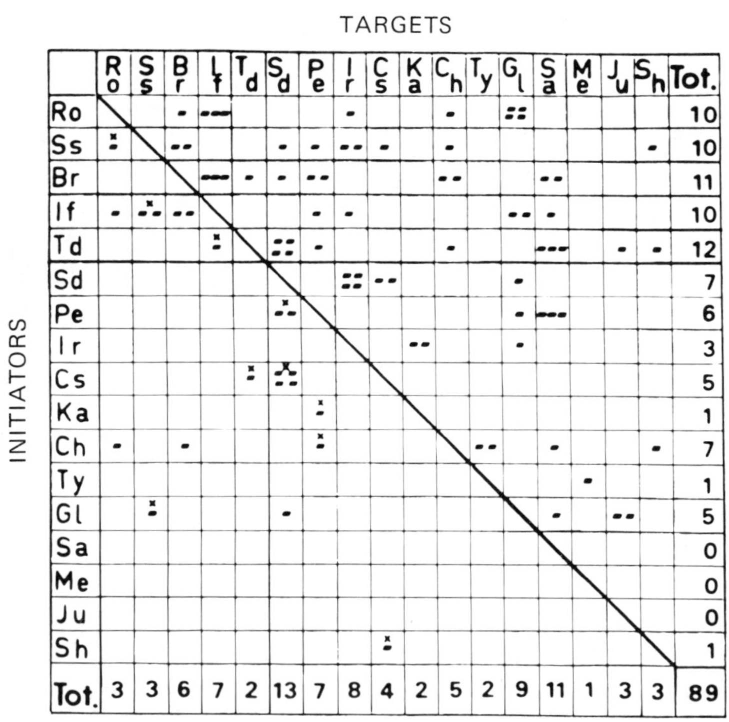

Figure 10. Initiated object/position struggles resulting in object/position loss. (X indicates dyadic reversal of linear dominance.)

violations of the linear transitivity rule and 21 episodic reversals within this class of interaction. The attack–threat dominance ranking predicted only 88 percent of object/position struggle relations, and 76 percent of outcomes in object/position struggles. Since there were 9 episodes which were contrary to observed object/position struggle relations, the rigidity index for this class of interaction was 90 percent. This level of rigidity differed from that found with either attacks or threats leading to submission.

Figure 11 shows threat interactions which led to submission by the target child. Although only 30 percent of the dyads engaged in this class of interaction, there were 6 violations of the linear rule and 9 episodic reversals. The attack–threat ranking would predict 88 percent of both threat relations and threat outcomes. However, given the 6 relational reversals in this matrix, the rigidity of threat relations is substantially higher than that of object/position relations. Since only 3 episodes violate observed relations in Figure 11, threat rigidity is nearly 96 percent. Finally,

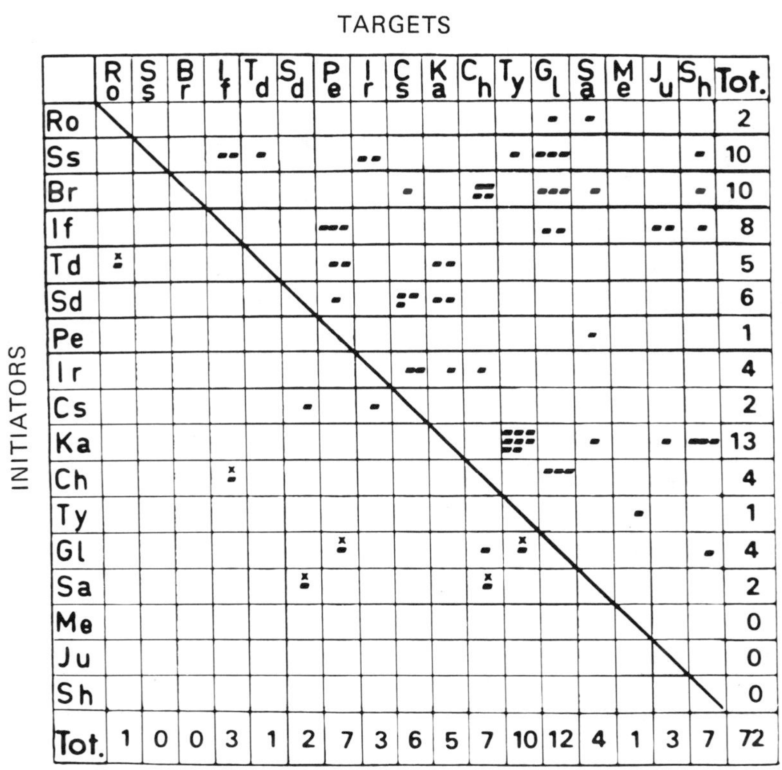

Figure 11. Initiated threats leading to submission. (X indicates dyadic reversals of linear dominance.)

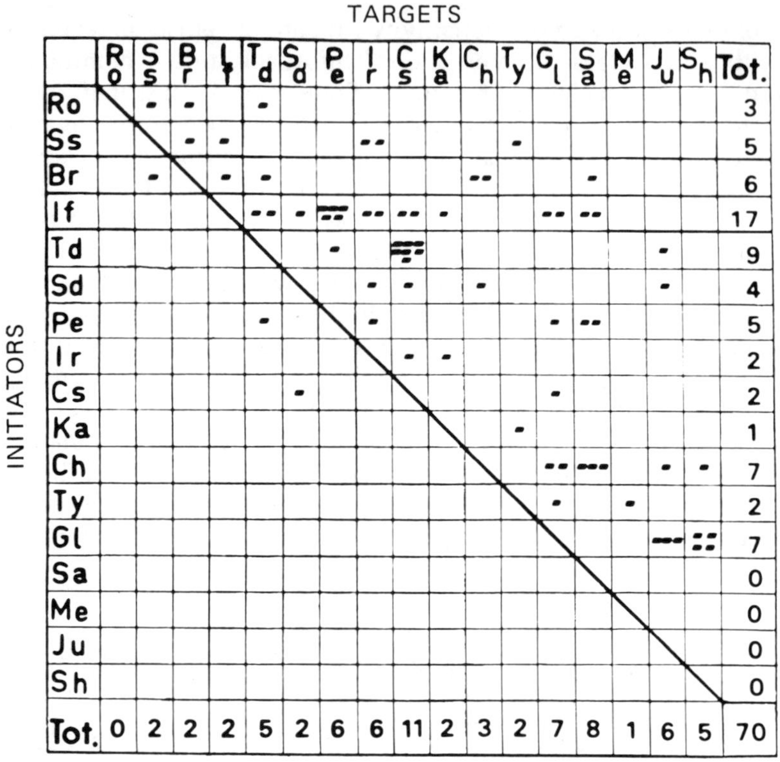

Figure 12. Initiated attacks leading to submission.

observed attack interactions resulting in submission are shown in Figure 12. Again, 30 percent of the dyads engaged in this class of interaction. There were no violations of the linear dominance rule for any of these 41 dyads. Thus, the attack ranking was 100 percent linear and in complete agreement with the original dominance assessment. Since there are 3 examples of episodic reversals, attack relations were also 96 percent rigid, or unidirectional.

Discussion

It is important to emphasize that each set of behavioral measures does not necessarily provide an identical view of either dyadic dominance relations or the group status hierarchy. The present analysis stresses the interactional nature of social conflict and requires that equal attention be given to the activity of both participants in the aggressive social exchange. This approach implicitly assumes that dominance between two individuals is established by a mutual agreement symbolized by the submissive response. Instances of self-assertion (i.e., initiated agonistic acts) that are not acknowledged by the social target do not necessarily provide evidence of dyadic dominance. The role of both participants is critical for an adequate assessment of a social relationship. This emphasis upon *submission* during social conflict is prerequisite to an empirical distinction between dyadic dominance and individual aggressiveness. For example, RO (the alpha girl in Figure 6) initiates dominance interactions with her peers on only five occasions. This contrasts sharply with the relatively greater initiation score of GL—a more submissive low-ranking boy.

The application of ethological concepts and methods to the analysis of preschool conflict and dominance relations resulted in the clarification of social constructs that have had a variety of meanings in psychological literature. The convergence of behavioral indices for dominance relations among the present preschool children corresponds quite well with dominance indices used to assess status with nonhuman primate groups. Furthermore, the low incidence of counterattacks (20 percent) suggests that the dominance structure of these preschool children may function to minimize intragroup aggression just as it does in other primate groups. Also, it is interesting to note that the two relational reversals in the dominance hierarchy involved a close friend of RO, the alpha member of the group. These two examples provide the only indication of possible interaction between affiliative and power dimensions in the present preschool sample. A more detailed ethological investigation of these two social dimensions, however, may reveal the existence of temporary alliances that could foster basic and contingent dominance relations among specific children. The possible existence of such alliances would likely result in context specific dominance reversals that would decrease estimates of rigidity and linearity of group dominance structure. Such cooperative interaction has been observed among nonhuman primates and should eventually play an important role in young children's dominance relations.

Contrary to earlier reports that the majority of conflicts among young children involved object struggles (Dawes, 1934; McGrew, 1972), only about one-quarter of the present observations were recorded in this category. This finding suggests a somewhat different developmental view of preschool conflict. Furthermore, differences in both linearity and rigidity of object/position struggle conflict compared with the primary dominance indices suggest possible ontogenetic trends in the development of dominance relations. Among younger children status rankings may be to some extent interaction specific. Only after more extended experience with the peer group would these different structures converge to produce a unitary group dominance hierarchy. The finding that object/position struggles seem to yield slightly different status rankings for very young children direction parallels developmental differences in status rankings recently reported among captive monkeys (Strayer and Cummins, this volume, Chapter 4). Young monkeys are able to displace and steal from more dominant group members with relative impunity. A decrease in the rigidity of dominance relations in order to permit such unorthodox behavior may well have important functional significance since it would facilitate exploration of the physical and social environment at a time when the young animal must acquire extensive practical knowledge. Perhaps the present discrepancy in status rankings reflects a particular time in development when similar impunity begins to wane among preschool children.

The presence of such developmental trends would imply that early dominance relations are modified through social learning. Further research should examine whether different types of reaction to initiated conflict function to strengthen or weaken specific patterns of initiated agonism. For example, submission could strengthen and counterattacks could punish aggression, while no response could function to extinguish it. Essentially such research could identify naturally occurring social reinforcers of aggressive behaviors.

The present analysis of social conflict also complements contemporary research on individual differences in aggressiveness (Feshbach, 1970; Maccoby and Jacklin, 1974). There were overall sex differences in frequency of initiated agonistic behaviors, with boys showing a higher rate of aggression than girls. Frequency of initiated aggression, though, does not determine dominance status. It is important to note that a relatively low-aggressive girl was the most dominant individual in this group. How individual differences interact with social organization is an important question. However, information concerning the relation of dominance to age, sex, and other individual "trait" concepts requires more general application of the present procedures to a variety of different peer groups. Some of these questions about individual difference are examined in the following chapters. The relation between behavioral dominance and other aspects of peer-group social organization, such as leadership and popularity, is an area amenable to investigation from ethological perspective. It seems probable that continued refinement of the present conceptual and methodological approach to children's social behavior will eventually provide more comprehensive insights into both the phylogeny and ontogeny of peer-group social structure. Such insights are essential for an adequate understanding of individual adaptation to social pressures which are provided by peers in a natural context.

Acknowledgments

The authors would like to express their appreciation to Ingrid Fischer who helped with data collecting, and to the children and staff at Langara Preschool Center, Langara College, Vancouver, British Columbia. Our special thanks to Teresa Blicharski for her illustrations of the conflict categories.

Chapter 8

Dominance Relationships in Preschool Children

Andrew M. Sluckin

The Concept of Dominance

Many everyday terms have been borrowed by psychology and given a specialized meaning. Whether this really enables the researcher to produce clearer scientific models is debatable, but it undoubtedly renders the subject even more difficult for everyone else to understand. The concept of dominance falls into just this category. Because it has been used in so many different ways, confusion has resulted in the minds of the laymen and between groups of professionals alike.

Much of this book has tried to clarify this issue. For our part, we should be clear that a score for "dominance" on some personality test means little unless it also predicts the behavior of that person in some novel situation. If it does, then we must be gaining some knowledge about the quality of his or her "dominance" relationships with other people. In particular, we would be finding out who is likely to be the dominant or subordinate partner under certain conditions.

Robert Hinde sums it up in the following way: "You cannot use dominance as though it were a thing. In my view, it can best be treated as a sort of intervening variable, useful when the various dependent variables are correlated, but not useful when there is only one dependent variable or when the dependent variables do not correlate" (personal communication). In other words, the term dominance may have descriptive value, but it has no explanatory value if one only observes one behavior, such as the number of times John hits Jim.

There is also an important distinction between discussing the dominance relationship between two children and whether these dominance relationships can yield information about the group dominance structure. To quote Robert Hinde again: "Thus, when one is dealing with a number of relationships, for each of which there is a dominant or subordinate partner, it may or may not happen that the various relationships can be arranged in a linear hierarchy."

With these points in mind, we shall first review the evidence for dominance relationships within pairs of preschool children. Next the range of interactions that dominance patterns may explain will be discussed. The third section will consider whether the various relationships can be ordered into a hierarchy, and, lastly, the implications that dominance has for a child's development will be examined. First, a few words about a methodology that may be appropriate.

An Ethological Approach

Ethology is particularly concerned with the observation of behavior in natural surroundings, the description of a wide range of behaviors in basic terms, and a wide (evolutionary) perspective on development and function of these behaviors (Smith, 1974b). Although it is not clear exactly what constitutes the "natural" environment for which preschool children are adapted, many ethologists have made observations while the youngsters are freely playing. Such studies have conveniently been carried out at preschool play groups or day nurseries.

As far as the concept of dominance is concerned, the observation and description of a wide range of behaviors should make it possible to delineate the range of encounters between children whose outcome can be predicted from a knowledge of the direction of their dominance relationship. As Hinde points out, this would be the way to test whether the concept of a dominance relationship has any explanatory value. Should this be the case, then an ethological approach would want to discuss dominance in terms of its possible phylogenetic adaptedness as well as the nature of its ontogenetic development in children.

Behavioral and Report Data

Even within an ethological framework the investigation of dominance relationships among preschool children could follow either or both of two very general procedures. On the one hand, it is possible to observe children's interaction and record the quality and quantity of their relationships, while another method is to ask questions of the children themselves about these relationships.

The distinction really concerns gathering behavioral data as against report data. Whereas most animal ethologists, with the notable exception of Dr. Dolittle, could not talk to their subjects, human ethology has this possibility. Indeed by gathering both these sorts of data it may be possible to gain a better view of "dominance relationships" than could be achieved from one source alone. Behavioral approaches will be considered first, however, since these provide the bulk of our data concerning dominance.

The Dominance Relationship

Evidence for the Relationship

What evidence is there for assuming that for some of a child's interactions with another child there is a consistent pattern of wins and losses? Researchers, in particular Krebs (1972), have observed what they have called "aggressive" encounters between children and conclude that stable patterns of wins and losses emerge. Thus, if two children, John and Tony, are observed on Monday and John is seen to win five times over Tony's one, then if we observe them again on Friday, we are likely to find a similar pattern. This consistency should also hold if John and Tony were each observed to win and lose an equal number of their encounters, and it is this stability over time that provides us with a means of prediction. We also need to know just exactly which behaviors and hence types of interactions are open to prediction from a knowledge of these win/lose patterns or dominance relationships.

The Range of the Relationship

The first observational studies with preschool children used very wide criteria for dominance. Thompson (1967), for instance, defines a dominance interaction as one in which a child "physically won a fight, displaced another child from a location, ended up with a mutually desired object, or obviously controlled another child's behavior, usually through verbal commands." The problem here is that we cannot know beforehand whether the same dominance relationships within a pair of children would be found were dominance separately assessed by reference to each of the various parts of this definition.

To answer this question, Krebs (1972) attempted to delineate a set of dominance behaviors. In all, she found 25 cases of behavior which predicted the same dominance relationships as those defined by hitting:

Behaviors Defining Dominance Relationships

Hit	Displace
Push	Throw at
Pull	Chase
Kick	Gain access to object
Hit with object	Cause to flinch
Fight	Raise hand to
Bear hug	Paint another's object
Punch	Push object at
Tread on	Go to hit with object
Command (obeyed)	Paint another child
Take	Cause to fend
Aggressive speech	Shake fist at
Prevent taking	

The importance of this list is that it enables the researcher to observe a far wider range of dominance behavior than might have been thought possible were the definition more arbitrary. For instance, common sense might tell us that painting another child says little about dominance, but quite a lot about that particular child's interests. Krebs' work shows us that it can also be used as an indicator of dominance. It is clear that it makes sense to talk of dominance relationships and such a concept may have some explanatory value, but the issue of whether these relationships may be ordered within the group has still to be resolved.

In Search of a Dominance Hierarchy

Two early studies (McGrew, 1972; Thompson, 1967) report that observations of aggressive encounters between children enabled them to construct a linear hierarchy for their groups. They did so by ordering the children within the win / loss matrix in such a way as to minimize the number of wins which went against the order. They did not have a large amount of raw data, and thus there could have been many such orderings which approximated their criterion.

A second problem is that this method does not really consider the question of a structure within the dominance relationships for the group of children. Rather, it arranges them in an order of those with the most "dominance" to those with the least amount of this magical quality. To consider relationships another approach is needed. In this respect, an experiment by the author will be presented to confirm the conclusions of Krebs (1972) that a dominance order can describe some aspects of children's behavior in a preschool play group.

Experiment I

In our Experiment I study (see Sluckin and Smith, 1977, for a fuller account) Child A was defined as dominant to child B if, within their encounters, he or she has two excess wins. Unlike Krebs' very systematic approach to what constitutes a win, Thompson's (1967) criterion for winning was used: "physically winning a fight, displacing another child from a location, ending up with a mutually desired object, or obviously controlling another child's behavior, usually through verbal commands."

Two independent play groups were observed. The first had 12 children (6 boys and 6 girls), while the second had 11 children (4 boys and 7 girls). These children were either 3 or 4 years old.

In all, 436 incidents were noted over a four-week period, and 79 percent of these could be scored as wins to a child. Incidents that could not be scored in this way included those in which more than two children were involved or where there was no clear winner. All wins were entered onto a matrix of wins and losses. From this it was possible to extract the dominance relationships. Whereas a criterion for dominance in this study was two excess wins, Krebs (1972) used a criterion of only one excess win.

THE IMPORTANCE OF TRANSITIVITY AND INTRANSITIVITY. The interesting question is whether the pattern of dominance relationships can be described by a linear dominance order. To answer this it is critical to understand the difference between three relationships showing transitivity or intransitivity. This can be best explained in the following way, where children are represented by letters and an arrow between them indicates a dominance relationship in that direction. Suppose it is found that A → B (i.e., A is dominant to B) and that B → C. If A → C is also found, the three relationships (A → B, B → C, A → C) show transitivity. If C → A is found, they show intransitivity. Thus, intransitivity occurs when the dominance relationships are nonlinear.

If the pattern of dominance relationships can be described by a linear dominance order, then we should expect to find only instances of transitivity. Depending on the number of cases of intransitivity, the evidence for a hierarchy becomes progressively weaker. Krebs (1972) argued for the existence of a linear dominance order since she found no case of intransitivity in one group of preschool children, and four cases out of another. Her statistical analysis showed these four cases to be insignificant in comparison to the evidence from the other transitive relationships which pointed to a linear order.

RESULTS. The dominance relationships of our study are displayed in matrix form in Tables 1 and 2.

Table 1. Observed Dominance Matrix for Group One

Boys [B] and Girls [G] separately numbered by dominance rank. A cross (x) indicates that child in side column is dominant over child in top row.

	B1	G1	B2	G2	G3	B3.5	B3.5	G4	B5	B6	G5.5	G5.5
B1		x	x	x			x	x	x	x	x	
G1			x	x	x		x	x	x		x	
B2						x						x
G2						x						
G3							x		x			
B3.5								x				
B3.5									x			
G4									x	x		x
B5											x	x
B6												
G5.5												
G5.5												

Table 2. Observed Dominance Matrix for Group Two

Boys [B] and Girls [G] separately numbered by dominance rank. A cross (x) indicates that child in side column is dominant over child in top row.

	B1	G1	G2	B2	G3	B3	G4	B4	G4	G6.5	G6.5
B1			x	x	x	x	x	x		x	x
G1							x	x	x	x	x
G2						x				x	
B2								x			
G3								x			
B3								x	x	x	x
G4											
B4										x	x
G4											
G6.5											
G6.5											

When these matrices are examined for instances of transitivity and intransitivity, the surprising result is that there is not a single case of intransitivity for either group—*all* of the relationships are *transitive*. Along with Krebs' study, this points very strongly to the existence of linear dominance orders at a very early age.

CONSTRUCTING A HIERARCHY. The path is not quite clear, however, for the construction of a hierarchical order. So far, there is information about a limited number of dominance relationships within the group, none of which are intransitive. But if the author had observed the children for even longer, it might have been possible to gain data about the other dominance relationships, and some of these might have shown intransitivity. Since there is no evidence to suggest this from all the other observations, it is possible to feel fairly confident in assuming that the unsampled relationships also show transitivity. It is this assumption that enables us to construct a hierarchy, and Tables 1 and 2 can be represented by the three forms of Figures 1 and 2.

These tree diagrams have the property of restricting the range of ranks that any particular child can occupy within a linear dominance order. Figures 3 and 4 show this range for each child, and Table 3 shows the final dominance orders for each group.

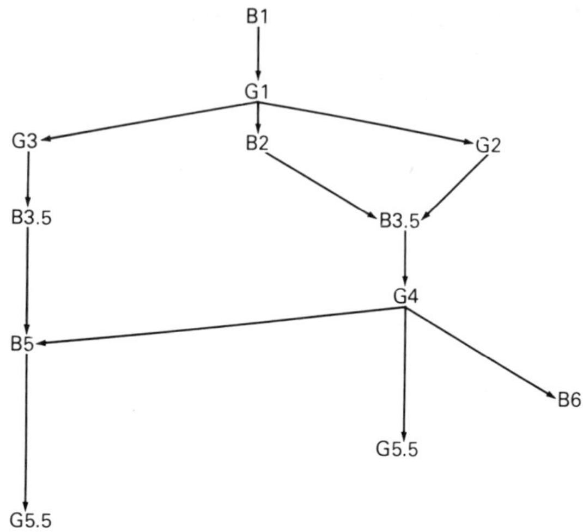

Figure 1. Dominance relationships in tree form for group 1.

It is clearly possible to construct a linear dominance order with some shared ranks for both groups. In Figure 3 each child in group 1 is restricted to a smaller range of ranks than the group 2 children of Figure 4, where, for instance, G1 could occupy any of the ranks from 1 to 6. For the mathematically minded, it should be possible to derive a numerical measure to describe this "average spread of ranks possible," and this could be used as an index of "orderliness" or "linearity" to show the extent to which the observed data really fix the hierarchy.

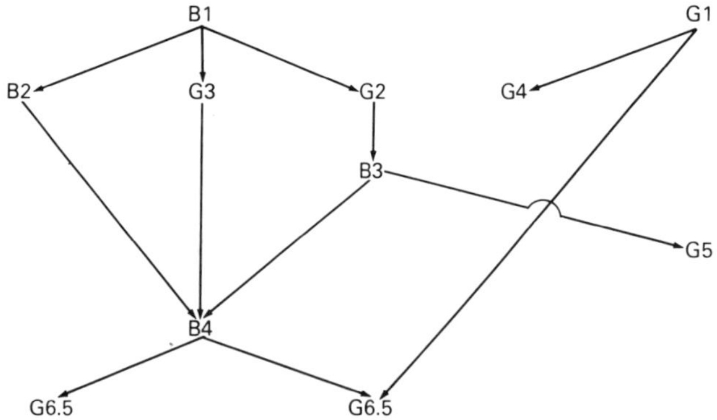

Figure 2. Dominance relationships in tree form for group 2.

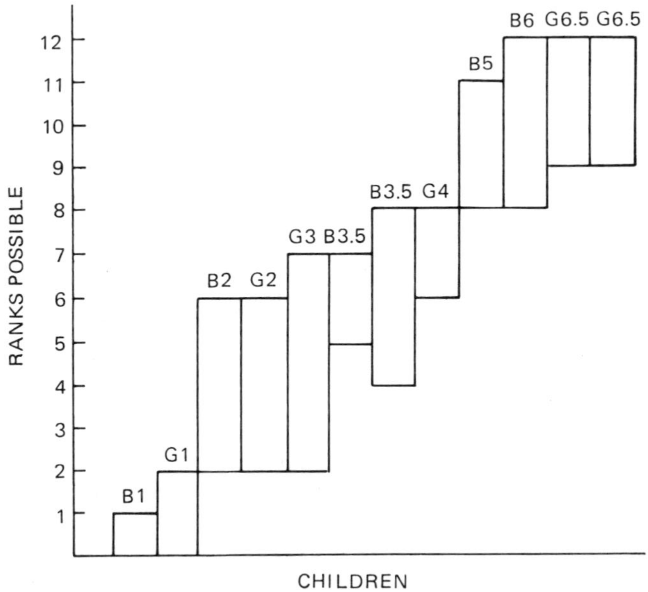

Figure 3. Possible ranks that any child can occupy in a dominance order for group 1.

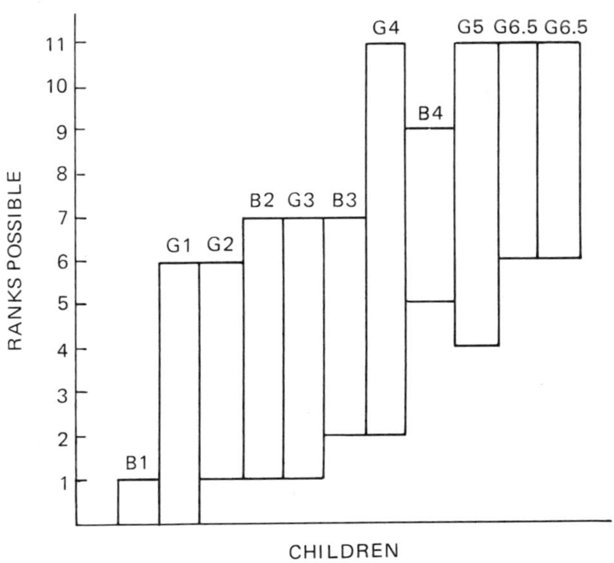

Figure 4. Possible ranks that any child can occupy in a dominance order for group 2.

Table 3. Observed Hierarchies in Two Groups of Preschool Children

Boys [B] and Girls [G] separately numbered by dominance rank in each group.

Group One	Group Two
B1	B1
G1	G1
(B2 G2)	G2
G3	(B2 G3)
(B3.5 B3.5)	B3
G4	G4
B5	B4
B6	G5
(G5.5 G5.5)	(G6.5 G6.5)

Report Data

Although it is obvious that any social structures must go hand in hand with a certain cognitive structure to enable an individual to perceive and act on the basis of that structure, it is still necessary to delineate the precise skills necessary for this. If a dominance structure is to have any effect on a child's behavior, then it could be argued that the child must have some idea of seriation. He or she must have, at least, some rudimentary notion that there is something predictable in his encounters with his playmates. If he or she had no idea whatsoever of this, it would be difficult to account for the observed stability in his dominance relationship.

This whole question of how a young child structures his perceptions of his peers is open to empirical investigation through asking the children questions. It is necessary, however, to be careful that the talks with children really do offer something relevant to the initial brief. With this in mind, we shall briefly look at what little work has been done in this field.

Edelman and Omark (1973, and also reported in this volume) asked a number of children between 4 and 8 years of age the question, "Who is the toughest in your class?" They suggest that children conceive of "the toughest" in the same way that primatologists mean the "most dominant." To see how 4-to-6-year-old children perceive their peers, they presented to each child all of the pictures of his or her classmates, with the snapshots laid out along a bench. The experimenter then said, "I'm going to ask you some questions about your classmates. The first question is about toughness. Now what is another word for tough? Now let us look at the first child in the row. If the child is tougher than you, turn his picture over." After the child followed these instructions, the experimenter again made sure that the child understood the question. The experimenter then repeated the question with each picture until he was confident that the child understood the task. Then he said, "Now continue on down the row, turning over the picture of each child that is tougher than you" (Edelman and Omark, 1976). When two children agreed that one was tougher than the other, then they called that pair a *dyad of established dominance*. From first grade through to fourth grade they found that between 62 percent and 73 percent of dyads agreed as to who was the toughest, and this was well above chance level. Other experiments showed that "along with a developing sense of a dominance hierarchy, children become better at reasoning about dominance relationships."

Edelman and Omark make the interesting suggestion that the developing dominance hierarchy in children may form part of the necessary social experience for this aspect of the child's cognitive development—the ability to seriate and make inferences. Each child has direct experience of his or her own encounters with others, and also indirect experience through observations of others' encounters, from which to make deductive inferences. Edelman and Omark, however, did not provide systematic evidence that the children's perceptions of toughness were actually concordant with observational measures of a dominance hierarchy. The next experiment examines the relationship between (perceived) toughness relationships and (behavioral) dominance relationships.

Is Toughness the Same as Dominance?—Experiment II

Although Edelman and Omark's work does show clear developmental changes concerning a child's conception of "toughness," it is not clear that the child necessarily has any conceptions about dominance relationships. There is no doubt that toughness is one aspect of a dominance relationship, but surely another aspect must be that of winning. It is easy to test whether toughness and dominance are really the same concept, and the experimental procedure described below does this.

Since dominance hierarchies had already been constructed from observational data of two preschool play groups, it was possible to compare them with hierarchies obtained by pooling the children's answers to questions about toughness. Should high correlations be obtained, then dominance and toughness can be thought of as the same concept. If, on the other hand, the correlations are low, it would be necessary to consider them as separate dimensions. The questioning was carried out in the following manner.

All of the children were individually shown photographs of themselves and their peers and then asked to point to the toughest child. Some children made absolutely no response, but made a choice when the word "toughest" was replaced by "strongest." This may be a cultural difference between British and American children. This photograph was then taken away, and the children were asked to point to the next toughest. In order to judge whether the children were using a concept and not ranking by chance, the photographs were rearranged and the task was immediately repeated.

RESULTS. Table 4 shows the correlation coefficients between the two orderings. In Table 5 there are two groups of children. Eight of the 20 showed that they could carry out the task reliably. Their average age was slightly older than the rest at 4 years 5 months; the average age of the others was 3 years 9 months. Since the mainly older group also showed substantial agreement on the order, it can be inferred that they showed some communality of awareness of an order for toughness. Consequently, a children's order of toughness for each group was obtained by combining the ranks of the mainly older children within each group.

In a sense, this was cheating, since the "group" hierarchy only reflected the older children's choices. Nevertheless, it is interesting to compare this hierarchy to the observed dominance order and to an order based on the amount of aggression initiated by each child. The Spearman rank correlation coefficients are presented in Table 6.

As expected, these children's conceptions of toughness related best to who initiates more aggressive acts and not to who is most dominant in a dominance and submission sense. Whether this is true for American children, is still an open question. Indeed, it may well be that "toughness" is a far wider concept in the United States than it is in Britain and has all the connotations of dominance that Krebs outlines as necessary. This is a question that can best be answered by empirical linguistic research of native speakers. No attempt to do so will be made here.

Table 4. Reliability Coefficients for Order of Toughness for Each Child in Each Group

Group One				Group Two			
Child	*Age* Yrs. Mos.	*Reliability* (Spearman r)	p *value* (df = 10)	*Child*	*Age* Yrs. Mos.	*Reliability* (Spearman r)	p *value* (df = 9)
B2	3 - 2	+.85	.01	G4	4 - 9	+.83	.01
B1	4 - 10	+.76	.01	B3	4 - 5	+.74	.05
G3	4 - 6	+.72	.05	G2	4 - 5	+.68	.05
G5.5	4 - 7	+.50	.10	B1	4 - 9	+.68	.05
B6	4 - 7	+.31	n.s.	G3	3 - 11	+.44	n.s.
B5	3 - 7	+.28	n.s.	G1	4 - 9	+.35	n.s.
B3.5	3 - 3	+.13	n.s.	G6.5	3 - 7	+.30	n.s.
B3.5	3 - 9	+.08	n.s.	G6.5	4 - 6	+.30	n.s.
G5.5	3 - 1	+.05	n.s.	B4	3 - 4	+.02	n.s.
G1	3 - 10	−.15	n.s.	B2	3 - 2	Refused	
G2	3 - 4	−.17	n.s.				
G4	3 - 6	Refused		G4	3 - 8	Absent	

Table 5. Mean Ages and Concordance Coefficients for Orders of Toughness for Reliable and Unreliable Children in Each Group

	Group One					Group Two				
	N	*Mean Age* Yrs. Mos.	*Kendall's*	W^2	p *value* (df = 11)	N	*Mean Age* Yrs. Mos.	*Kendall's*	W^2	p *value* (df = 10)
Reliable children	4	4 - 4	.61	26.9	.01	4	4 - 7	.67	26.8	.01
Unreliable children	7	3 - 8	.15	11.6	n.s.	5	4 - 0	.30	15.0	n.s.

Table 6. Correlation Coefficients Between Observed Hierarchy, Perceived Hierarchy, and Aggression Initiated for Each Group

	Group One		Group Two	
	Spearman r	p *value* (df = 10)	*Spearman* r	p *value* (df = 9)
Observed hierarchy and Perceived hierarchy	+.36	n.s.	+.20	n.s.
Aggression initiated and Perceived hierarchy	+.51	.10	+.47	n.s.
Aggression initiated and Observed hierarchy	+.46	n.s.	+.67	.05

Dominance Relationships and Information about Social Structure

Much of the interest in dominance relationships has arisen from trying to describe the real social structures that exist in society. Since adult societies are so amazingly complex, the first speculations by ethologists seem to have often been about the nature of preschool groups. There is nothing necessarily wrong with this except that if one were to judge from the number of conference papers on dominance, one might be forgiven for thinking that all that children do is squabble for dominance positions.

There are, of course, all sorts of other interactions that occur between children—playing together and talking, to name but two. While Krebs' list of dominance behaviors is extensive, it is also true, at least in the present study, that some 20 percent of dominance accounts ended up in wins against the dominance order. No one would expect that knowledge of relationships should completely predict the outcome of an encounter; yet, if we want to describe real social structures, then we should try to account for these occasional anomalous wins. At a more general level Hinde and Stevenson-Hinde (1976) discuss at length this whole issue of how one might proceed to gain information about social structures through observations of a range of interactions. As they point out, it is not an easy task.

The results from this study, however, do confirm ethologists' beliefs that a linear dominance hierarchy is a useful approximation in describing the outcome of conflict situations in preschool children. Amongst individual children *no* intransitive dominance relations were found, and at the level of individual incidents the outcome of 80 percent of all encounters was consistent with the observed hierarchies. No clear sex differences were obtained for either group. This too is consistent with the earlier work on preschool children, although Edelman and Omark suggest that there is a fairly marked sex differentiation in the perceived hierarchies of older, grade-school children.

Dominance Relationships and Cognitive Abilities

The rank ordering of classmates in a reliable and meaningful way shows some ability to seriate. Piaget (1952) distinguishes three stages in the development of this ability. During the first stage the child always fails to make a complete series, while a child on the second stage produces a complete series by trial and error but cannot systematically correct mistakes. It is only in the third and final stage that inference procedures are used. The ability that some of the children showed to order their peers for toughness need not necessarily imply mastery at the level of the third stage of seriation, which involves using transitive inferences. Recent work by Bryant (1974) and others, however, has shown that in experimental situations children as young as 4 years can indeed make transitive inferences provided that experience with the test materials is sufficient to overcome memory difficulties. There is some evidence that these abilities can be put to active use in cognitive tasks (Bryant and Kopytynska, 1976).

On the basis of the present findings it seems that the children's social experience in the preschool also provides a suitable experiential matrix for the development of transitive inference procedures, at about the age at which this ability first becomes manifest in experimental situations. Indeed, the evidence for this cognitive ability is not in conflict with earlier findings by Edelman and Omark (1973), and Omark and Edelman (1975), that nursery-age children are inaccurate at agreeing on *dyads of perceived toughness*. Like these researchers, the present author also found below chance agreements (31 percent and 46 percent in our two groups) on this statistic. Edelman and Omark's finding that preschool children tend to overrate their own status vis-à-vis a peer is confirmed. It now appears, however, that some 4-year-old children, at least, can fairly accurately perceive and communicate rank orderings of dyads not involving themselves.

The Importance of Social Experience—An Evolutionary Perspective

What aspect of the child's actual social experience is being utilized in the growth and organization of his or her cognitive abilities, as manifested in the report data on toughness? Since it is clear that a behavioral dominance hierarchy is a valid description of the social structure that the child is experiencing, it might be supposed that the children's perceived toughness hierarchy is derived from experience of, and attention to, the observed dominance outcome in the group. However, the correlations between the *perceived hierarchies* and the *observed hierarchies* are, for both groups, small and nonsignificant. Rather, there are more substantial correlations between the *perceived hierarchies* and the rank orderings for *aggression initiated*. It may be that the initiation of agression by peers is a more salient feature to the preschool child than the actual outcome of the aggression. In further work it should be possible to examine more thoroughly the nature of the preschool child's perceptions by asking a wider range of questions, including ones specifically about who wins conflicts. It may turn out that for older children a perceived toughness hierarchy will correlate more closely with an observable dominance hierarchy. There is the complication, however, in the present study, as well as in earlier reports by McGrew (1969), Krebs (1972), and Smith (1973), of finding moderately high correlations between dominance (defined as winning encounters) and the initiation of aggressive acts and dominance and initiation of social interaction generally.

Omark and Edelman (1975) view aspects of social structure as having importance at the phylogenetic as well as the ontogenetic level. An increasing ability with age to accurately perceive peer relationships can be seen as useful for the formation of a cohesive group. It may also be adaptive for a young child to overrate his position within a group and help him or her to face the initially strange and possibly frightening world of peer-group relations. Paying more attention to the initiation of aggression, rather than its outcome, would be consistent with the young preschool child overrating his ability to win encounters while still developing accurate ranking inferences as to his peers' social and aggressive behaviors. These speculations certainly paint a plausible picture!

The results also suggest the importance of certain aspects of the child's social experience for cognitive development. It does seem that the child's attention to aggressive encounters between peers is one of the first instances of an active use of developing inferential abilities. In a much wider context, it would seem that the use of verbal reports and ethogenic inquiry (Harré, 1974), combined with observational and ethological methodologies and perspectives (Smith, 1974a, b) may throw much light on the little explored area of interplay between cognitive and social factors in child development. Any approach that ignores this interplay can only have a restricted contribution to make.

Acknowledgments

The author is indebted to Dr. Peter K. Smith for his comments on issues raised in this chapter and to the Social Science Research Council, London, for financial support.

Chapter 9

The Structure of Young Children's Authority Relationships

Carol Barner-Barry

Authority is a concept that is central to the study of political behavior. Yet, among empirically oriented political scientists there has been little interest in the study of authority until quite recently. One possible reason for this relative neglect is the difficulty of conceptually separating authority in what have traditionally been regarded as political settings from authority in traditionally nonpolitical settings.

It is usually in the ostensibly "nonpolitical" setting of the early childhood peer group that humans have the earliest opportunity to explore their abilities to deal with authority relationships in terms of the balance between authority and acquiescence that is characteristic of the behavior of mature adults functioning in political settings. In relationships with parents, teachers, and other adults, the emphasis is on the child's skills in acquiescence. It is only with peers that children have real opportunities to test their capacity to exercise authority or influence over others (Garvey, 1974). There are two central issues in this area. The first is the way in which some children at an early age develop more skill in authoritative behavior than others. The other is the effect this skill differential has on group structure. It is the latter issue that will be explored in this chapter.

Dominance and Authority

Human ethology, as noted throughout this volume, is a discipline which is in its infancy. Human ethologists have emerged from a diversity of traditional disciplines. They have been inspired and heavily influenced by the mainstream of ethology, in which research has an evolutionary perspective and focuses on the functioning of groups of nonhuman animals (Smith, 1974b). This influence has led human ethologists in three directions which are of interest in relation to the research reported here. First, like other ethologists, human ethologists analyze groups and group structure. They have spent much time studying groups of children in nursery schools and similar settings. These groups have the advantage of being stable over time and composed of relatively unself-conscious and, thus, easily observed subjects. The basic and relatively uninhibited behaviors exhibited by children in such settings lend themselves more readily to analogies with animal behavior than would the behaviors of human adults in those stable group settings to which a researcher might obtain relatively easy access.

Second, direct access to stable groups has been important to human ethologists because their method of research has emphasized the observation of human behavior in natural settings. The focus has been on what humans *actually* do. Spontaneous physical behavior is recorded as completely as possible; verbal behavior is given secondary emphasis, if it is recorded at all. Thus, human ethologists—under the influence of a tradition where the subjects do not communicate by means of a language that is similar to human language—follow the old adage that enjoins us to attend to what people do rather than what they say. This is in sharp contrast to the tradition followed by empirically oriented political scientists who, using survey research, tend to focus on what people say rather than what they do.

Finally, because dominance and submission has proved to be a useful variable in comparative research, human ethologists have tended to analyze authority and power in human groups in terms of dyadic dominance. Since dominance is usually measured by the analysis of incidents involving agonistic behavior, such research has yielded information that seems more relevant to theories about social power than to theories about authority.

When one is considering the entire complex behavior system of a group of human children, the most useful way of separating authority from power seems to be to impose some standard for legitimacy (Bell, 1975). In the study reported here, the criterion for legitimacy was the reaction of the acquiescent child to the demands of the authoritative child. Thus, the children's own reactions to an event were used to interpret the significance of the event to the children themselves (Garvey, 1974). The characteristics of the actual standard used are specified below. Where the standard is met, the behavioral interaction is classified as one involving authority; where it is not, the interaction is classified as one involving power. Such a distinction results in the classification of most agonistic interactions as power interactions. However, once the focus is broadened to include nonagonistic behavior in which individual A induces individual B to do what individual A wants B to do, a range of behavior which is much more rich and diverse is opened for exploration. This latter class of behavior seems more relevant to the complexity of human political behavior.

Ethological studies of peer-group dominance relations based on agonistic encounters have quite consistently reported very hierarchical dominance structures. An important question concerns whether the authority structure in similar groups would be equally hierarchical. In the following research, information was gathered on authority structure. Methods similar to those used in recent studies of dominance hierarchies were used to explore the structure of authority in a group of young children. The data base encompassed a broader range of asymmetrical interactions than just agonistic encounters and used a participant-indicated criterion of legitimacy.

Data Collection and Analysis

The data to be analyzed were gathered over a four-week period by means of nonparticipant observation of interaction events in focal subgroups (Altmann, 1974; Wright, 1960). The 38 subjects ranged in age from 3½ to 6½ and were attending a summer program at a combined kindergarten and nursery school. Observations took place five days a week between 9:00 A.M. and 2:30 P.M. Additional data were collected on the children's background and their families' socioeconomic characteristics by means of a short questionnaire administered to their parents at the conclusion of the observation period. The parents of 33 of the children (87 percent) returned completed questionnaires. The questionnaires indicated that the children's families were predominantly middle class and that most of the children had had previous nursery school experience.

The set of decision rules which governed the observer's choice of whom to observe was based on two assumptions. The first assumption was that authority is a relational phenomenon. Thus, the basic unit of analysis was an interaction episode between two or more children, not the behavior of individuals. In order for the child to be authoritiative, at least one other child had to be acquiescent. The second assumption was that authority relations between peers are the most important factor deteriming the peer authority structure of the group. The authoritative behavior of the adults present was regarded as a secondary, though not theoretically unimportant, influence on the information of the peer authority structure. The resulting set of decision rules follows:

1. Choose groups of two or more children.
2. Choose groups without adults.
3. If more than one group has no adult present, choose the group farthest from any adult.
4. If two adult-absent groups are equidistant from the nearest adult, choose the group which is larger.
5. If the two equidistant groups are of equal size, choose the group which is engaged in the least passive form of play.
6. If all groups contain adults, choose the group in which the present adult has assumed the least active role.

Utilizing these rules, the observer would choose a group to observe. Once having established herself in a location from which she could see and hear all that went on within the group, she would begin choosing events to observe and record. The recording was done either by writing into a notebook or by speaking softly into a tape recorder. The event sampled was any asymmetrical interaction between two or more children. An attempt at authoritative behavior was scored as successful (S) if either of two things occurred:

1. The target child(ren) accepted the intiating child's decision without question, or
2. The initiating child managed, physically or verbally, to persuade questioning child(ren) to accept the decision *without alientating* the questioning child(ren).

The latter distinction was made in order to operationalize the concept of legitimacy on the assumption that the key element which differentiates authority from power is legitimacy. Here acquiescence without subsequent alienation was interpreted as a behavioral indicator of the acceptance of the authoritative child's legitimacy or right to exercise authority in that situation.

Three factors were taken into consideration to determine individual levels of authoritativeness. One was the frequency within each dyadic context of attempts at authoritative behavior by child *j* with reference to child *i(T)*; another was the frequency of dyadic successes *(S)*; the final factor was the number of children over whom a given child had been authoritative. On the basis of these three measures, an authoritativeness score was computed for each individual *(J)* in the group using the formula shown in Figure 1.

Authoritativeness scores range from 0 to 1. A score of 1 indicates a high level of authoritativeness and a score of 0 indicates a lack of authoritativeness. Figure 1 shows the distribution of authoritativeness scores for the children in the group under consideration.

Once these authoritativeness scores were computed, they served as the basis for rank ordering individual children to

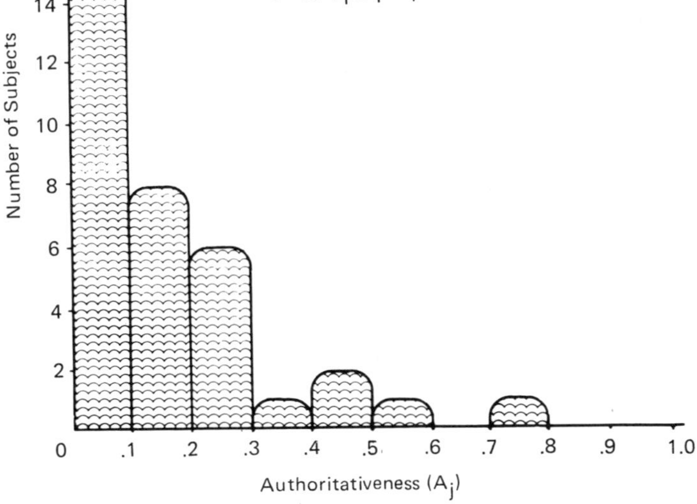

Figure 1. Distribution of authoritativeness scores ($N = 38$).

examine the authority structure of the peer group. One of the first determinations made was the extent to which the authority structure for this peer group corresponded to the linear transitivity rule.

Whether or not a dyad had an authoritative relationship (AR) was determined by counting the number of successes each member had in relation to the other and scoring the child who had more as authoritative. In cases where both children had an equal number of successes, the dyad was judged to have a mutual relationship (MR).

The linearity of the authoritative relationships in Figure 2 was evaluated by taking the number of ARs falling below the diagonal and substracting them from the total number of ARs plus

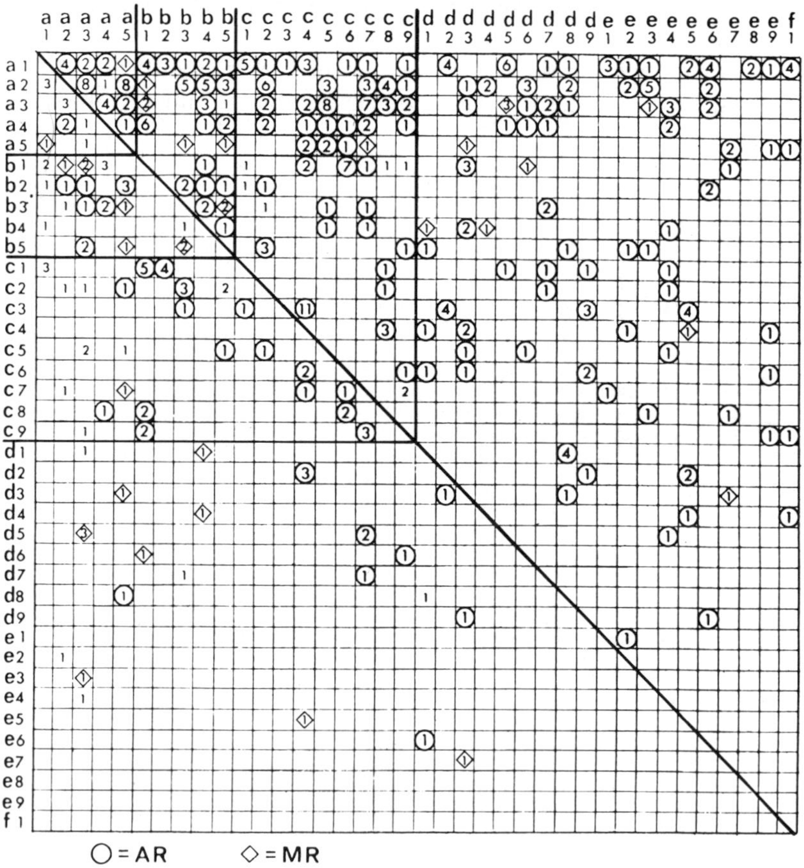

Figure 2. Dyadic matrix of observed authoritative interactions.

MRs for the group. The result was used to compute the percentage of dyadic relationships which follow the linear transitivity rule:

$$100 \times \frac{\text{Total Dyads} - \text{Relational Reversals}}{\text{Total Dyads}}$$

Where total dyads equal the number of observed dyads (ARs plus MRs) and relational reversals equal the number of observed dyads violating the linear transitivity rule (Strayer and Strayer, 1976). Using this procedure, the percentage of linearity obtained for the dyadic matrix in Figure 2 is 87 percent. This figure indicates that there was a high level of linearity in the authority structure for this group of children.

It seemed possible that the observed transitivity of authoritative relationships might be accounted for by the fact that 27 of the 38 members of the group were highly acquiescent, having authoritativeness scores in the range of .00 to .20. In order to explore the effect of increasing "eliteness" on the linearity of the authority structure, the matrix in Figure 2 was progressively decomposed into three successively smaller matrices by taking the upper left quadrant of each matrix to form a submatrix. Thus, the upper left quadrant of the complete dyadic matrix (38 by 38) was designated as Submatrix 1 (19 by 19); the upper left quadrant of Submatrix 1 was designated as Submatrix 2 (10 by 10) and the upper left quadrant of Submatrix 2 was designated as Submatrix 3 (5 by 5).

When the percentage of linearity was calculated separately for each submatrix, an interesting pattern emerged. Submatrix 1 with 76.5 percent linearity was substantially less linear than the overall matrix. Submatrix 2 with 82 percent linearity indicated a movement back toward increasing linearity. Finally, the greatest linearity was evident for Submatrix 3—90 percent. Visual inspection of Submatrix 3 shows that were it not for one reversal and one mutual relationship, the linear hierarchy would be perfect. This analysis of successively smaller matrices seems to indicate that while the authority structure is quite linear for the group as a whole, there is a slight tendency toward less linearity among the children who occupy positions in the upper midportion of the authority structure. However, at the very top, or "elite" level, linearity again is quite high.

A similar mathematical formula can be used to test for rigidity in the authority structure. This measures the extent to which there have been incidents in which a member of the group exercised authority over someone who was generally more authoritative. A high level of rigidity indicates that this seldom happened; correspondingly, a low level of rigidity would indicate that such incidents were frequent. The level of rigidity in dyadic relations was computed using the following formula:

$$100 \times \frac{\text{Total Episodes} - \text{Episodic Reversals}}{\text{Total Episodes}}$$

Where total episodes equal the number of observed dyadic interactions and episodic reversals equal the number of dyadic interactions in which a member of the group exercised authority over an individual who was more authoritative for the group as a whole (Strayer and Strayer, 1976). If only one member of each dyad assumed an authoritative role, the authority structure of the group would be totally rigid. Departure from this situation indicates relatively less rigidity. Although previous empirical findings indicate that linearity and rigidity are usually correlated measures of asymmetrical social relations, "it is theoretically possible to find a perfectly linear status hierarchy with relatively fluid" authority relations (Strayer and Strayer, 1976, p. 984).

Comparison of the levels of linearity and rigidity in Table 1 indicates that while the overall levels of both linearity and rigidity are high, there are frequent challenges to dyadic authority relationships which do not result in an alternation of the relative authoritativeness for the members of the dyad. This, in turn, seems to point to a conscious or unconscious reluctance, especially at the upper middle level of the structure, on the part of the acquiescent children consistently to play the acquiescent role.

The overall levels and patterns of linearity and rigidity suggest that the children who are close to, but not in, the most elite group were most apt to jockey for position and test the limits of their ability to be authoritative with children relatively close to

Table 1. A Comparison of the Percentages of Linearity and Rigidity for the Whole Group and Successively More Authoritative Subgroups

	% *Linearity*	% *Rigidity*
Complete matrix	87	76
Submatrix 1	76.5	71
Submatrix 2	82	70
Submatrix 3	90	75

them in the authority structure. Qualitative evidence supported this interpretation and also indicated that such jockeying for position did not seem to be the result of a pure situation of competition for authority. Rather, it seemed more the result of a lack of knowledge on the part of the children involved as to exactly where they stood vis-à-vis their close neighbors in the hierarchy and a corresponding relative lack of restraint in attempting authoritative behavior during interactions with children close to them in the hierarchy. On the other hand, the children closer to the bottom of the hierarchy tended to cluster more with regard to level of authoritativeness and seldom attempted authoritative behavior in interactions with anyone, even those below themselves in the hierarchy. Thus, sheer lack of interest in attempting authoritative behavior tended to insure that these children would maintain their low ranking position. Inspection of the lower right quadrant of Figure 2 reveals a rather small number of ARs (12 as compared to 91 in the upper left quadrant of the same matrix). In contrast, the most authoritative children, especially the top three, attempted authoritative behavior substantially more often than any of the others and attempted it with a much wider range of children. However, they tended to avoid each other most of the time and direct authoritative confrontations between any two of the top three were quite rare, considering their overall activity level in this area.

There were no instances when two of the top three children put an acquiescent child in a position where the acquiescent child had to follow one or the other, and there were several instances in which two of the authoritative children cooperated to solve a problem. The observer was left with the overall impression that they simply accepted where they were without thinking a great deal about it, but that they were clearly aware of their relative positions and little inclined to put much effort into changing them. The one exception was the most authoritative child, who occasionally seemed to invite a confrontation with another high-ranking child to reinforce her position, but these incidents were infrequent.

Comparisons and Conclusions

The findings with regard to the hierarchical nature of this authority structure parallel recent ethological findings concerning dominance structures in groups of young children. Both dominance and authority show a high level of linearity and rigidity in the structure of dyadic relationships. The patterns, however, differ somewhat. The present results can be most closely compared to those of Strayer and Strayer (1976), who used similar techniques for data collection and analysis. Although the Strayers did not divide their dominance matrix into submatrices, it is possible to do so. Given the size of their total matrix (17 by 17) and the fact that the total number of interactions is smaller, the most valid procedure would seem to be to compare their dyadic data with the submatrices which were abstracted from the present study. The result of this comparison can be seen in Table 2. An inspection of Table 2 indicates that the analysis of submatrices from the Strayers' dominance data yields information which suggests a higher level of linearity and rigidity than was found in the study reported here. This difference may be due to the fact that the Strayers used a method for ranking their subjects which maximized linear transitivity rather than ranking in terms of an individual assertiveness measure. When the subjects in the present sample were ranked according to this maximization criterion, however, the results seemed to misrepresent important differences in both the frequency and scope of initiated activity. The present authoritativeness index was designed to provide a means

Table 2. Comparison of Authoritativeness Submatrices with a Dyadic Dominance Matrix

	Matrix Size	*Observed Dyads (N)*	*% Linearity*	*Observed Episodes (N)*	*% Rigidity*
Authority submatrix 1	19 x 19	98	76.5	277	71
Dominance matrix 1	17 x 17	61	98	142	95
Authority submatrix 2	10 x 10	39	82	122	70
Dominance submatrix 1	9 x 9	23	100	55	93
Authority submatrix 3	5 x 5	10	90	44	75
Dominance submatrix 2	5 x 5	9	100	12	100

of ranking which emphasizes both of these latter variables. Because of this difference in the definition of linearity, results in Table 2 should be interpreted cautiously.

The matrix constructed by the Strayers was composed of "agonistic" episodes involving physical attacks and threat gestures leading to submission. From the details given in their article as to the specific acts involved (biting, kicking, showing intent to hit, and the like), it is clear that few, if any, of the interactions studied by the Strayers were likely to have involved any question of legitimacy. Thus, the comparison of the data from the two studies can be interpreted to suggest that both linearity and rigidity of hierarchy may be more pronounced where pure power (defined as physical force or the threat of physical force) is involved. While the levels of linearity and rigidity in the authority structure were high, they were not as high as those in the dominance structure. Examination of the patterns of linearity and rigidity in the dominance matrix show perfect (100 percent) linearity and rigidity at the elite level. The linearity pattern is consistent throughout, while the rigidity pattern shows a slight dip at the upper middle level which is suggestive of the pattern in the authority matrix, but considerably less pronounced.

The results reported in this chapter point to the need for more extensive work that would examine systematically the problems of empirically distinguishing between social authority and social power. These two concepts must be differentiated operationally and then studied with reference to each other. A solution to the problem of operationalization would necessarily involve distinguishing between these two types of social interaction on the basis of behaviors which are reasonably unambiguous and congruent with traditional concepts of power and authority. The experience of the human ethologists in studying agonism indicates that reasonably unambiguous operational definitions of power are possible if power is regarded as an agonism-based phenomenon. Traditional political science treatments of power have emphasized the importance of coercion and sanctions, but they have generally not conceptualized power as narrowly as is implied by an agonism-based approach. What would be lost in conceptual richness, however, would be offset by gains in precision, which, in turn, could facilitate badly needed basic empirical research in this area.

Approaching the empirical study of authority using a legitimacy criterion also seems feasible and not unduly at odds with traditional notions of authority, if one makes a distinction between *de facto* and *de jure* authority. Since this distinction is made either explicity or implicity in much of the traditional literature, it is quite reasonable to regard studies such as the one reported here as studies of *de facto* authority. This criterion also carries the implication that the acquiescent child is an active, rather than a passive, actor. By, in effect, legitimizing the attempts of certain others to regulate or control their behavior, the acquiescent children play a part in regulating or controlling the situation. To regard acquiescence as necessarily a negative or passive stance is to assume that all persons are similarly motivated to seek authority. The reality of human diversity argues against this interpretation. Control and regulation are but two aspects of social and individual behavior. They do not necessarily interest every child equally. In fact, a child who is both independent and, say, deeply interested in cognitive tasks may choose minimal involvement in control and regulatory activities. On the other hand, many authoritative children tend to be quite peer-dependent.

Thus far in the discussion power relationships and authority relationships have been treated as separate sets of social interactions which create separate social structures. This is useful for analytic purposes, but reality is seldom so neatly and clearly defined. Thus, analysis must be accompanied by synthesis. One must not only consider the distinctions between authority structure and power structure, but also the ways in which they intermingle or blend. Do power relationships predominate among children at earlier stages of social, cognitive, and emotional development? Do authority relationships become more important along with the maturing process? In the group studied here, as well as in other groups observed by the author, there has seemed to be some correlation between development and increasing reliance on authoritative, as opposed to power, interactions. Given the fact that authoritative behavior is more complex and requires higher levels of skill, this is not surprising. The issue, however, is important in and of itself and needs to be addressed explicitly rather than noticed as a by-product of other research.

While the preliminary comparative efforts in this chapter have a certain intuitive appeal, more systematic scrutiny may reveal more important empirical distinctions between these two political concepts. Current research suggests, however, that when power and authority relations are measured in terms of dyadic behavioral interactions, the structure of power relationships in children's peer groups is less egalitarian than the structure of authority relationships. There is nothing in this work which indicates what influence the superimposition of a formal authority or power structure might have on these patterns of social organization. Thus, while research on groups of very young children avoids "contamination" from such external influences, when well developed it could conceivably generate a basic or "ideal" type of social organization. Later researchers could use this model to develop ways of estimating the extent to which a formal authority network tends to modify the authority network which might have arisen in the absence of rules and regulations which enhance the authority of persons who do not have well-developed authoritativeness skills.

These, however, are long-run research objectives. In the short run there is much to be learned from work which concentrates on the extent to which such patterns are characteristic of various types of stable childhood groups—and there is considerable evidence that they are. The nature of politically relevant behavior patterns arising in peer groups needs much more study and refinement. Once we know more about this, we can begin to ask questions about the implications for political development of the experiences children have in such structures, as well as the implications of such experiences for a more adequate theory of political socialization and political psychology.

Acknowledgments

The collaboration of Linton Freeman in the development of an adequate descriptive statistic for authoritativeness (Aj) is gratefully acknowledged. In addition, the author would like to express her appreciation to both the children and staff from the summer program, as well as acknowledge the technical assistance in the preparation of graphics provided by Teresa Blicharski. A preliminary report of this research was presented at the annual meeting of the American Political Science Association, in Washington, D.C., during the fall of 1977.

Chapter 10

The Perception of Social Power Relations among Preschool Children

F. F. Strayer
Janet Strayer
Thomas R. Chapeskie

It is only during the past five years that both concepts and methods from animal social ethology have been applied directly to the study of human social interaction. Most of these studies have focused upon preschool behavior, particularly upon preschool social dominance. Two quite different approaches have been used to investigate social dominance among young children. A number of researchers have examined naturally occurring episodes of social conflict in order to determine dominance relations and status structures within stable peer groups. Abramovitch (1976) and McGrew (1972) both found nearly linear status structures using "wins" during object conflict as a criterion for assessing social dominance. Similar dominance hierarchies have been described by Missakian (1976) and Strayer and Strayer (1976), who examined all social exchanges involving attack and submission. A second approach to dominance assessment relies upon analysis of consensual agreements in children's verbal reports concerning dominance relations within their group (Omark, Omark, and Edelman, 1975). According to this latter method, when both members of a dyad agree concerning who is "tougher," the two children are considered to have an "established dominance relationship." All such established relationships are examined to determine the group dominance hierarchy.

Although both approaches to the assessment of dominance

have led to the identification of linear status structures for young children's peer groups, potential discrepancies between the two procedures pose an important methodological question. To what extent do the two procedures provide convergent information about the same structural aspects of the peer group? A second, and perhaps more psychologically interesting question, concerns individual differences in agreement between statements concerning social relationships and a behavioral assessment of social interaction within the peer group.

At a methodological level there are a number of potential problems inherent in use of the verbal report for the assessment of dominance relations. It is conceivable that child A could overrate his dominance in relation to child B, while the latter could underrate his dominance in relation to child A. If the verbal report method is used, the dominance relation for this dyad would be considered consensually "established" by dyadic agreement. This would result in unavoidable mistakes in the derivation of group dominance structure because there is no way of independently assessing instances of reciprocal errors. A second problem occurs, particularly in developmental research, using verbal reports. Since the derivation of a social dominance structure depends upon verbal agreement between a number of dyads within the group, it is impossible to derive a hierarchy if there is insufficient or only chance agreement. In the past, it has often been difficult to derive social dominance hierarchies from verbal reports by preschool or nursery school children (Omark, Omark, and Edelman, 1975).

Even when the verbal report method is used primarily to assess the accuracy of children's perceptions about dominance relations, problems concerning the adequacy of the "criterion for accuracy" must be resolved. If accuracy is judged in terms of verbal consensus alone, it seems necessary at least to know how the consensual status structure obtained from children's reports compare with an independent assessment from teachers or adult supervisors of the peer group. Perhaps the most important question, given the divergences of methods for the assessment for social dominance, is how well verbal reports from both children and teachers reflect the group dominance relations derived from behavioral observation in intragroup conflict.

The present investigation was designed to determine the extent of agreement between the verbal consensus and the behavioral assessment procedures. One objective was the replication and comparison of previous findings concerning preschool social dominance behavior (Strayer and Strayer, 1976). Following this replication, our major interest concerned the degree to which preschool children and teachers were accurate in their perception of dominance relations and if children's accuracy of perception were related to their position in the social dominance hierarchy. To answer these questions data from both behavioral and verbal report methods were compared.

Behavioral Assessment of Social Dominance

Method

Two groups of preschool children were observed. Only children who were present for at least 75 percent of the sampling period were included in the present analysis. The first group attended the Langara City College preschool in Vancouver, British Columbia. This group included 17 children (7 girls, 10 boys) whose mean age was 53.4 months (*SD* = 7.6). The second group was observed at the University of Waterloo Early Education Center, in Waterloo, Ontario. This sample contained 18 children (10 girls, 8 boys) whose mean age was 51.4 months (*SD* = 7.1).

A total of 30 hours of videotaped social interaction were obtained during free-play periods for each of the groups. Records were collected, daily for six weeks using a matrix completion sampling method (Altmann, 1974). Episodes of conflict were analyzed using the behavioral inventory for preschool agonism reported by Strayer and Strayer (1976). In each agonistic interaction the initiator, agonistic action pattern, the target child, and the target's response were identified. Social dominance hierarchies were constructed on the basis of conflict episodes that terminated with a submissive response by one of the children involved in the dyadic exchange.

Results

Comparison of observations for the two groups showed that while most of the specific behavior patterns were evident at both locations, there were important between group differences in the relative frequencies of the conflict categories. Table 1 summarizes the general forms of behavior which began conflict episodes in each of the two groups. Both the total frequency and estimated rate of occurrence of each category of action are shown. Children in the

Table 1. Summary of Initiated Agonism at Two Preschool Centers

	LANGARA		WATERLOO	
Behavioral Category	*Total Frequency*	*Rate per Hour*	*Total Frequency*	*Rate per Hour*
Physical attack	173	5.8	150	5.0
Threat gestures	146	4.9	158	5.3
Object/position struggles	124	4.1	489	16.3
Total Initiations	443	14.8	797	26.6

Waterloo group were involved in four times more object/position struggles than the children in the Langara sample. This might be attributable to differences between the two environments. While the Langara children were observed for the most part in an open playground, the Waterloo study was conducted inside a building that contained many toys and recreational materials to occupy the children. The larger number of toys and the relatively closer proximity of other children may have contributed significantly to group differences in the frequency of object/position conflict.

Table 2 shows the distribution of response to initiated conflict. A comparison of these data with the earlier report again indicated that the same action patterns occurred with approximately the same relative frequencies in the two groups. There was one significant difference, however. The children in the Waterloo sample responded to initiated agonism by counterattacking much

Table 2. Summary of Responses to Agonism in Two Preschool Centers

	LANGARA		WATERLOO	
Behavioral Category	*Total Frequency*	*Rate per Hour*	*Total Frequency*	*Rate per Hour*
Help seeking	4	.1	10	.3
Submission	173	5.8	225	7.5
Counterattacks	95	3.2	345	11.5
No response	171	5.7	217	7.2
Total responses	443	14.8	797	26.6

Video records of responses coded as object/position loss in the earlier report (Strayer and Strayer, 1976) were recoded into one of the above categories.

more often: 43 percent counterattacks were observed at Waterloo, as compared to 21 percent at Langara ($x^2 = 11.9$, $p < .01$). Again this difference seemed attributable to differences in the observation settings.

Since the coding procedure identified both the initiator and the target in each conflict episode, data can be presented in a dyadic matrix format. This representation of observed conflict is shown in Figure 1.

Individual cells across the rows indicate a particular initiator and the frequency with which he initiated agonism toward particular targets, who are identified by column headings. Row totals indicate the number of times each child initiated agonism. The number of times each child was the target in dyadic conflict can be reviewed by examining individual cells down the columns in the matrix. Column totals report how often agonistic acts were directed toward each individual during the observation period.

Several findings are indicated by the Figure 1 matrix. First,

TARGETS

INITIATORS	Ma	Ch	Ja	Bl	Tl	Jo	Cm	Je	El	Ef	Bk	Em	Eh	Pa	Rs	Rm	Sh	Bp	Ca	Total
Ma		2	4		4	1	2	8	1			1	1		1		6			31
Ch	3		27	3	3	8	2	29		7			2	1	1		1			87
Ja	2	16		6	5	7		14		4	1		26				4			85
Bl	1	2	5		18	5	3	1		3		2				2		2		44
Tl	4	3	5	14		6	8	10	3	20	12	2	10	1	1	10	1	10	6	126
Jo		4	6	3	4		6	4	1	10			2				1			41
Cm	1	1	1	12	4	9		2		7			1		2	1			1	42
Je	9	20	19	1	3	19	3			2			4		1		2	1		84
El				1		1								1	1		2		1	7
Ef		7	2	2	11	7	8	3			15	2	7	1				2		67
Bk					14					6		2	6				3	7		38
Em				1	1				1	2	1						1		1	8
Eh	3	2	18		4	1		2		7	3			1		1		3		45
Pa					1		1		1	2					11	9				25
Rs	1	2			1		1							5		4	2		1	17
Rm					3		1					2	1	5	1		2		1	16
Sh	4	1	2		2			2			2				2	2			2	19
Bp					7			1		2	3		3							16
Ca					2										3	1	1			7
Total	28	60	89	43	87	64	35	76	7	72	37	11	63	15	24	30	26	25	13	805

Figure 1. Dyadic representation of total agonism observed at the Waterloo preschool center.

approximately 65 percent of all observed agonistic interactions are located in the upper left of the matrix. This indicates that higher status children both initiated and received the major proportion of observed agonistic behaviors. Second, as reported previously for the Langara sample, the use of all agonistic interactions provided a poor index of social dominance at either the level of group structure or dyadic relations. This is evidenced by the fact that approximately 42 percent of the observations fall below the diagonal in Figure 1. If the combined use of all initiated agonism produced a good fit to the linear dominance model, virtually all observations would fall above the diagonal of the interaction matrix. In general, research with primates suggests that the best index of social dominance includes only those interactions leading to social submission (Rowell, 1974).

The set of agonistic interactions leading to submission in the Waterloo group is shown in Figure 2. In this more restricted class

TARGETS

INITIATORS	Ma	Ch	Ja	Bl	Tl	Jo	Cm	Je	El	Ef	Bk	Em	Eh	Pa	Rs	Rm	Sh	Ba	Cp	Total
Ma		•	2			1		2	1				1		1		1			9
Ch	•		10	2	1•	1	1	8		1			1		1		◆			26
Ja		1		4	2	2	1	3		2			10							25
Bl			1		1•	1	1•									1				5
Tl		1•	1	1•		1		5	2	8	2		3	1		3		1	2	31
Jo			1				1			2			1							5
Cm				1•				•	•	1			1		1					4
Je			1				•		•	2	2		2							7
El							•	•		•		1								1
Ef					1	1			•		1	1•						1		5
Bk												3	1					1		5
Em										1•			•							1
Eh					1							•		1						2
Pa															1			2		3
Rs																1	◆			1
Rm																	1			1
Sh		1◆													1◆			•	1	3
Bp																	•		•	0
Ca																		•		0
Total	0	3	16	8	6	7	4	18	3	17	5	5	20	2	5	5	2	5	3	134

Figure 2. Dyadic dominance matrix for the Waterloo group. (Diamonds indicate relational reversals of linear dominance, and dots indicate tied dyadic dominance.)

Table 3. Fit of Dominance Rankings for Two Forms of Conflict

	LANGARA (%)	*WATERLOO* (%)
Attack/threat interaction		
Linearity	98	96
Rigidity	93	94
Object/position struggles		
Linearity	82	81
Rigidity	90	94

of social conflict approximately 95 percent of the observations fall above the diagonal of the interaction matrix. These data correspond well with expectations concerning social dominance derived from the primate literature and the earlier Langara sample.

The degree of similarity between the two samples was also evident in the comparison of linearity and rigidity of initiated conflict. Table 3 shows rigidity and linearity indices for aggressive conflict and object/position struggles. Linearity of dominance represents the degree to which the rank ordering conforms to the rule of transitivity; that is, if A > B, and B > C, then A > C. Since only two relationships contravene this rule, the Waterloo dominance structure is 96 percent linear. The eight observations below the diagonal in Figure 2 that are not designed by a diamond indicate episodic role reversals, that is, a particular episode in which the usually less dominant child won. The frequency of episodic reversals provides an indication of the rigidity of dominance relations within the group. For the Waterloo sample the rigidity index is 94 percent. It is evident from Table 3 that the coherence of social dominance is quite similar in both of the preschool groups. Despite the reported differences in the rates and forms of social conflicts, both preschool groups were similar in the rigidity and linearity of their dominance structures.

The Perception of Social Dominance

Method

At the close of the six-week observational period both children and teachers were interviewed individually to assess their perceptions concerning dominance relations within the preschool group. At Langara 14 children and 3 teachers were present for interviews; at Waterloo 15 children and 3 teachers were present. Two different approaches were used to assess the perception of social dominance.

The first approach was designed to assess accuracy of dominance judgments in dyadic comparisons of peer-group members. There were two versions of this procedure, an other–other comparison in which the subject was not included in the stimulus pair, and a self–other comparison in which the subject was one member of the stimulus pair. Children at each preschool, as well as teachers, were tested with other–other comparisons. During each testing session the subject was presented with from 10 to 16 stimulus pairs and asked to select the "tougher" child in each pair. Eight stimulus pairs were presented in the self–other comparison to children at both preschools. Individual photographs of peers were used during testing of children in order to prevent confusion about which classmates were being compared. In addition, children were asked to give another word for "tougher" prior to testing in order to insure that they understood the term. Since some children did not respond to all stimulus pairs, a percentage correct score was obtained for each subject by dividing the number of correct responses by the total number of judgments made.

The second approach to assessing the perception of social dominance emphasized judgments concerning children's position in the dominance hierarchy. Teachers at both preschools were given a list of children's names and asked to rank order the entire group according to "toughness." Since in preliminary work preschool children had difficulty rank ordering a large group, the subjects from the Langara sample were asked only to rank order a subgroup of nine children (eight randomly selected classmates and themselves). This simpler task still proved quite difficult. Consequently, children at Waterloo were shown photographs of all children in the group including themselves. They were asked to pick the child who was the most "tough." Their first choice was then eliminated and the question repeated, yielding their second- and third-ranked classmates. A composite score was given to each child based on the number of times each was ranked first, second, or third yielding a consensus score for 16 children in the Waterloo sample.

Results

Perception of Dyadic Dominance

Since eight self–other comparisons were presented to children from each group, it was possible to compare reciprocal judgments concerning relative dominance for 28 dyads in each sample. At Langara 14 of these pairs of children agreed that one member of the dyad was tougher; at Waterloo, 16 pairs agreed. Thus, using the method of Omark, Omark, and Edelman (1975), there was 50 percent and 57 percent agreement on verbal reports of dyadic dominance. Although these figures are slightly higher than the level previously reported for preschool children, they confirm the earlier finding that preschooler's agreement on dyadic dominance differs little from the percentage expected given random choosing. Consequently, as in the earlier study (Omark, *et al.*, 1975), it was not possible to construct a consensual status ranking based upon verbally "established relationships" which could be used to assess preschoolers' accuracy of dominance perceptions.

However, accuracy of dyadic dominance judgments could be assessed relative to the behavioral observations. To facilitate these comparisons, children from both groups were classified into categories of high, medium, and low dominance on the basis of their ranking in the behavioral dominance structure. For the Waterloo sample five children were assigned to each status level. For the Langara sample five children were assigned to the high- and low-status groups and four to the medium-status group. Figure 3 represents the mean accuracy of both children and teachers in these dyadic comparisons. Inspection of other–other comparisons in both Langara and Waterloo samples shows that high- and medium-status children are more accurate than the low-status group, but not as accurate as the teachers. A two-way analysis of variance in accuracy on other–other comparisons as a function of preschool group and status level indicated a significant main effect only for status ($F(2.23) = 6.64$, $p < .01$). Figure 3 also shows that mean accuracy in self–other comparisons appears to be related to the child's status position. Statistical analysis, however, indicated that there was considerable variation in these latter accuracy scores within each of the status levels and that the differences between levels were not statistically significant. Mean accuracies for the three subgroups in self–other comparisons reflect

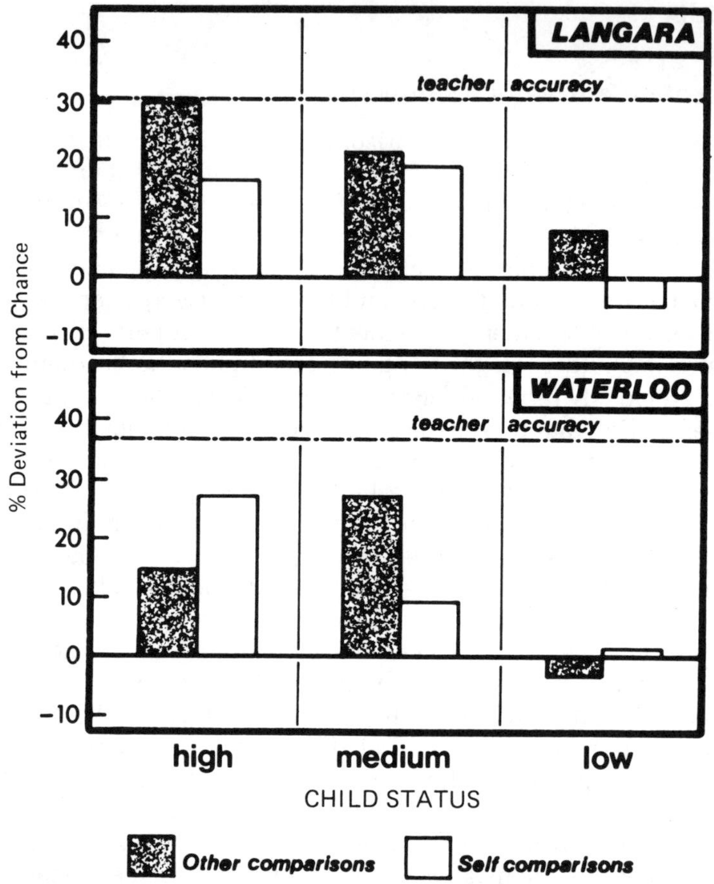

Figure 3. Accuracy of dyadic dominance judgments as a function of group dominance status.

the effects of both under- and overrating of their dominance by some children in each status subgroup.

Perception of Group Status Rankings

Although the children's dyadic comparisons did not produce sufficient agreement to generate a single status ranking from verbal reports for either group, the results from the rank-ordering procedures provided information which permitted rank ordering of a subgroup of children in each sample. By weighting individual scores given to each member in these subgroups, we were able to

Table 4. Rank Order Correlations between Status Rankings Based on Behavioral Observations and Verbal Reports

	TYPE OF DOMINANCE ASSESSMENT		
	Behavioral Observations	*Teachers' Reports*	*Children's Reports*
LANGARA (N = 8)			
Teachers' reports	.66*	—	
Children's reports	.52	.69*	—
WATERLOO (N = 16)			
Teachers' reports	.81**	—	
Children's reports	.51*	.44*	—

* = $p < .05$
** = $p < .01$

derive a single status ranking that reflected the general relative position assigned to the children by their peers. Table 4 shows the degree of association between the status ranking obtained from children's and teachers' verbal reports and the behavioral observations.

While it is clear that the teachers' verbal reports (especially at Waterloo) agreed to a significant extent with the behavioral assessment of the dominance structure, the children's reports appear much more discrepant. Even if one takes into account the cognitive difficulties inherent in the task of rank ordering a group, and uses the teacher's score as a "ceiling," the children's accuracy in both samples was at most marginal. Their judgments provided the poorest consensual index of behavioral dominance ranking for both groups. Although teachers proved to be better judges of observed status ranking, it should be emphasized that even the adult judgments provide only 50 percent to 65 percent accuracy in the prediction of individual differences in status position. Thus, complete reliance upon teachers' reports would lead to errors in the ranking of from one-third to one-half of the children. This level of accuracy does not seem sufficiently high to warrant use of teacher judgments rather than behavioral observations for derivation of peer-group dominance hierarchies.

Discussion

In conjunction with our earlier report the present behavioral analysis indicates that the typical primate linear dominance model adequately summarizes the organization of aggressive relations within the preschool group. These results provide an excellent replication of our earlier findings. Attack and threat patterns leading to submission patterns again produced the best fit to the linear dominance model assumptions concerning asymmetry and transitivity of power relations. The Waterloo observations also concur with the earlier report that the organization of social relations defined in terms of object and position conflict differs from that defined only in terms of aggressive interaction. The slight discrepancy between these two classes of conflict suggests that dominance relations do not necessarily determine the outcome of all social conflict. There are probably other contextual factors, such as prior possession of a play area, or ownership of an object, that influence the expression of dyadic social dominance in a particular situation.

A child's position in the group dominance hierarchy did not predict how accurately he perceived his own dominance status. Regardless of individual status, perceptual accuracy scores were uniformly poor when children made judgments that involved themselves. However, in making judgments about others, the perceptual accuracy of low-status children was significantly poorer than that of other group members. Since these low-status children were less frequently involved as participants in conflict episodes, their lack of accuracy in the perception of other children's relations may reflect their lack of exposure to, or interest in this particular type of social interaction.

That children were more accurate at assessing dominance relations among other children than when they themselves were a member of the dyad in question may be related to cognitive egocentrism. One aspect of egocentrism is the difficulty children have in "decentering," or placing themselves in various vantage points. Children might be expected to perform better in judgments involving themselves since they need only coordinate their own viewpoint with that of one other person relative to them. On the other hand, problems in perspective taking would make any judgment of relative dominance difficult but might make judgments involving the self particularly difficult because of an egocentric response bias. Since higher status children approached the accuracy level of teachers in other–other comparisons, the latter interpretation of errors in self–other comparisons seems more plausible. A profitable direction for future research would entail simultaneously examining social status, perspective taking skills, and accuracy of dominance perception.

The extent to which people agree in their perception of social relations is an issue which is of interest, aside from the question of whether or not they are correct in their agreement. As reported earlier by Omark, Omark, and Edelman (1975), preschool children did not agree in dominance comparisons of themselves and classmates. Furthermore, their verbal reports differed from those of teachers in the preschool. Children often gave idiosyncratic reasons for choosing one member of a dyad as tougher (e.g., "He's tougher 'cause he's my friend" or "She's tougher; she bites!"). The particular behaviors to which teachers and children selectively attend in order to judge social dominance remain to be investigated. Since verbal interviews alone provide a relatively poor index of actual dominance relations, future research should include some form of naturalistic behavioral assessment. The strongest conclusion from the present research is that behavioral observations provide a more complete picture of peer-group relations. Consequently, in future research such observations should be used to assess developments in social perceptiveness.

Chapter 11

A Longitudinal and Cross-sectional Study of Dominance in Boys

Glenn E. Weisfeld
Donald R. Omark
Carol L. Cronin

Stability of Dominance Ranks

Animals usually determine hierarchical position by fighting. Following this model, Omark, Omark, and Edelman (1975) asked boys to rank themselves and each other as to "toughness" (see Parker and Omark, this volume, Chapter 24). They found great concern with being tough and high agreement among the boys regarding their respective toughness—higher than for other traits: smartness, niceness, and popularity. Omark and his co-workers subsequently found agreement on and concern with toughness in Swiss boys. For both the Swiss and American boys toughness was salient from about 3 to 10 years of age.

Some of the boys from the American sample have now reached high school. In order to see if dominance ranks showed any stability from grade school to high school, those boys who had been tested previously and who were still in the same school were tested again. This time, dominance was measured by means of two questionnaire items. Classmates ranked the subjects, in groups of 4 to 7, on one item asking, "How dominant are these boys? Rank them on the basis of how easily each gets his way when people want to do different things." This item was designed to represent general ability to prevail in disputes in the high school setting. Boys who were ranked high on this item were also ranked high on a second item trait purporting to reflect dominance: being respected by others. The two items gave very similar results.

In measuring dominance ranks in terms of prevailing in disputes, we were employing a highly general and versatile measure, one that not only has been used in animal studies, with reference to disputed resources, but can also be used on adult humans, even when material resources are not directly under dispute. In contradistinction, measures of dominance that refer only to aggression are not relevant to nonviolent competitive encounters, such as those prevalent in human adults and, possibly, girls. Furthermore, prevailing in disputes seems to be the ultimate function of being dominant rather than a means to or a by-product of that end (such as winning direct encounters, signaling one's rank, or gaining attention). Being dominant allows an animal to gain prerogatives of rank, which are resources or social opportunities that enhance survival and reproduction.

Other items asked for rankings on leadership and popularity, traits that have been shown to bear a relationship to dominance. Questionnaires were completed by 25 to 60 classmates for each high school class of subjects.

Table 1 shows the results of this attempt to relate toughness rank in grade school to dominance, leadership, and popularity in high school. Two cohorts were followed: the graduating class of 1980 and the class of 1979. The former was tested for toughness on three occasions: first, second, and fifth grades. As can be seen, toughness ranks at all three times were significantly correlated[1] with dominance, leadership, and popularity when the boys reached ninth grade. Similarly, the class of 1979 showed a posi-

Table 1. Relation of Toughness in Grade School to Dominance, Leadership, and Popularity in Ninth Grade, as Judged by Boys

Class	*Grade When Toughness Measured*	N	*Dominance*	*Leadership*	*Popularity*
1980	5th	20	.79(.01)	.69(.01)	.56(.02)
	2nd	20	.72(.01)	.66(.01)	.52(.02)
	1st	8	.74(.03)	.74(.03)	.68(.04)
1979	3rd	7	.69(.05)	.83(.02)	.58(.10)

All figures are Spearman rank-order correlation coefficients; levels of significance in parentheses.

Table 2. Correlation of IQ and Achievement Test Scores in Third Grade with Dominance, Popularity, and Leadership in Ninth Grade ($N = 14$)

	Dominance	*Popularity*	*Leadership*
IQ	−.06	−.19	−.05
Achievement test	.08	−.29	.20

tive relationship between toughness in third grade and dominance, leadership, and popularity in ninth grade. On the other hand, IQ and achievement test scores, recorded when members of the class of 1979 were in third grade, were not correlated with ninth-grade dominance, leadership, or popularity (see Table 2).

Determinants of Dominance

The literature on the determinants of popularity and leadership in children and adolescents is inconsistent (Mussen, Conger, and Kagan, 1974). Some studies (e.g., Hahn, 1970; Marks, 1957) show physical traits as being crucial, whereas others emphasize mental traits. Part of this inconsistency is probably due to the use of different measures of popularity and leadership. It has been found, for example, that when a child is asked to name his friends, he often names a popular boy whom he would like for a friend but who would not necessarily accept that child's offer of friendship (Gronlund, 1959; Petry and Omark, n.d.). Because of this inconsistency and these methodological pitfalls, the area could benefit greatly from more behavioral studies of actual friendship choices, including detailed observations of friendship overtures and their acceptance or rejection.

The information available thus far indicates that mutual friendships usually involve individuals who are similar to each other, including similarity in social status (Byrne, Clore, and Worchel, 1966; Hodges and Byrne, 1972; Neugarten, 1946). A great disparity in rank apparently precludes or undermines a friendship. Popularity, on the other hand, is a unilateral attribute that usually characterizes high-ranking individuals. Self-esteem and self-confidence may be compounded of both popularity with peers and respect from adults (Coopersmith, 1967).

In order to determine if physical traits were related to dominance in high school, as Savin-Williams (1976, 1977b, this volume, Chapter 12) had shown for adolescent boys at a summer camp, questionnaire items were included that asked for rankings on athletic ability, physical appearance, intelligence, and fairness. As Table 3 shows, athletics and appearance were salient throughout high school, especially in ninth grade, whereas intelligence showed salience only later in high school. Similar findings held for leadership and popularity. For the class of 1979 in ninth grade popularity was actually negatively correlated with intelligence ($-.30$, $p < .005$). To control for the effect of the setting in which the questionnaires were completed, they were administered in a gym class, an academic classroom, and the school cafeteria, with comparable results.

It is interesting that, for the tenth-grade boys, dominance showed some correlation with intelligence as perceived by peers. However, objective measures of intelligence (grade-point average, achievement tests, and aptitude tests) revealed no substantial correlation with dominance (.28, −.26, and −.10, respectively). This suggests that the raters overestimated the intelligence of dominant boys, in accordance with the reported finding that successful individuals are overrated by others, as well as overrating themselves, even at tasks unrelated to the task of their original success (Harvey, 1953; Sherif, 1956). This halo effect occurred even though the precaution was taken of having no rater judge the same boy twice; i.e., judge him on more than one trait. One might say

Table 3. Correlations of Dominance, Leadership, and Popularity with Other Factors in Ninth-, Tenth-, and Twelfth-grade Boys, as Judged by Boys

	Grade	*Class*	N	*Athletics*	*Appearance*	*Intelligence*	*Fairness*
Dominance	9th	1980	20	.82(.01)	.56(.02)	−.12	—
	9th	1979	50	.73(.001)	.62(.005)	.01	.01
	10th	1979	12	.20	.40	.54(.08)	—
	12th	1977	16	.49(.06)	.41	.40	—
Leadership	9th	1980	20	.90(.01)	.82(.01)	−.07	—
	9th	1979	50	.70(.001)	.56(.01)	.22	−.11
	10th	1979	12	.56(.07)	.78(.02)	.34	—
	12th	1977	16	.27	.14	.69(.01)	—
Popularity	9th	1980	20	.71(.01)	.71(.01)	.13	—
	9th	1979	50	.78(.001)	.80(.001)	−.30(.005)	.04
	10th	1979	12	.87(.01)	.81(.01)	−.18	—
	12th	1977	16	.55(.04)	.48(.08)	.20	—

that this was a true halo effect as opposed to an experimental artifact. Lest too much be made of this finding, however, it should be noted that judgments of intelligence in the same boys one year before and in the other two cohorts accurately reflected objective measures of intelligence and were not correlated with dominance (Weisfeld, 1978).

Despite the importance of athletics and appearance as demonstrated here, it is significant that the study was conducted at a university laboratory school that places great emphasis on scholastic achievement. When informed of the results of the study, all 12 sophomores of the class of 1979 who volunteered to be individually interviewed stated that they believed that academic success was, and ought to be, more important as a means of gaining respect than athletic prowess. Several of them rose to the defense of their school by contrasting it with other, less academically oriented schools.

A Prerogative of Dominance: Access to Females

One of the principle resources for which males of almost all mammalian species contend is females. Accordingly, the high school girls were asked to rank boys in their class on desirability as a date and as a party guest (see Table 4). These items were designed to reveal, as discreetly as possible, males' differential access to females. For both items the respondent was asked to assume that she had no personal attachment or commitment to any boy.[2] These measures may merely have reflected popularity with both sexes, however, since boys who were ranked as desirable were also rated as popular by male classmates. In addition, boys who were rated as desirable dates were also rated as desirable

Table 4. Correlations of Desirability as a Date and as a Party Guest in Ninth-, Tenth-, and Twelfth-Grade Boys, as Judged by Girls

	Grade	*Class*	N	*Athletics*	*Appearance*	*Intelligence*
Desirability as a date	9th	1980	20	.64(.01)	.81(.01)	−.17
	10th	1979	12	.87(.01)	.93(.01)	−.02
	12th	1977	16	.67(.01)	.71(.01)	.23
Desirability as a party guest	9th	1980	20	.54(.02)	.56(.02)	.15
	10th	1979	12	.85(.01)	.78(.02)	−.08
	12th	1977	16	.60(.03)	.77(.01)	.03

party guests. Girls and boys agreed closely as well in their judgments of boys' athletic ability, appearance, and intelligence in all three high school classes (see Table 5).

Table 5 indicates that the girls were concerned solely with boys' physical traits, even in the senior year. Intelligence played no demonstrable role, even though marital partners are known to be very similar in IQ. Presumably, intelligence becomes salient in mate selection later on, although physical attractiveness continues to operate (Stroebe, Insko, Thompson, and Layton, 1971).

It is possible that the girls followed the boys' lead in selecting traits on which to judge the boys or, alternatively, that the girls responded to the outcome of the boys' dominance struggles without being aware of the determinants of dominance among the boys. Either explanation would account for the girls' delay in appreciating intelligence in boys.

Militating against the idea of girls' dependence on boys' perceptions of each other is the fact that the girls showed great ability to reach judgments similar to the boys' on the subjects' intelligence, athletic ability, and appearance. Moreover, when the sexes disagreed on these assessments, the girls followed their own judgments, not the boys', in rating the subjects' desirability. For example, the senior girls' judgments of athletic ability were correlated with their judgments of desirability as dates by .67 and with desirability as party guests by .60 ($p < .03$), whereas the correlations of athletic ability as judged by boys with desirability to girls were only .33 and .41 (neither significant).

For both sexes, boys who were judged to be physically attractive were also identified as athletically skilled. This suggests that natural selection has occurred for an ability to estimate a given individual's physical condition on the basis of physiognomy. Such a capacity would be advantageous to males in a potentially violent confrontation and to females if they were attracted to vigorous males.

Next, the stability of girls' judgments of boys' dominance was

Table 5. Correlations between Boys' and Girls' Judgments of Boys' Athletic Ability, Appearance, and Intelligence

Grade	*Class*	*Athletic Ability*	*Appearance*	*Intelligence*
9th	1980	.97(.01)	.79(.01)	.60(.01)
10th	1979	.83(.01)	.76(.02)	.83(.01)
12th	1977	.76(.01)	.64(.02)	.68(.01)

Table 6. Relation of Toughness in Grade School to Desirability as a Date and as a Party Guest in Ninth and Tenth Grades, as Judged by Girls

Grade	*Class*	*Grade When Toughness Measured*	N	*Date*	*Party Guest*
9th	1980	5th	20	.37	.39(.09)
		2nd	20	.62(.01)	.46(.05)
		1st	8	−.05	.43
10th	1979	3rd	4	1.00(.04)	1.00(.04)

investigated. Table 6 presents the data for the same two cohorts of boys discussed previously. As can be seen, the toughness of grade-school boys, as judged by girls, was correlated with their desirability to girls in high school. Taken as a group, the six correlations for the class of 1980 were significant at the $p < .005$ level (Jones and Fiske, 1953). Evidently, physical dominance struggles in grade school carry implications for access to females seven or more years later.

Observational Measures of Some Correlates of Dominance

Lastly, an attempt was made to demonstrate some behavioral correlates of dominance and hence to validate our questionnaire measures. Ninth-grade boys were observed playing volleyball in gym class. This setting offered several noteworthy methodological advantages. First, athletic competition is a relatively natural, spontaneous, and unsupervised activity. Second, the volleyball court is small enough to allow the players to interact closely and to permit an observer to overhear their remarks. Third, the rules of the game equalize the players' roles almost completely, so that the game itself does not confer status differentials on the participants. Finally, team membership is stable, with no substitutions.

Two observers stationed themselves in the stands adjacent to the court and tape recorded their observations of each evaluative comment between teammates. The identity of the evaluator and that of the recipient were recorded, along with the nature of the evaluation: praise (containing positive reinforcement), criticism (containing negative reinforcement), or instruction (neutral).

Teammates also made quasi-observational (retrospective)

judgments of each other's volleyball ability and the frequency with which each player hit the ball. This was done after each of two tournaments, new teams being chosen for each tournament. The 6 or 7 boys on each of the four teams per tournament also ranked each other on leadership.

As Table 7 shows, players high in ability got to hit the ball frequently (rho = .88), this behavior being taken as the principal prerogative of rank in this setting and, consequently, a measure of dominance. Their teammates also judged these high-ranking players to be leaders. Further, these boys gave more instruction, praise, and criticism than their teammates and received less. In other words, evaluation flowed down the hierarchy. This effect was even more pronounced than the data indicate because dominant players by definition hit the ball more frequently and hence were more often in a position to have their actions evaluated by their teammates.

The finding that evaluation flowed down the hierarchy validated and extended results obtained in social psychological research on small-group processes (Bales, 1953; Homans, 1974). Most of these studies have been done on adults performing an intellectual task under rather controlled conditions, whereas the present study involved adolescents participating quite spontaneously in an activity that they obviously enjoyed. One might say that where spontaneity and affective involvement are high, so is ecological validity.

Table 7. Relation of Teammates' Judgments of Dominance (Frequency of Hitting the Ball), Leadership, and Ability at Volleyball (as Judged by Teammates) to Various Observational Measures

	Ability	*Dominance*	*Leadership*
Giving instruction	.55(.005)	.66(.001)	.80(.001)
Giving criticism	.46(.015)	.53(.005)	.60(.001)
Giving praise	.26	.31(.05)	.44(.025)
Receiving instruction	−.56(.005)	−.40(.005)	−.53(.005)
Receiving criticism	−.18	−.01	.09
Receiving praise	−.25(.10)	−.13	−.21
Posture rated by teammates	.51(.01)	.48(.011)	.45(.05)
Posture rated by observers	.39(.10)	.47(.025)	.47(.025)
Ability	—	.88(.001)	.83(.001)

Subjects were 25 ninth-grade boys (class of 1979) engaged in two tournaments with 6 or 7 boys per team.

Erectness of posture, an element in the dominance displays of most mammals, was also correlated with dominance in the volleyball setting. Posture was judged by the observers and the subjects, with comparable results. Posture is only one indicator of dominance or submission (Maclay and Knipe, 1972). Presumably, the operation of multiple cues—including direction of gaze, muscular tension, and "nervous" movements—renders the communication of rank highly reliable and, at the same time, difficult to falsify.

Dominance interactions in the volleyball setting may be taken to represent the overall high school setting in microcosm. Boys competed with each other for prerogatives, including access to a limited resource and the opportunity to exercise leadership. At the same time specific adjustments in dominance behavior took place in the volleyball setting. Posture in the volleyball setting more closely followed the dominance order in this setting, as measured by frequency of hitting the ball, than the dominance order overall, as measured by ability to get one's way in general. This suggests that posture is adjusted to reflect momentary triumphs and failures rather than being quite invariant, as Maslow (1937) contended. Similarly, behavioral measures and judgments of leadership at volleyball followed ability and dominance at volleyball more closely than they did overall athletic ability and dominance (see Table 8). It seems appropriate, therefore, to apply the dominance paradigm to peer interactions in general (*absolute rank*) and in specific settings (*relative rank*).

Table 8. Correlations of Dominance and Ability at Volleyball and in the Overall High School Setting with Manifestations of Dominance in the Volleyball Setting

	Dominance		*Ability*	
	Volleyball	*Overall*	*Volleyball*	*Overall*
Posture	.48(.02)	.38(.08)	.45(.02)	.50(.01)
Leadership	.92(.001)	.64(.001)	.83(.001)	.70(.001)
Giving instruction	.66(.001)	.40(.03)	.55(.005)	.41(.02)
Giving praise	.31(.05)	.19	.26	.15
Giving criticism	.53(.005)	.22	.46(.02)	.32(.05)
Receiving instruction	−.40(.005)	−.48(.005)	−.56(.005)	−.46(.005)
Receiving praise	−.13	−.21	−.25	−.13
Receiving criticism	−.01	.02	−.18	−.20

Conclusions

In summary, the study indicated that dominance in boys (1) is based mainly on physical traits, (2) is highly stable, (3) leads to access to resources (including the dating of females) and the opportunity for leadership, and (4) is expressed by erectness of posture, among other possible physical attributes. These results are consistent with ethological findings on other mammals in which individuals have been described as competing for prerogatives of rank and communicating their status by stereotypic displays.

What are reported here are some questionnaire data, some interview data, and a limited set of observations within a single school. Such studies need to be replicated but, more importantly, future studies have to dig deeper and examine more factors. One obvious factor is that adolescent girls tend to date older boys (a typical primate pattern). By limiting itself to particular classes, the present study did not comprehensively investigate the topic of high school dating patterns. The next study that someone does will hopefully be broader (an entire high school and beyond?) and will include detailed behavioral and questionnaire data.

What does seem to be of prime importance in the present study is that males of our species negotiate a critical, or sensitive, period from about 4+ (see Sluckin, this volume, Chapter 8) to 7 years of age. During this period they (1) become combative with their peers, (2) become more muscular, (3) acquire the cognitive capacity for transitive reasoning necessary for comprehending rank orders, and (4) begin to experience pride or shame when they succeed or fail. The outcome of these emergent dominance struggles predictably affects social interactions through adolescence and possibly beyond. One might almost say that an individual's rank is imprinted at this time.

The stability of boys' ranks through childhood can be explained as follows. A primacy effect may operate, such that the outcomes of early dominance encounters influence later encounters. This "conditioning to success" phenomenon appears to be a general characteristic of dominance hierarchies (see Weisfeld, this volume, Chapter 15). An important antecedent factor in this developmental progression may be physical maturity. Bigger, stronger preschool boys tend to dominate their peers (McGrew, 1972; Omark and Edelman, 1973). A dominant, habitually successful boy and youth would be expected to be relaxed, confident, and outgoing. Therefore he should excel in competitive situations. Many investigations have in fact revealed that adolescent boys who mature early perform well at a variety of social and cognitive measures and retain this advantage into adulthood (reviewed by Conger, 1977). Savin-Williams (1976, 1977), on the other hand, has found no relationship between pubertal maturity and dominance.

In late adolescence or early adulthood mental traits may also influence dominance. Savin-Williams (1979) has now reported that intelligence but not athletic ability was related to dominance status in 16-year-old boys. If so, this may be a case of ontogeny recapitulating phylogeny. That is, as brains came to rival brawn as an adaptive quality in human evolution, mental prowess may have come to inspire respect. A shift in the criteria of dominance of this sort would have been accelerated by the added advantage conferred in reproductive fitness:

> The more intense this competitive behavior is, the more it is correlated with a breeding premium, which means a very rapid rate of selection for the characteristics that lead to dominance . . . (Chance, 1961, p. 29).

To the extent that the importance of physical traits in peer relations comes as a surprise to us, we may adopt two caveats for the future. One is to beware that children do not necessarily have the same concerns that adults do. We cannot assume that just because we have tried to instill a given value in a child and he or she understands it, the child will respond to our admonitions in all settings. The second warning is that adults as well as children are a continuing product of the interaction between their hereditary endowment and their environmental experiences. Contrary to what many psychological textbooks imply, we are not emancipated from our genes at the age of majority. For example, we should not be surprised to find that adults tend to favor the attractive child (Dion, 1972) as well as the attractive adult (Farina, *et al.*, 1977). We adults may evaluate people by more complex standards than do children, but we seem to retain our evolved responsiveness to human physiognomy.

Notes

1. All correlations in this chapter are Spearman rank-order correlation coefficients (rho).

2. Instead of duly ranking the subjects from 1 (most desirable) to 6 (least desirable), several of the girls felt compelled to express their disenchantment with their male classmates by resorting to the use of larger integers, on the order of 10^4 in some cases. These data were discarded.

Chapter 12

Dominance and Submission among Early Adolescent Boys

Ritch C. Savin-Williams

This chapter summarizes a series of studies (Savin-Williams, 1976, 1977a, b) which extends the investigation of dominance and submission behavior to developmentally more advanced groups of children. These studies demonstrate the existence of dominance–submission hierarchies in male adolescent groups, as well as their stability over time and settings. Characteristics of various status positions and of the hierarchy itself are presented. Dominance and submission behavior for adolescent female groups will be published elsewhere (Savin-Williams, in press).

The setting for the studies was a boys' summer camp in the northern section of the United States. The camp has over 400 acres of natural forests, sand dunes, and lakes, encouraging a number of recreational and interest activities. The camp's program emphasized opportunities for leadership development, balanced activity, friendly competition, and an acceptance of oneself and the groups in which one lives. The campers, over 200 in number and ranging in age from 10 to 16, attended a five-week session. They were divided into four age units; each unit contained eight groups of six boys and a college-age counselor who lived together in a cabin. The daily schedule included such group events as athletic games, cabin discussions, religious services, rest periods, meals, crafts, and outcamping.

During each summer between 1973 and 1975 one of these cabin groups was extensively studied throughout the five-week camping period. Each of the three groups contained six boys between the ages of 12 and 14 years who came from similar suburban, upper-middle-class socioeconomic backgrounds. They had been placed in cabin groups by the camp administration under the general policy of randomly assigning boys to groups, considering diversity in geographic representation and athletic ability, and not placing campers in cabin groups with "old friends."

Procedure

Observations

The author, as cabin counselor, recorded all episodes of social dominance which occurred between members in each cabin group during each regular day of camping activity. Observations entailed the use of an event sampling technique, "all occurrences of some behaviors" (Altmann, 1974). Records of behavior were obtained for three hours during five different activity periods: (1) rising from and going to bed (bedtimes), (2) meals, (3) cabin clean up and rest hour, (4) cabin discussions and meetings, and (5) athletic activities. Eight general categories of behavior were used as measures of the dominance of one individual (X) over another (Y):

1. *Verbal directive:* X verbally communicates to Y what to do and Y complies
   a. Direct order or request.
   b. Indirect directive or suggestion of behavior.
   c. Giving unsolicited advice or information.
2. *Verbal ridicule:* X raises his status or lowers Y's status by verbally abusing Y or by putting himself in a good light at the expense of Y. Y does not contradict, usually withdrawing from further interactions.
   a. Name calling, teasing by direct confrontation
   b. Talk about, verbal put down, gossip, cattiness.
   c. Bragging or boastful behavior.
3. *Physical assertiveness with contact:* X pushes, shoves, kicks, or hits Y. Y takes a submissive posture, flees, or, if asserts self in turn, loses.
4. *Recognition:* Y acts in such a way as to place X in a more powerful position. X becomes a social monitor for Y
   a. Imitation or modeling behavior, appearance, or speech; agree with.
   b. Ask for approval of behavior or appearance; apologize.
   c. Give compliments; give favors; ask where is; defend.

   d. Ask or solicit information or advice; divulge information; wait for.
5. *Physical or object displacement:* X takes an object away from Y, or X approaches and Y moves away.
   a. Direct removal or supplantation.
   b. Indirect control; not asking to borrow; moving into space.
6. *Verbal or physical threat:* X asserts verbal or physical authority over Y with Y not countering.
   a. Verbal challenge, usually with threat of bodily harm.
   b. Physical challenge without making actual physical contact; glaring.
7. *Counterdominance:* X, commanded by Y, assertively or passively disobeys, and Y does not pursue the demand.
   a. Ignoring a direct order or request.
   b. Ignoring the other; shunning; spatial exclusion.
8. *Verbal control:* X verbally argues or battles with Y and gets the last word in or monopolizes the content and structure of the verbal interaction
   a. Argument or battle; direct refutation.
   b. Monopolizes a conversation; interrupts the other's speech.
   c. Contradictions without anger; corrections.

Sociometric Data

In order to provide a more diversified approach to the evaluation of social organization, a questionnaire was given to each camper during the first and last weeks of camp, asking him to list all members of the cabin in order of toughness or dominance. In addition, twice during the camping session each boy was asked to rank the other cabin members according to how much he liked them. Finally, midway through the camping period, each boy completed a group characterization exercise which entailed selecting the one member of the group (including himself) who was best described by a list of 40 adjectives taken from junior and senior high school personality inventories, e.g., most stubborn, shortest, funniest, most feminine, bully.

Other Testing Procedures

In order to examine other physical, social, and behavioral attributes which might characterize different status positions within a group dominance hierarchy, several general assessments were made during the course of the five-week camping session. These measures included the following:

1. *Physical fitness:* The speed by which pulse rate recovers to its normal level following exercises.
2. *Creativity:* Fluency, flexibility, originality, and elaboration in the usage of words and ideas (Torrance, 1966).
3. *Intelligence:* Utilizing the general information, similarities, and digit span subtest of the Wechsler Intelligence Scale for Children (Weschler, 1965).
4. *Physical size:* Body mass derived by multiplying weight x height x a constant (Gallagher and Brouha, 1943).
5. *Pubertal maturation:* Based on Tanner's stages of pubic hair and genitalia development (Tanner, 1962).
6. *Chronological age:* Birth dates as reported by parents on camper application forms.
7. *Hiking position:* The predominant position of the boys as the group hiked single file during the five weeks of camp.
8. *Bed position:* The distance from each boy's bed to the counselor's bed.
9. *Leadership:* From the observational record, the percentage of times an individual's suggestions for group activities was accepted by the group.
10. *Athletic ability:* Assessed verbally by asking each camper to rank the entire cabin for each sport played at camp and then averaging the rankings of each for each sport and assessed behaviorally by compiling a conglomerate ordering based on athletic achievement during camp (the two ranking were always identical).

Results

Assessment of Dominance Structures

Since status ranking based upon the various behavioral indices showed a significantly high degree of interindex similarity (W = .69 for Group 1; W = .62 for Group 2; W = .68 for Group 3; for all, $p < .01$—Kendall's coefficient of concordance, Siegel, 1956), the eight behavioral interactions were combined to yield a general score describing dyadic dominance. The dyadic matrices in Table 1 indicate that a systematic rank ordering of campers for all ob-

Table 1. Summary of Group Dominance Hierarchies for the Entire Camping Session (in Percentage)

Group 1 (1973)	Dominant (X)	Dominated (Y)					
		Ad	Be	Ge	Di	En	Ol
	Adam	X	77	93	100	82	100
	Ben		X	90	87	100	100
	George			X	43	81	86
	Dick			57	X	100	57
	Enos					X	86
	Oliver				43		X
Group 2 (1974)	**Dominant (X)**	**Dominated (Y)**					
		Al	Bo	Ga	Do	Ed	Os
	Alan	X	66	79	81	88	85
	Bobby		X	76	86	86	88
	Gary			X	86	62	92
	Don				X	48	53
	Ed				52	X	49
	Oscar				47	51	X
Group 3 (1975)	**Dominant (X)**	**Dominated (Y)**					
		Ar	Bj	Ge	Do	Er	Om
	Ara	X	67	94	83	100	80
	Bjorn		X	80	63	91	90
	Gene			X	65	71	67
	Doug				X	74	90
	Ernie					X	58
	Omar					42	X

Dyads with percentage entered below the diagonal did not differ significantly ($p > .06$) in the direction of observed dominance.

served dominance–submission behavior was evident in each group. The table summarizes the percentage of times that any particular boy was dominant in his dyadic interactions with all other cabinmates. These behavioral dominance orders were derived for each group by arranging members according to the number of cabinmates significantly dominated. Significance was determined using the binomial probability distribution formula; if the percentage were significantly different from chance, it was assumed that one boy dominated another. Entries below the diagonals in Table 1 reflect instances where the dyadic relationships were not significantly unidirectional.

Of the dyadic pairings (15 for each group), 38 were significantly unidirectional. The exceptions involved low and closely ranked group members. In some cases these ties seemed to reflect slight shifts in relative status during the camping session; in other cases the ties were associated with a low frequency of observed behaviors for the dyad. Ties in the number of cabinmates significantly dominated were broken by placing the pair member who had the higher percentage over the other.

Onset and Stability of the Dominance Hierarchy

A separate analysis of dominance behaviors that occurred during the first three days of camp indicated that an ordered hierarchy was evident quite early for each group. For example, in Group 3 the alpha individual was dominant in 82 percent of his encounters with cabinmates during the first three days, while the beta member was dominant in 67 percent of dyadic interactions with other group members. The percentage of behavior during the first three days which violated hierarchical relationships was only slightly higher than the percentage for the entire five weeks of observations (Group 1—23 percent versus 18 percent; Group 2—23 percent versus 20 percent; and Group 3—28 percent versus 27 percent).

The stability of the hierarchies generally increased during each camping session. In all three groups the level of hierarchical reversals decreased from the first to the last third of the observation period. Table 2 shows the frequencies of dyadic dominance interaction for each of the three groups during the beginning, middle, and end of the camping session. It is evident in these matrices that the majority of instability in the overall organization of dyadic conflict resulted from occasional shifts in relative status between adjacently ranked boys after the initial period of group experience.

Regardless of the specific behavioral setting (athletic games, discussions, and so on), cabin members ranked the same in all three groups ($W = .81$, $W = .71$, $W = .91$, respectively; for all, $p < .01$—Siegel, 1956). The group hierarchy was most stable while the boys ate and the least stable during rest hour, the setting in which the highest rate of conflict was noted (13.7 episodes per hour). Mealtimes proved to be the most peaceful period (5.5 episodes per hour). For both Group 2 and Group 3, however, consistent setting differences were apparent in the occurrence of some behavioral measures. Thus, although the dominance hierarchy remained intact from one setting to another, each setting apparently elicited a different mosaic of behavior.

Sociometric Assessment of Dominance

The sociometric dominance orderings were calculated twice for each group—once on day 3 and again on the next-to-last day of camp. A composite group ranking was obtained by averaging the individual rank orderings, including self-rankings, of the boys. Within each group the boys significantly agreed among themselves on relative dominance status—xr^2 ranging from 14.10 to 26.42; for all, $p < .01$ (Friedman's two-way analysis of variance by ranks, Siegel, 1956). There was nearly perfect agreement between these perceptions and the observer's recording of the daily

Table 2. Frequencies of Dyadic Dominance Behavior during Early, Middle, and End of the Observational Period

Group 1	*Dominant (X)*			*Early* *Dominated (Y)*				*Total*
		Ad	Be	Ge	Di	En	Ol	
	Adam	—	24	8	4	6	5	47
	Ben	5	—	11	5	3	1	25
	George	0	1	—	5	10	10	26
	Dick	0	1	9	—	4	4	18
	Enos	1	0	3	0	—	0	4
	Oliver	0	0	1	2	0	—	3
				Middle				
	Adam	—	2	4	4	2	1	13
	Ben	0	—	2	3	1	3	9
	George	1	0	—	3	4	0	8
	Dick	0	0	2	—	0	0	2
	Enos	0	0	1	0	—	1	2
	Oliver	0	0	0	0	0	—	0
				Late				
	Adam	—	8	2	1	2	0	13
	Ben	5	—	5	5	5	6	26
	George	0	1	—	1	3	2	7
	Dick	1	1	1	—	0	0	3
	Enos	0	0	0	0	—	5	5
	Oliver	0	0	1	1	1	—	3

Table 2 (Continued).

Group 2	Dominant (X)	Early Dominated (Y)						Total
		Al	Bo	Ga	Do	Ed	Os	
	Alan	—	68	38	46	21	15	188
	Bobby	46	—	45	27	30	10	158
	Gary	14	13	—	32	26	3	88
	Don	7	3	5	—	11	3	29
	Ed	1	7	21	38	—	5	72
	Oscar	2	2	1	10	0	—	15
		Middle						
	Alan	—	40	26	18	52	16	152
	Bobby	21	—	12	13	77	13	136
	Gary	7	5	—	9	25	8	54
	Don	7	2	3	—	40	6	58
	Ed	5	13	9	29	—	8	64
	Oscar	7	2	1	12	23	—	45
		Late						
	Alan	—	39	20	9	6	20	94
	Bobby	8	—	8	8	15	13	52
	Gary	2	3	—	10	5	12	32
	Don	3	3	0	—	20	24	50
	Ed	5	0	4	10	—	10	29
	Oscar	0	1	0	7	1	—	9

behavior of the campers in all three groups at the beginning and closing of camp (r ranging from .90 to 1.00; for all, $p < .05$—Siegel, 1956). Junior counselors, who assumed leadership over the cabins on the author's days off, independently reported the same hierarchy as did the boys and the observer (r ranging from .83 to 1.00; for all, $p < .05$—Siegel, 1956).

Overall, there was a consistent tendency for the adolescent boys to express egocentrism rather than self-abasement. In the three cabin groups the boys overrated themselves by at least one position in 40 percent of the 40 sociometric self-ranks. By contrast, 10 percent underranked themselves and the remaining 50 percent were accurate in self-perception of dominance rank. This egocentrism is further indicated by the fact that of the 20 dyads who disagreed on relative standing, 85 percent were ones in which both dyadic members claimed that they were more dominant than the other. In only three pairs was the disagreement in

Table 2 (Continued).

Group 3	Dominant (X)	Ar	Bj	Ge	Do	Er	Om	Total
				Early Dominated (Y)				
	Ara	—	20	2	14	17	41	94
	Bjorn	19	—	8	95	16	37	175
	Gene	0	2	—	1	5	1	9
	Doug	4	53	4	—	15	17	93
	Ernie	0	3	3	10	—	6	22
	Omar	9	6	3	2	8	—	28
				Middle				
	Ara	—	29	22	4	12	5	72
	Bjorn	11	—	46	96	12	14	179
	Gene	0	11	—	4	23	28	66
	Doug	0	55	3	—	10	15	83
	Ernie	0	0	10	5	—	8	23
	Omar	3	1	9	0	2	—	15
				Late				
	Ara	—	21	8	2	4	10	45
	Bjorn	5	—	23	56	11	19	114
	Gene	2	6	—	8	18	28	62
	Doug	0	39	0	—	21	13	73
	Ernie	0	1	6	1	—	1	9
	Omar	2	1	16	3	1	—	23

the opposite direction, each boy reporting that the other was more dominant. These dissenting dyads were usually composed of middle-positioned and closely ranked individuals.

Predictors and Characteristics of the Dominance Hierarchy

In all groups athletic ability, bed position, and leadership were significantly correlated with position in the behavioral hierarchy (Table 3). Although measures of peer popularity failed to predict dominance status, it is interesting to note that best friends were often individuals immediately adjacent in the dominance structure (average distance between best friends was ±1.7 status positions).

The most frequently occurring dominance behaviors were, in

Table 3. Rank-Order Correlations of Trait Measures and Sociometric Choice with Status in the Dominance Hierarchy

	Group 1		*Group 2*		*Group 3*	
	r	*p* <[a]	*r*	*p* <	*r*	*p* <
Traits						
Leadership	.83	.03	.88	.02	1.00	.00
Athletic ability	.94	.01	.71	.07	.89	.02
Bed position	.83	.03	.83	.03	.89	.02
Physical fitness	—	—[b]	.66	.09	.66	.09
Pubertal maturation	−.31	ns	.09	ns	.89	.02
Physical size	.09	ns	−.31	ns	.77	.05
Hiking order	.26	ns	.71	.07	−.09	ns
Chronological age	.31	ns	.60	ns	−.09	ns
Intelligence	−.14	ns	.14	ns	.06	ns
Creativity	—	—[b]	−.37	ns	.31	ns
Sociometrics						
Popularity (early)	.03	ns	.26	ns	.80	.03
Popularity (late)	−.31	ns	.60	ns	.66	.09

[a]Spearman rank correlation (Siegel, 1956).
[b]Trait not measured.

order, verbal ridicule, verbal directives, and physical assertiveness; the least frequent were displacement and verbal/physical threats. Various status positions, however, had unique characterizations in regard to the number and kind of dominance behaviors employed. For example, beta-ranked individuals were most and omega boys least likely to be involved in dominance interactions. In all groups the gamma-ranked boy was the "bully," successfully employing an unusually large percentage of physical assertions and threats.

Alpha group members ranked first in athletic ability and physical fitness. In addition, they were pubertally advanced for their age, slightly older than cabinmates, and more intelligent. During group meetings, alpha boys proposed the most ideas for group activities and their ideas were almost universally accepted by other group members (90 percent acceptance rate).For example, in Group 2 no activity proposal was passed by the group unless Alan suggested or strongly seconded it. During hiking trips one could usually find alpha near the middle of the single-file order; when he slept in the cabin, his bed was closest to the counselor.

While the group overwhelmingly attributed positive adjectives to the most dominant individual in all three groups, he was the only group member who underestimated his own positive characters. In all groups the alpha was seen as athletic, dominant, strong, masculine, serious, mature, popular, rich, and handsome and as the leader of the group. Alphas accurately perceived other group members' relative status (average correlation, .88). They engaged in dyadic dominance interactions as often as the average group member (245 versus the group average of 249 per individual).

Omega individuals, by contrast, were poor athletes, physically unfit, pubertally immature, and younger than cabinmates. During cabin meetings the omega boy suggested few ideas for cabin activities, and less than 10 percent of his suggestions were accepted by the group. He slept far from the counselor and hiked at the front or the end of the group during hiking expeditions. These boys overestimated their sociometric dominance rank by one or more places. Cabinmates described omega campers as possessing more negative (60 percent) than positive traits—as a follower, smiler, and cabin "brown-noser" as well as most religious, feminine, submissive, and quiet. Finally, omegas had the lowest rate of involvement in dyadic dominance interactions (mean number of episodes was 168).

Discussion

The present findings add to the increasing evidence that man shares with other primates a means of structuring dyadic power relations within a group dominance hierarchy. This stable and ordered—but not invariant—structure is evident not only through observation of behavior, but also in the verbal reports of group members. While it is not possible to determine how early in the history of the group that group members accurately perceived the hierarchical structure, it is likely that some preliminary organization emerged within hours of meeting. Certainly, the dominance structure was established at both a verbal and behavioral level by the third day of camp, gaining in stability as camp progressed.

Callan (1970) points out that it is extremely difficult to give a simple definition of dominance and submission that is valid for all species. Even though roughly two-thirds of the dominance interactions recorded in these studies involved language, other behaviors frequently employed to assess dominance and submission relations among nonhuman species (e.g., winning fights, supplanting others, displaying threat gestures, gaining priority to resources) were evident in all three groups.

Many individual characteristics distinguish alpha group members from omegas, but only three individual variables—athletic ability, leadership, and sleeping position—significantly predicted position within the entire group dominance hierarchy. While the first of these variables may be a determinant of status, the other two are probably the result of being dominant.

Pubertal development, chronological age, and body mass were less highly related to dominance status than athletic ability and physical fitness. This suggests that psychological and sociological studies of adolescents which have traditionally examined early and late maturation, relative age differences between subjects, and weight variables to determine important predictors of group structure and interpersonal behavior have perhaps missed important determinants of early social relationships: coordination, physical fitness, and athletic ability. According to Omark, these latter factors begin to distinguish young children as early as kindergarten (personal communication).

The consistently significant correlation between bed position and the dominance hierarchy is not easily interpretable. It was previously suggested (Savin-Williams, 1977b) that for the early adolescent to sleep next to the adult counselor is to associate and perhaps to identify with the ultimate source of power in the cabin. In this sense, bed position may imply nothing more than a preferred sleeping position, an advantage afforded to those with high status—a common primate pattern (Aldrich-Blake and Chivers, 1973; Carpenter, 1964; Jay, 1965). Group 2 is a good example of the prevalence of this sleeping status relationship. During the course of the five weeks the position of beds within the cabin changed three times. Despite these changes, the relative position of each camper's bed to the counselor's bed remained constant.

It was noted that most boys had a reasonably accurate view of their relative ranking within the cabin group. This is in marked contrast to preoperations-age children (Savin-Williams and Freedman, 1977), but not to older children in same-sex groups (Omark and Edelman, 1976; Parker and Omark, this volume, Chapter 24). Thus, while an adolescent boy can accurately report where he ranks in relation to others in the group, he tends to overrank rather than underrank himself. This may reflect the social and personal desirability of high status within the adolescent group.

The relationship between best friends and similarity in rank is also difficult to interpret. Perhaps two boys, because of their friendship, form a coalition which results in the two maintaining close ranks within the hierarchy. From another perspective, because two closely ranked individuals are likely to be competitors within the group, it may behoove an individual to make friends with those directly below and above himself in order to reduce tension, to gain favors, to decrease the likelihood of physical harm, and to inform himself of what competitors are thinking and acting.

The social roles of alpha and omega individuals were considerably different. The alphas initiated and determined group activities while the omega's role entailed accepting ideas, doing the necessary work for their implementation and completion. While specific high status prerogatives of nonhuman primates may not be applicable for high status in a human group (food, water, sex), there are similarities: sleeping sites, preferred sitting and marching positions, and desirable foods—the scarce resources at summer camp. Perhaps the most salient motivating force behind the seeking of high status is internal or, as Barkow (1975b) has labeled it, "prestige-striving." Washburn and Hamburg (1968) write, "Being dominant appears to be its own reward—to be highly satisfying and to be sought, regardless of whether it is accompanied by advantage in food, sex, or grooming" (p. 473).

The function of a dominance hierarchy in human groups is perhaps similar to its effect for other primates: it adds stability and predictability to social living and thus contributes to a cohesive group. One mechanism for achieving these ends is to reduce the amount of intragroup conflict. In Group 2 and Group 3 there was a sharp decline in rate of dominance behaviors during the course of the camping session. There were only two exceptions to this general decrease. On day 7 Gene entered the group and on day 16 Doug was unexpectedly elected as a compromise candidate by the "brother" cabin to be a unit chief, a position of formal leadership established by the camp administration. In both cases the rate of physical assertiveness and verbal/physical threat increased rather sharply. Perhaps these increases in conflict activity are best viewed as a group response to external factors which disrupt social stability. From this perspective, it is noteworthy that the increment in the initiation of conflict behavior corresponded within the bounds of established patterns of dyadic dominance rather than being reflected in rate increases of violations to the existing dominance structure. Such data begin to address the functional significance of dominance structures and will hopefully provide the direction for future research on human social dominance.

Chapter 13

The Umwelt and Cognitive Development

Donald R. Omark

> . . . there is no fundamental difference between man and animals in their mental faculties . . . as man possesses the same senses as the lower animals, his fundamental intuitions must be the same.
>
> (Darwin, 1871[1])

In this chapter an overview will be presented of the developmental studies of dominance relationships that appeared in this section of the book. This overview will be presented, insofar as possible, in terms of the child's possible conception of his or her world. A lot is known about the child's cognitive perceptions of the physical world, and some suggestions will be made about what our conceptualizations of the child's social cognitive world should consider. The notion of the Umwelt will be used as an heuristic device for such conceptualizations.

Jakob von Uexküll (1926, 1957) introduced the concept of the Umwelt and defined it as the perceived world of the organism. Since every organism has its own particular set of sensory organs, the perceived world will be slightly or greatly different from that of other organisms. For example, imagine seeing the world only in black, white, and grey as many predators do, or being able to "hear" how far away from objects you may be as bats do, or being able to smell a thousand more smells as dogs do. This perceived world is that portion of the total environment to which the organism can respond; it is that part of the world which is necessary for its survival in terms of food, shelter, mating partners, danger, and so on.

The notion of a feedback system was probably first applied to animals by von Uexküll in the 1920s. Figure 1 illustrates his idea about the organism perceiving its Umwelt (receptor), processing this information, and then responding to it (effector). The environment was then affected in some way, leading to new perceptions, and so on. It should be noted that not all of the environment is perceived and that selective forces may act upon the organism to its advantage or disadvantage with the organism being only peripherally aware (e.g., feeling sick) or even totally unaware of such forces (see Alexander, 1971, pp. 109–110). Imagine what a world it would be if germs could be seen as bright yellow dots and the evolutionary advantages to the organism that could do that.

But, you say, humans are aware of germs, and they have developed instruments that can hear like a bat, smell like a dog, see in the dark, and so forth. It is for these reasons that von Uexküll's notion of the Umwelt has to be updated to include not only the perceived world, but also the conceived world of the organism (Freedman and Omark, 1973; Omark, Omark, and Edelman, 1975; see also Boehm, 1978). The conceived world is that part of the external world which comes to be represented within the organism, in forms peculiar to each organism. It is this data which the organism then uses as it confronts its environment. Both learning and thinking, as concepts, require that something happens within the organism. This "something," which will be different at different phylogenetic levels of organization, is the active conceptual processes which the organism is capable of performing. Whatever produces these conceptualizations will be left to others

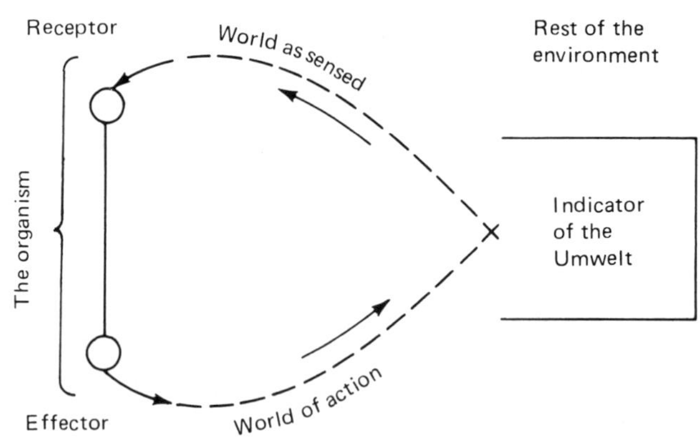

Figure 1. The organism in relationship to its Umwelt (after von Uexküll, 1926).

to discuss. What is important here is that these conceptualizations lead to observable behaviors that can have evolutionary consequences for the organism. For example, if an organism can conceive, or think, that it is important to imitate the actions of another, then observational learning can begin. For the organism to conceive of danger, as yet unseen, when hearing an alarm call means that communication can serve a group survival function, and when communications are exchanged, as with Kortlandt's chimpanzees (1962; Korlandt *et al.*, n.d.) about where danger once occurred, then both mythology and science can begin.

The perceived world coupled with the conceived world, stored in relationship to available levels of cognitive structures, constitute the Umwelt of the organism—its knowledge about its environment. This knowledge may be true or false (to an outside observer); it may be good or bad in terms of survival of the individual; and when transmitted to others, it may be good or bad depending upon the end result of the group's responding to and incorporating it. For example, the Japanese at their various primate centers have been provisioning these centers with a variety of foodstuffs. At some centers all of the food is eaten, while at others some types of food are ignored (Miyadi, 1959). Hence, because some of the food has not yet been perceived as good within the Umwelten of some of the troops, these troops are depriving themselves of possible provisions.

Cognitive development has been studied in humans, and more recently in other organisms, and the really surprising finding that Piaget and others have made is that the young child is not just a blank slate *(tabula rasa)* to whom information is provided following the rules of learning theory, but that the young organism has a different Umwelt. What is perceived and what is conceived appear to follow fundamentally different rules for the child than for the adult. The child obviously learns facts about its environment, but the way the child structures these facts may be very different from the way an adult would structure the same information.

Much of the cognitive developmental research has focused upon how children conceptualize relationships between physical objects (which is heavier, longer, taller, and so on). Only recently have attempts been made to extend the idea of stages in cognitive development to social situations, e.g., infants' object permanency as related to the mother (Bell, 1970), later changes in attachment as a function of declining egocentrism (Marvin, 1977; Marvin and Greenberg, n.d.), and seriation and transitivity as factors in the establishment of social hierarchical structures (Omark and Edelman, 1976; Omark, Omark, and Edelman, 1975). From an ethological, or evolutionary, perspective the changes that occur in cognitive development should be equally, or even more, important for relationships in the social sphere than they are in the physical sphere. For instance, is the concept of object permanency more important to a primate when it is attempting to follow a bug which runs under a leaf or when it is able to turn its back to its mother while hunting the bug? It will, after all, be fed by its mother even if it does not catch the bug.

Size and weight probably play a role in adult–infant/juvenile interactions, since a good deal of punishment and reinforcement come from the adult, but what in the physical world requires similar responses? The largest leaf is not eaten first, followed by the next largest, and so on down the line. One eats what is at hand, literally. Contrasts between the social behaviors of adults, juveniles, and other infants present distinctions that should be important for cognitive development. As suggested in Parker and Omark (this volume, Chapter 24), some other distinctions between peers may be a necessary part of cognitive development. Being able to make finer and finer distinctions between others and between the self and others is, of course, one of the bases upon which hierarchical social relationships can develop. It also is a necessary part of the cognitive stages involving seriation and transitivity.

Genetic Perspective

The Umwelt, as an heuristic concept, is the internal representation of the external world. Evolutionary theory (and genetic theory) is predicated upon variation in the population. First, because without variation in the population, there can be no evolution, no progressive shift in the genetic base, and thus no change in the phenotypes toward a more adaptive norm. Second, there are always mutations that introduce variations into the structures and behaviors of populations. Third, and perhaps most important, is the fact that most structures and behaviors derive from the actions and interactions of more than one gene (a pleiotropic effect). In addition, every individual can be expected to be genetically unique (see Chapter 1, this volume). Hence, no behavioral, structural cognitive, communicational, physiological, or other system should be expected to be invariant across any population (see

Scott, 1976). This means that the Umwelt of any member of a species should be expected to be slightly or drastically different from that of every other species member, except for the case of identical twins reared together.

For example, in Figure 2 we consider two organisms with different genetic, and hence different biological conditions (B_1 and B_2) living in the same environment (E), but each having that environment enter consciousness, or the Umwelt, slightly differently (U_1 and U_2). There are some portions in common (U), which if common throughout the species might come to be thought of as "universals" or "species-specific characteristics." Depending on the species, these overlapping areas might be broad or narrow. One can certainly have bird songs which show local dialects but with the song from a different dialect similar enough so that mating can occur across the boundaries. When Umwelt differences are too great (when the "other" is not recognized as a conspecific), each popuation is on its way to becoming a separate species.

A different kind of universal may also be operating in Figure 2, and this universal is the way in which information is processed. When going to a foreign country, one can recognize that a language is being spoken or that certain nonverbal gestures must mean something. Anthropologists manage to survive in the field because what they encounter is not gibberish, but part of a process which they have experienced before. Some cognitive processes may then be more universal than the products which they produce. Ethologists and linguists have both considered universals at other levels (e.g., genetic similarities, neurological commonalities, "deep structure," and so on. Human ethologists should not get lost in similarities and differences at the observable behavioral level without also considering what the Umwelt of the

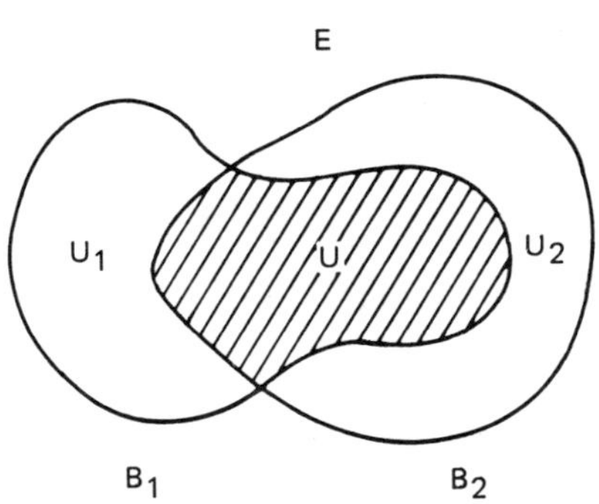

Figure 2. Each organism has both distinct and common portions to its Umwelt.

species may be. Barkow in Chapter 18, this volume, discusses one such universal that may take many different forms in adult life.

The Umwelt is that part of the total environment which becomes part of the organism between the reception and production aspects and is presumably closely linked to them. Taking communicatory structures (the structures needed to produce, and the sense organs needed to receive messages, as well as the underlying neurological structures) as one example from the many related to an Umwelt, a number of the chapters in Hahn and Simmel (1976) consider the genetic relationship between transmission and reception. For example, among the arthropods crickets have been extensively studied, and both parts of the communicatory structures have been found to be under the control of the same genes. Ginsburg (1976, pp. 59–80) traces this genetic control upwards through man, with more "degrees of freedom" appearing at each level (i.e., environmental factors, or possibly other genetic factors, having an influence without destroying communication).

A problem continually mentioned throughout the Hahn and Simmel book seems to be whether the same genes continue to control both the receptive and expressive sides of the process. This appears to be an unnecessary restriction, even though Ginsburg suggests that some human communicational problems suggest such a link. However, once communication has started (and, as we suggested earlier, this should occur simultaneously with bisexuality), then it appears immaterial whether both sides of the communicatory process are controlled by the same genes. That both sides are is probably the simplest beginning point. That the process continues to be governed at both ends seems superfluous since the total communicatory process can probably be better enhanced by genes which contribute to either the production or reception side, basically very different physiological mechanisms. The limiting and controlling factor, then, is that this total complex of genes designed to establish and (perhaps) maintain the group is subject to the forces of natural and social selection. In other words, once communication has begun the same genes need not continually be involved in both transmission and reception as long as communication is possible.

The ontogeny of our species presents similar problems. If one reads Piaget and the cognitive developmental literature, one wonders how these young organisms could possibly develop into the adult organisms that they become. While Piaget presents a nice progression of stages toward the adult, the difference between children's Umwelten and that of adults can be so different that an observer from Mars could say that children must be a different species, smaller in size. It will become apparent as this chapter continues that genetics, through the action of regulatory genes which control when particular structures (hormones, and so on) will appear, obviously plays a role in such ontogenetic changes.

Social Perspective

The Darwinian model of evolution is directed toward the reproduction of the species. Sociobiology, as a subdiscipline within evolutionary theory, similarly focuses upon reproduction. Adulthood is, of course, the period when reproduction occurs and the point where the final determination is made of what the next generation will be like. Much attention has been focused upon this point in development. Even Piaget describes his stages as leading toward an adult model (see Flavell, 1963). But ontogeny is as major a focus for ethologists (and others) as is the final end product. What must be kept in mind is that the organism has to survive to the point of reproduction for reproduction to occur at all. One should not expect that an organism remains biologically invariant throughout its life. Different sets of genes may be operative for different portions of the life cycle, and with different consequences since the organism within which they are operating is also changing.

If children, and adults, are seen as being fundamentally different at different points in their life, then the question, How does this behavior seen in a child lead toward adulthood or serve adult functioning? becomes supplemented with a question that takes a "stage-in-life" focus. This second question is, How does this behavior (cognitive process, physical structure, and so on) enable the child to survive *now*, and what function does it serve? Play is one set of behaviors that is frequently investigated in terms of its potential effect on adult life (for example muscles are developed, coordination is enhanced, fighting techniques are learned, communicative signals are refined). Many of the above factors also enhance the chances of survival of the young organism while it is young and might *also* appropriately be viewed as having immediate benefits. Play is much more common in the juvenile than in the adult stages of life, and its function in the juvenile stage may be as a strategy for acquiring information in that, at least some, "play" behaviors have "exploratory" and "manipulative" phases where one first uses an object, partner, and the like to see what he or she can do, and then one acquires (some) mastery over the object (see Hutt, 1966; Brenner and Omark, 1979).

Similarly, cognitive stages may not *just* be an invariant pathway to an adult model. If Piaget's cognitive stages are relatively invariant and universal (cf. Ember, 1977) then one should question how these stages promote survival at the ages when they appear. To be able to turn away from mother in order to explore the environment, "knowing" that mother will still be there if needed, is an important early stage. Being able to enter a peer group and suffer defeat in a fight without that first, or tenth, defeat forever after determining one's place in life is an obviously important later stage. To fit finally into a social group, while still maintaining some confidence in the self, is yet another stage. Some of these points will be elaborated below and in other chapters. Piaget's cognitive stages are rich in detail, and the student is invited to explore what they may mean in terms of social interaction, with an evolutionary view always being kept in mind (a beginning point might be Piaget and Inhelder, 1969, and then Flavell, 1963, for more detail). What developmental psychologists, human ethologists, sociologists, and anthropologists should be interested in exploring is the broader question about the way the environment and the developing organism interact to promote the process of surviving to adulthood (for examples see Marvin *et al.*, 1977; Omark and Brenner, 1976).

Part of children's entrance into the adult world depends upon their actions and thought as they reach an age where they become young adults. Savin-Williams (this volume, Chapter 12, and 1977a, b) and Weisfeld, Omark, and Cronin (this volume, Chapter 11) examined subadults (in primate terminology) or young adults (as children beyond 12 or 13—the age of puberty—have been considered by many religious and tribal groups over at least the last 2,000 to 3,000 years) as they interacted within their peer groups. Barkow (this volume, Chapter 18) considers the fully adult individual. Ethological studies of humans are needed to examine the shift from just within the former peer group to interactions with adults. This examination has to include movement into work settings (where hierarchical relationships will certainly be present), into courting situations (but do, please, permit them some privacy so that next generation can be started), and into all the other areas of social interaction that affect our species.

Many primate studies, like Kummer (1968), present an overview of all the age-sex classes as they interact and can serve as heuristic models of what may occur during this transition phase from adolescence to adulthood in humans. Hunter-gatherer groups probably provide the best analogy, and perhaps homology, to primate groups, as do anthropological reports of small villages, nomadic tribes, and the like, because they examine groups which include all age and sex classes. It is left to the student to develop the researchable questions needed to explore ethologically this transition period.

But let us return to an earlier age where data is available for reexamination. At some age children in every culture, as do young primates, move away from the mother, parents, or home, and confront their peer group. Why they do this is not directly explainable in a cause–effect fashion from the ethological, or any other, perspective. There is the obvious evolutionary reason in that those of peer-group age may provide a mate for the next generation.[2] Incest (copulation with a parent) is as taboo in human groups as it is in primate groups (Sade, 1968), so it is to the peer group that the juvenile must first turn if it is to establish a future societal position and reproductive relationship.

The chapters of the preceding section present a number of interesting phenomena that appear to be related to age. As with primate groups, human children's groups are initially behaviorally undifferentiated in sexual composition (Chance and Jolly, 1970, p. 161). Missakian (this volume, Chapter 23) emphasizes this point in her presentation, and it is apparent in the data in the Strayers' and Sluckin's chapters in this volume. As the children grow older (by first grade, Omark and Edelman, 1975), their play groups become differentiated by sex, with some exceptions, and remain that way for many years. Within many societies groups undifferentiated by sex, when there is a choice, may be relatively rare (see Draper, this volume, Chapter 25).

The first pattern of group formation then is boys and girls playing together (3 to 5 years of age) followed by a second pattern where they basically play separately (6 years on, Edelman and Omark, 1973). These patterns fit well within a primate model that would include many primate species. The earlier period corresponds with the beginnings of symbolic play where the child initiates control of his/her world and starts to develop a sense of self. The latter corresponds with the concrete operational abilities of performing seriated and transitive thought operations (among others) as attention is directed toward similarities and differences within the peer group (cf. Piaget and Inhelder, 1969). Questions need to be raised and explored about the earlier shift from orientation to the mother to orientation toward peers, and then about the second shift toward the same sex portion of the peer group (see Parker and Omark, this volume, Chapter 24, for possible ecological/social effects upon cognitive development during this latter shift).

The first shift, toward the peer group, seems to resemble a phenomenon that might be labeled "species recognition." If you own a dog or a cat, you know that either can spot a conspecific two blocks away, and long before you are aware of what it is that is attracting the animal's attention. A similar phenomenon, perhaps better labeled "age class recognition" seems to occur in children; they can similarly spot another child, even when there are far more numerous adults present. This part of their Umwelt is an "attention structure" that precedes the one Chance and Jolly (1970) and Chance and Larsen (1976) illustrate, but one which serves the purpose of moving the child or young primate toward and then into the peer group.

This "earlier" form of an attention structure, hypothesized here, is in obvious need of further research. That young children from the age of 2, if not earlier, will gravitate toward another child, or other children, at picnics, parties, outings in parks is a fact that can be readily observed by parents and others. That young primates also leave their mothers to enter peer groups is documented in numerous primate studies. In both of these examples it is the youngster who makes the choice of moving toward peers, and the mother continues to be present as a secure base. Being "sent" to nursery school, where the mother immediately leaves after bringing the child, both removes the element of choice and the secure base. As we have observed in various nurseries and as Waterhouse and Waterhouse (1973) document, such an introduction to a peer group can, at least initially, be traumatic to the child. What the long-term effects may be are not known; they would require detailed comparisons between children introduced with and without the mother present.

Given a choice, however, even infant primates (human and nonhuman) will move away from their mothers in order to approach their peers and interact with them. The young at this point begin interacting for the first time with similar-size organisms. They also gain a little more control over their choice of interactants. In numerous primate and human groups they, prior to this age, had little direct control over who would handle or interact with them outside of their mother. Finally, they can pick and choose among their playmates, and they have taken the initial step toward creating the next generation's group.

In these initial asexual groups individual differences will express themselves. Some individuals will be more agile, more graceful, bigger, smarter, faster, and so forth. There should be no lack of opportunity for the youngster to perceive differences between the abilities of others. The youngster of 3, 4, and 5 years of age is then encountering social differences well before the ages of 6 or 7 where the child can readily respond to questions about seriated differences between physical objects. Within a Piagetian assimilation/accommodation model the youngster should be provided with numerous perceptual examples from the social world of differences in behavioral attributes before he or she has to deal with the physical attributes of objects with which experimenters present them. Once the cognitive structure has been developed for dealing with differences per se, and the following step of ordering these differences in a serial fashion has been developed, then the structures should also be available for further social as well as physical tasks.

This line of reasoning suggests that any research comparing a child's abilitiy to respond to both the physical and social world should not simply see, for example, whether a child can both seriate physical and social objects in some hopes that the social results will come out earlier (see Edelman and Omark, 1973; Omark and Edelman, 1976). By the time children are 6 or 7 years of age the above argument would suggest that they can handle equally well both the physical and social aspects of seriation. Such were the results of the Omark and Edelman studies cited.

What really needs to be examined in more detail are the social and physical experiences that children have long before they are capable of demonstrating that they can perform certain cognitive tasks. Where the child or young primate actually gains the experience necessary to move from one cognitive stage to another needs more detailed research efforts, and these efforts should initially be directed at the real world of the child rather than at experiemental manipulations of that world based upon *a priori*, but not necessarily true, assumptions about that world.

Learning That There Is a Hierarchy

The acquisition of particular patterns of behavior has long been of interest to learning psychologists, but from the stance of their being able to control the acquisition process. The factors operating in the real world that "teach" the developing organism something are more difficult to detail because they occur spaced in time (rather than in discrete trials) and these factors are embedded within a variety of environmental stimuli (rather than as specifically controlled experimental stimuli). One prime example of an external agent presenting particular stimuli to a developing organism occurs during the weaning process in many primate species. At some age, depending on the species, the mother avoids, hits, or pushes away the youngster when it tries to suckle. Human mothers may or may not be as severe, depending on the group norms, but still the child is weaned to solid foods somewhere between 1 and 4 years of age. Of course, even where numerous studies would suggest that it is the mother who weans the infant, counterexamples can be found, as in our own experience, where the infant decides that it no longer needs or wants the breast.

The studies done by Piaget and his followers suggest that what the child "knows" may be more age-related than due strictly to any particular set of reinforcing stimuli. Many parents of very young children (under 3 years of age) who have attempted to punish their children so that they will not do certain things know how futile such punishments can be. The child does have experiences which it can remember and respond to, but such experiences seem to have different behavioral effects at different ages. Attempts by learning psychologists to speed up the Piagetian stages by intensive "learning" have been notable for their lack of success.

In the evolutionary scheme where a particular Umwelt of the world may be taken as important for the survival of the organism at a *particular point in time,* such failures of the learning psychologists to control developmental processes should not be surprising. This does not mean, however, that the principles of learning theory must necessarily be abandoned because a variety of the techniques seem applicable, from time to time, to species ranging from rats to man. Rather, one must ask where and when such techniques naturally appear and may be responded to by organisms living in their own world.[3] One must also question the relevance of what is learned for the life of the organism. Washburn and Hamburg (1965, p. 613) suggested that "evolution, through selection, has built the biological base so that many behaviors are easily, almost inevitably learned." This statement may be true at the behavioral level without necessarily being true (from the organism's own perspective) at the cognitive level. As Strayer and Strayer, Missakian, and Sluckin (this volume) indicate, a child of preschool age and even younger can "learn" a position in a dominance hierarchy. If the children's behaviors were random (producing dominating or submitting behaviors willy nilly), an observable dominance hierarchy with a marked degree of linearity could not be formed. If the children had some built-in predisposition to act in just one way or another (say, always aggressive or always submitting), again there would be no hierarchy. Instead, there would be one cluster of aggressive children and another of docile children. It is possible, though, for the observer to construct a dominance hierarchy, and, hence, it is apparent that the children are acting in a differential fashion toward their peers. In essence, the children have "learned" a set of dominance relationships between themselves and their peers.

What does this differentiated form of interaction mean? It means that the children are perceiving and conceiving of their peers as being different from each other. While this fact may appear to be obvious, when reading the experimental psychology or educational literature where children are paired to perform some task (modeling, coaching, and so on), this fact is generally not considered. Frequently some hoped-for results are sought by "randomly" pairing individuals. This randomness is only in the experimenter's head if the children know each other.

What the children are perceiving and responding to is an open question. Sluckin (this volume, Chapter 8) and Strayer and Strayer (this volume, Chapter 7) suggest that it is initiations of aggressive actions rather than completed dominance/submissive interactions that are important. Height and/or weight may or may not be factors (Edelman and Omark, 1973), and responses to these physical characteristics may shift with age and can be different for initial and repeated encounters. Overall physical coordination may be important at an early age and certainly seems to be so later (see Chapters 11 and 12, this volume). For instance, the smallest boy in the first-grade class was beautifully coordinated, being able to swing across a suspended ladder like a gibbon. Even though he was the smallest boy, he was ranked third in the hierarchy by almost all of his peers. Similarly, girls who are ranked high by both girls and boys are those who are equally proficient in both the boys' and girls' games. Leadership versus followership is not really examined in the chapters of this volume, but this is an obvious area that needs more research. In other words, what are the characteristics of self and others to which children come to pay attention?

The children learn about others by interacting with them and by observing them. As should become evident when reading this book, not everyone interacts with everyone else. A sociometric matrix (Altmann, 1974; Omark, Fiedler, and Marvin, 1976) generally cannot be completed. Some dyads never interact, and this is as true for human children as for any primate group. In fact, from nursery school on, in at least three cultures (United States, Switzerland, and Ethiopia) not only particular dyads but each sex generally does not interact with the opposite sex (Omark, Omark, and Edelman, 1975). However, in at least the American sample each sex was as accurate in its perceptions of the *opposite* sex on four dimensions (Who is the toughest, smartest, nicest, and has the most friends?) as was the opposite sex in perceiving itself (Omark and Edelman, 1975). At some phylogenetic levels it may not be necessary for every organism to interact with every other because observations of others replace direct experience. As a hypothesis, the particular phylogenetic level at which this occurs may be closely linked with the point where imitative learning becomes important. Both forms of learning require that attention be paid to others who are not directly interacting with the self.

That observational and verbal learning occurs need not be argued at the human level. For example, a new boy came into a third-grade class in Switzerland. On the way home at lunchtime on his first day in class he had a fight with the alpha boy (who lost), and this fight was witnessed by about one-third of the class. The fight was reported by the children, with a great deal of excitement and even glee, to the teacher and the other children that afternoon. In the "Who is the toughest?" sociometric test (Omark, Omark, and Edelman, 1975) given a few days later all of the children ranked the new boy as alpha (first in the class), except for the former alpha who placed himself first and the new boy second in rank.

If "learning" one's place in a hierarchy can be shown to occur behaviorally in children of nursery school age, what does this say about cognitive development, and what may be the function of both what is learned and what cognitive developmental stages the children exhibit during their early preschool and school years? Preschool-age children (4 and 5 years of age) were asked to rank order their peers on the "Who is the Toughest" test in an attempt to see if there was any agreement about the seriated order among peers (Edelman and Omark, 1973). What was found was that they were significantly *below* chance level, with most of the children ranking themselves first or second and the rest of the children were apparently randomly ordered. Sluckin's report (this volume,

Chapter 8; see also Sluckin and Smith, 1977) also supports this finding. An examination of *how* the children performed the seriation task, across nursery school, kindergarten, and first grade, basically followed the developmental pattern described by Piaget and Inhelder (1969) for children's remembrance of the seriation of physical objects (Omark and Edelman, 1976).

In contrast, preschool-age children seemed to learn their place in a hierarchy and "pay attention" more to initiations than outcomes. At the same time, when asked about the hierarchy (the total set of dominance relations) children with an average age of 4 years, 5 months (Sluckin, this volume, Chapter 8) seemed reasonably accurate when viewing those interactions that did not involve the self. This set of findings suggests some hypotheses for further testing and some speculations about the biological function of learning versus cognitive development.

First of all, Chance's (1976; Abramovitch, this volume, Chapter 22) notion of attention structures is important because if "initiations" of aggressive interactions indicate the dominance relationships to a preschool-age group, then this means that the group members have to be attuned to the rest of the group (as much as is an observer who is deliberately observing the group). Since preschool play is so frequently described as being "parallel play" but not "interactive" play, this is a surprising finding, as much as the finding that older children, girls as well as boys, were able accurately to describe the dominance relations of the other-sex group as was that sex able to describe its own dominance relationships despite the fact that the groups basically did not interact. This means that children from a very early age are continuously attuned to the social events and the persons.

Why initiations rather than outcomes are focused upon by young children may have a number of explanations. First, considering a cause (the initiation) and the effect (the outcome) that followed may be beyond the cognitive abilities of preschoolers. Second, that cause–effect relationships are even an important factor to consider may be a cultural artifact that takes some time to acquire (Ember, 1977). The mythologies that explain the universe are certainly as, or more, frequent a phenomenon among human cultures as is a more "scientific" cause-and-effect view. Even chimps display superstitous behaviors, and that children do need not come as a surprise. Perhaps the amazing thing is that some children do shift from initiations to outcomes, and a number of studies might be done about when, and if, such a shift occurs in naturalistic settings.

Third, from a purely practical point of view, it might be better to learn which other individuals tend to start fights than whether they ultimately win or can be beaten. Knowing the first means that most damage to the observing individual can be avoided, while knowing the latter two outcomes suggests that some individuals should either definitely be avoided or that injury to the self may or may not occur. In the early stages of entrance into a peer group it might be more important to move into that group successfully and form bonds with group members then to establish one's position high in the hierarchy and attempt to maintain that position until reproductive age is reached. A repeatedly reported phenomenon among newly formed primate groups is that aggression is initially high, but it quickly drops to a lower rate. The hypothesized premise for this shift in behaviors is that an initial rapid assessment is made about others' abilities and a slower working out of eventual position can then follow. If aggression is to serve a stabilizing rather than a disruptive function in group formation, then attention to initiations may facilitate this process.

Sluckin in Chapter 8 (this volume) suggests that around 4½ years of age children can verbalize about dominance relationships between others. However, his study and our earlier ones (Edelman and Omark, 1973; Omark and Edelman, 1975) indicate that problems exist when the self is included in such verbalized reports. The cognitive development of children in the sense of what they "know" about their social world and their own position within it may not necessarily match with how they have "learned" to act within that world. These factors will be explored in the next section.

Learning One's Own Place

While young children appear to learn how to respond to others such that linear relationships can be established for aggressive/submissive encounters, with the children paying somewhat more attention to who initiates such encounters, a major question should be what the child thinks about him- or herself during such encounters. Such a question is necessary if behavioral actions toward others, or manipulations of physical objects, are to be directly related to the cognitive structures that are being formed within each child. In other words, how is the child per se involved in what it is doing, and what in this process is observable to an outsider. If what is observed does not match up with what is said by the child, while both physical and verbal behaviors are being produced by the same organism, then questions need to be asked about those discrepancies which appear.

Language acquisition is a continuous process, and a continual problem for researchers working with young children. Even when a word has been acquired, some period of time follows before a firm meaning may be attached to that word. Little of a child's vocabulary may be acquired with a dictionary meaning attached to it; instead the vocabulary is gained through experiencing its usage within various settings (Hanson, 1978; Miscione, Marvin, O'Brien, and Greenberg, 1978). For example, children were asked, "Who is the smartest?" (Omark and Edelman, 1976). There was little agreement among the children in first grade but significant agreement in second grade. One explanation for this shift would be that it took more than one school year for the children to attach meaning to this more-or-less scholastically related concept and then to be able to use it when making comparisons between peers. The word and its meaning appeared to be the problem rather than the children's abilities in seriating the ranks of their peers because agreement significantly above the chance level was found in first grade for the concept "Who is the toughest?"

Being tough is a concept that appears to have been acquired much earlier. The concept is not unidimensional, even for American, English, or Swiss 4-year-olds; some of them conceive of it as having aggressive elements, others see it as relating to strength, and many see it as involving elements of both. In fact, an American dictionary gives twelve definitions for the term, and Chapter 11 of this volume suggests a thirteenth (for young children) that may consist of some components of coordination or smoothness of behavioral actions.

The concept is not unidimensional, but this need not be disturbing because even the observable behavioral actions that describe the initiation of aggressive acts are not unitary. Krebs (1972 as noted in Sluckin, this volume, Chapter 8) found 25 behavioral elements that "predicted the same dominance relationships as those defined by hitting." If "aggressive speech" were expanded upon by a psycho- or sociolinguist, Krebs' list could be increased enormously. The first interesting set of points about what is happening is that somewhere after 4 years of age the children are able to combine their experiences under a single concept, to show some agreement across a class full of peers, and this concept is beginning to show some relationship to behaviors that can be observed by an adult.

The other interesting point discernable at this early age is where and when disagreements occur. Cognitive development for humans is, at least in part, related to verbal development. One gains knowledge about the external world, and part of this knowledge is stored in verbal form. As the above discussion of "toughness" indicates, much of it may be retrieveable in verbal form—at least grosser concepts like being tough. Such methods of retrieval are obviously not possible in primates, but since primates too produce hierarchical structures related to a variety of experiential events, they should have similar cognitive ways of combining events. Hence, what can be learned from human verbal reports might be reflected back upon primate behavior rather than simply being dismissed as beyond a primate's cognitive abilities because we are verbal and they are not.

The verbally reported disagreements with peers about hierarchical status is one area where research on human children's cognitive development might be explored in more detail among primates. We leave it to the primatologists to develop the appropriate experiments.

Before 4 years of age children cannot vebalize a linear or hierarchically seriated structure for their peer group. Their linguistic abilities and their amount of experience with peers may be a problem when confronted with the tasks of the experimenter. A salient point, however, is that the children's responses to experimenters' questions are almost invariably favorable for the self. If a child is asked, in contrast with another, "Who is tougher, taller, shorter, faster, slower," and so on, the child will respond "Me!"

When constructing a hierarchy from pictures of nursery school peers, children overrate their own position, even when they could fairly accurately rank their peers (Sluckin, this volume, Chapter 8). In our experience, children of nursery school age would place themselves first or second in the ranking and then deal with the rest of the class. This misperception of the self would mitigate against a strong correlation between hierarchies perceived (by the child) and observed (by the adult). Since the older children could agree on the placement of others in Sluckin's data, this means that the children were capable of observational learning but were denying or cognitively falsifying experiential learning. At the same time they were behaving in such a way as to suggest that they had experientially learned their position in the hierarchy. As mentioned earlier, if they can observe and learn (and accurately report these facts), and if they can experience and learn (and behave appropriately), why do they "tell a lie," as it were, about themselves?

Overrating

The next piece of developmental data about how children in the lower grades both perceive their peers and behave toward them is missing, but something is known about their perceptions of peers. With increasing age the children show more agreement between each other on their relative ratings of peers, and there is more agreement between their own placement of the self and where their peers place them (Omark and Edelman, 1976; Omark, Omark, and Edelman, 1975).

Where disagreements do occur between the relative placement of the self and another child, the child almost invariably ranks the self higher. For example, Suzy and Mary are ranking each other (on their individual tests), but if they disagree on their positions, then Suzy will rank Suzy higher and Mary will rank Mary higher. Out of four grades (K–3) tested in both Switzerland and America (36 classes) this result occurred, except for four classes of first-grade American girls.[4]

From the above set of results on children's perceived hierarchies the derived hypothesis would be that children will cognitively skew their experiences in favor of the self until they are absolutely convinced about a relationship. The limited set of experiences in nursery school does not convince any of the children, but they do become convinced of certain relationships as they get older. As Omark and Edelman (1976, pp. 139–141) illustrate, this skewing for the dimension "toughest" may be more problematic for boys than for girls.

From an evolutionary perspective where man, the primates, and perhaps other species are seen as group-dwelling species and where dominance relationships are also seen as serving a stabilizing function, the above set of findings may begin to make sense. The developing organisms act in such a way that their peer group is momentarily stable; they "learn" a position within the group. At the same time they are moving toward adulthood, reproductive age, and the probable necessity of leading the group. Hence, it seems important that the child may cognitively be viewing the self as being *more* and, in part, different from what is expressed behaviorally.

An egocentric view of the self may never leave us, even as we become more realistic in judging the external world and assessing our own capabilities. Such a view may be very important for group maintenance (see Barkow, this volume, Chapter 18). If a leader is lost, someone has to take his or her place (depending on the species). If the number 2 (beta) individual always saw itself as number 2, even in the absence of number 1, then groups should disintegrate frequently when the leader and focus for attention is lost. There has to be a continuous readiness to move up. This readiness appears as the continual testing of higher members, although not necessarily the top member, in numerous species of primates and man (see Savin-Williams, this volume, Chapter 12). This means that a group with a hierarchical structure is a dynamic entity, momentarily stable but poised to change and reconstitute itself into another stable structure (cf., Hinde, 1974).

Underrating

The converse of seeing the self as *more* than behavioral actions or others' perceptions would indicate is a downward placement of the self within the hierarchy. It has been our experience, in numerous classrooms, that when children seriously *underrate* their own positions in relation to how their peers perceive them, other social, emotional, or academic problems are indicated. When such children have appeared on tests, both in the United States and abroad, and the teacher is questioned about them, almost inevitably a problem exists. This caveat is inserted to alert future researchers about anomalies in their data. When and why a child may begin to undervalue the self has been explored in numerous research efforts, but tests of the nature of "Who is the toughest, and so on" do seem to indicate which children might be withdrawing into themselves. Even children at the absolute bottom of any perceived and verbalized hierarchy will overrate their position; contrary findings should be investigated further, or the teachers should be alerted to the findings.

In a recent study (Petry and Omark, in prep.) an entire culturally distinct sex group was found to underrate themselves vis-à-vis other members of this same cultural/sex group. These children were Chicano (Mexican-Americans raised in the United States) girls in a fifth-grade class of 10 Chicano girls, 9 Anglo

(Caucasian, non-Chicano) girls, 10 Chicano boys, and 1 Anglo boy. The Chicano girls underrated their positions on "Who is the toughest," which suggests that they are following their cultural stereotype of girls being humble and meek, at least in terms of aggression. The Chicano boys and Anglo girls and boy overrated their position. When an individual underrates the self drastically, then problems are indicated; when a group does this, the group may have problems or other factors (such as cultural factors) may be operating.

Related to underrating is a phenomenon suggested by Barkow (this volume, Chapter 18, and personal communication). This is the problem of those who find themselves at the bottom of a social caste or class system. As one becomes an adult, and generally more realistic in the assessment of self and others, it just may happen that one finds oneself at the bottom of the totem pole. Such individuals may copulate and reproduce, even in primate groups and certainly in human groups, so the hypothetical, strict dominance-related matings proposed by the sociobiologists are not of interest here. What is of interest is how such individuals relate to the total group or societal structure. If dominance relationships occur, in any species, then low-ranking individuals are inevitable. Their escape from the group is not a viable alternative path (cf. Wilson, 1975) because this simply moves the problem up the hierarchy until the group per se ceases to exist. Such is not the pathway for most primate/human groups. The "low" members continue within the group.

Group Identity

An alternative approach is to permit the group to enter the Umwelt or cognitive awareness as an entity—"This is the group that I belong to"—and to draw boundaries that exclude nongroup members. This is the typical pattern found among primate groups where nongroup members have to win their way into a group by fighting lower ranking individuals before they confront the higher individuals. If the nongroup members survive the lower ranking confrontations, they will probably be permitted to remain with the troop or group. What seems to be happening is such instances is that there is a definite group boundary, with the lowest ranking male marking the boundary (females are different and their entrance might depend on estrous condition). If the new male passes the boundary by defeating one or more low males, he then is, in essence, within the group. The hypothesis here would be that new males cannot enter a group by becoming the new omega (lowest) male; they would continually be attacked until they left.

An Umwelt capable of perceiving group/nongroup boundaries may be important for continued group membership. Even an individual's Umwelt that remains mildly egocentric in adulthood is still continually confronted with reality. To be at the bottom of the group would seem to carry some psychological/physiological weight in terms of stress, as numerous writers have suggested. Hausfater and Watson's data (1976) indicates that, for at least one species of baboon, this may not be the case. A basic research question would then be whether individuals at the bottom of the hierarchy are more egocentric (e.g., place themselves higher in a perceived hierarchy) than the average group member, even if behaviorally they do not manifest these perceptions of the self.

If the low-ranking individuals do not differ from the average group members (of their own sex) in their amount of expressed egocentricity, then questions arise as to whether x amount of egocentrism and/or identity with a group is enough to maintain mental stability (i.e., a "good" feeling about the self) and continued group membership or whether something else is required.[5] The something else may be a combination of the above *plus* an identification of one's own group as being better than another group, the extreme of which would be xenophobia (ethnocentrism). The research questions become more complex here because they involve contrasting groups which have contact with other similar groups, versus groups which lack such contact, while measuring amounts of egocentrism (or overrating of hierarchical status), and other group identify factors.

The only data that we have to begin to illustrate what may occur when groups meet are the above mentioned confrontations of Chicano/Anglo girls in a bilingual classroom (i.e., Spanish and English were used on alternate days). While the Chicano girls underrated the self vis-à-vis other Chicano girls, they overrated themselves vis-à-vis the Anglo girls (as did the Anglo vis-à-vis the Chicanos). The apparent acceptance of a cultural norm (being humble) obviously stopped at the cultural boundary and noncultural members were placed lower in the hierarchy than were cultural members. Any number of factors may be operating in the development of group boundaries, and these factors need to be explored if desegregation or bilingual classrooms are to meet their supposed objectives. One needs to know more about how children structure their groups and how boundaries are developed and successfully passed.

Whether one feels that one is a member of a group or not has not really been explored in the early school years when peer groups are first formed. A few notes on what has been observed but not reported elsewhere may be useful for future researchers for the questions that they might suggest.

The boys' pattern of group formation starts with small groups, 3 to 6 children, playing together and responding to the commands of one or more individuals. By kindergarten, if a leader has emerged, groups will coalesce around him and follow his lead or imitate his actions. There did not appear to be much stability in these groups, outside of the identifiable leader. In first grade leaders and groups were apparent, but group members could travel back and forth between groups without the leader being aware that they had left. By third grade the groups appeared to be more stable, large enough to include most of the boys and sometimes a few girls, with certain individuals actively being excluded from group membership or excluding themselves (Ginsburg, this volume, Chapter 20).

Girls' groups across this entire age range were typically smaller (3 or 4 children), less active in the sense of covering less distance (e.g., jump rope instead of dodge ball). Whether or not some of the girls were acting as leaders in their groups is not known because the observations occurred at a distance. By third grade, and definitely in fourth grade, cliques had formed with definite exclusion of others (e.g., who could or could not join a group for lunch). Most of these statements are based upon incidental observations that need to be made more systematic. The statements also refer to within-class groups; no observations were made on between-class interactions during free play time.

At the other end of this developmental process of group formation are the adolescent and adult structures that are created. Weisfeld *et al.*, (this volume, Chapter 11) suggests that there is some continuity, at least among boys, between their earlier—in first and third grades—and later perceptions of other group members—in ninth and tenth grades (14- and 15-year-olds). Hence, despite how the individual perceives the self, his Umwelt may perceive the group in fairly constant terms. As further illustration of this constancy, Savin-Williams (this volume, Chapter 12) infers that when new groups are formed among adolescents, dominance relations appear by the third day of interaction (a period similar to experimental manipulations of primate groups) and these groups remain stable thereafter. Both of these studies imply that group stability is as important a feature in human groups as it is in primate groups. In essence, the Umwelten of involved individuals have been selected to need some group focus. This focus either maintains or quickly establishes a group with a leader as its center of attention. This would suggest that perhaps most humans and primates would prefer to identify with a group and know their place within that group than to act as free agents. The data are slim and need to be expanded upon, but the suggestion is there.

We will not attempt to explore here the relationship between various adult human groups and the possible cognitive processes necessary for dealing with such relationships, interchange between groups, and so on. Such an exploration would require a book (but see Mazur, 1973, for a cross-species comparison of small established groups as a starting point).

Instead, let us consider briefly what may happen cognitively to adult group members who are at the bottom of the hierarchy and for whom egocentricity per se may not be sufficient mechanism for maintaining mental stability. What human groups create are other groups to which *the group* is superior. This means that every group member is superior to every nongroup member. This is reflected in the anthropological translation of numerous tribal names as "The People" or "Man" with the implications being that any other tribe or group must be "nonman," and hence of lower status.

Such a philosophy, or construction of "Man's" Umwelt, appears humorous when one examines primitive tribes or college fraternities; the thought that they would label other humans as nonhuman seems a bit ludicrous. Any reading of the daily newspaper, however, shows that such attitudes are ever present in modern societies, and active attempts are continually being made to exclude various classes of humans from access to human institutions. Scientists are no more objective! Following from Descartes and his dualistic view of the mind and body, most of social science has accepted a philosophical model which separates the human from the nonhuman—e.g., man can do this but apes, or monkeys, or rats, or flies cannot. Rather than viewing "Man" as a currently examinable point in the evolutionary process, social scientists continually stress the difference between Man and nonman, on the one hand, or deny that man and others may from time to time *think* about what is happening to them, on the other hand. A truly evolutionary perspective on human behavior has yet to gain a foothold (Freedman, 1979; Freeman, 1977; Omark *b*, in press).

If the typical human response is to consider nongroup members as being of lower status, then even bottom-ranking individuals have someone to whom they are superior.[6] This implies that, in general, it is better to be in a group rather than to be freely wandering between groups, even if one is of low status within that group. That a group is really better than another group, to an outside observer, is probably immaterial; other groups need only be conceptualized within the individuals' Umwelten as inferior for the individuals to have a perceived status higher than omega (the bottom of the group).

As far as we know, the dissolution of human groups and the forming of new groups has not been studied developmentally in naturalistic settings. The first-grade example mentioned earlier is one instance of group members moving between groups, but when and why they do so remains an open question. Insufficiencies in the abilities of the leaders as well as limits in the cognitive abilities of the followers are two possible explanations. However, what is and is not being incorporated into the cognitive or conceptual portion of the Umwelten of the individuals involved is still unknown. When a group is first seen as an entity perhaps happens at about first grade. The age when a group is also adopted as that entity with which an individual identifies him- or herself as a member needs to be explored by political scientists as much as by other social scientists.

A Final Hypothesis

Combining the above set of statements with those statements in Chapter 2, "The Group," we would like to present one final hypothesis. If the group is a viable entity upon which natural and social selection can operate and if group members have been selected to function in terms of the group (such that they maintain its form, as they appear to), then one has to consider how group structures are continually re-created. Hinde and Stevenson-Hinde (1976) suggest that the interactions between individuals combine, per their model, into more complex and abstract entities (e.g., relationships, structures). They approach the problem of how groups are created by using a learning theory paradigm.

Piaget's work (and our own) suggests that a recognition of

relationships (seriation, transitivity, and so forth) has been selected by evolutionary pressures to be part of the Umwelt of each individual organism (certainly at the primate and human level and possibly at the carnivore and other levels). The individual then comes prepared to deal with relationships—both between himself and others and, observationally, just between others. Males and females may differ on the amount of experiential information they require, but they both seem to be able to process observational data. This suggests that a reasonable hypothesis might be that we (and other species) are not just a product of relationships developing ontogenetically through time, but that our Umwelt was organized phylogentically in a way which facilitates group membership. The ready acquisition of language, or conspecific signals, being but one example of our predisposition to group life and maintenance of group structure.

Conclusion

The development of perceptual and cognitive structures for handling information gained from the environment is as important at other levels of phylogenetic development as it is at the human level. The nature of the information received and what is then done with that information will, *per force*, be different for different species because of their particular evolutionary encounters with specific habitats (as are individuals' responses to their environments, Omark *b*, in press). Still, the Umwelten of related species should be expected to exhibit some similarities, particularly to the extent that ecological and social environments produce pressures for structures which function similarly. This is not to say that the physiological/neurological structures of different species need be the same. Rather, it is that these various structures process similar information to similar end results, even as species that are more related phylogenetically should be expected to exhibit more nearly identical bodily structures.

Once a species has crossed the line from being basically solitary living to group living (or, perhaps, just being a bisexual species), the Umwelt has to be expanded to include the social world. This expansion may include perceptual changes, but it should certainly include the necessity for cognitive changes. Being able to handle communicatory information and being sensitive to group relationships are two of the more obvious basic changes. The more complex the social structures become, the more complex will the cognitive structures need to be. No single cause-and-effect relationship need be isolated here because it is the genetic variability of a population that both permits it to produce more complex structures while also responding to such structures. For example, one need not understand the complexities of religious, military, or political structures for one to be able, nonetheless, to play a role within these structures.

Over time, those social (or cultural) structures that are detrimental to the group in terms of the broader environment will be selected against, as will the groups, by natural and social selection. Those individuals within the group who do not respond to the created social structures will either be ostracized by the group, exiled, ignored, or isolated. In turn, the individual can rebel, form new groups, and so forth. No social structure need be taken as an absolute given. Again, this is because genetic variability will continually produce new, absolutely unique, individuals who will respond to and create social structures in their own way.

As Piaget has argued, it is the ontogeny of the individual and the experiences that occur during this ontogeny that interact to produce the cognitive structures which an observer can discover. As has been presented in this chapter, and previously, there probably are phylogenetic aspects to cognitive development that should also be considered. Within our own human phylogenetic line, stretching back through the primates, important aspects of change in cognitive structures should be found which relate to increasing abilities to (1) handle socially communicated information, (2) develop relationships with others along a number of pathways of social interaction, e.g., attraction, affiliation, dominance, submission, and so forth, and (3) recognize the total group and respond differentially to subclasses within that group. Within the general range of knowledge that the Umwelt may be genetically programmed *to receive* should lie most of the particular cultural or intragroup linguistic, communicative, and group structural signals which the organism is apt to receive (with the caveat noted above that a place can be found for even those who do not fully comprehend).

What we have attempted to convey here is that the world is a complex entity. That is both a truism and something that is generally ignored in most social science research. As Tinbergen (1973, pp. 138–139) suggested, experimental models which only consider immediate *cause-and-effect* relationships between stimuli and behavior are probably missing a "truer," or at least broader, picture of what the organism is, in fact, experiencing and why it is responding the way it does. Human ethology is probably the only discipline that permits investigations ranging from the genetic and physiological structures to the behavioral and cultural aspects of man. A human ethologist should not be limited to the strictly observable, nor to the "more animal-like" infancy or early childhood periods. Human ethologists should be interested in humans in their full complexity—be it linguistic, cultural, nonverbal, or otherwise. This book presents a smattering of data related to a particular focus—dominance relations. The life of humans and other animals is certainly broader than that, and we invite you to begin the explorations of other areas. We expect that numerous books from an evolutionary perspective will follow.

Notes

1. Quoted from Ginsburg (1976, p. 61).
2. In humans a small peer group per se, if started at a very early age, will probably not provide a mate. Children from kibbutz groups tend not to marry within the group (Tiger and Shepher, 1974, 1975), and the prediction would be that the Synanon group's children (Missakian, this volume, Chapter 23) will not. The apparent desire for an out-group member as a mating partner lowers the chances of "lethal genes" (Ginsburg, 1966) appearing; hence, it is to the group's advantage. This desire also introduces variability into the group. What the limits are on "familiarity" versus "availability" needs further study.
3. Learning techniques such as those employed in "behavior modification classes" as applied to specific behavioral problems may have specific results. What is being considered in the foregoing discourse, however, is the normal course of events experienced by most children.
4. Why an entire cohort of girls one year should significantly underrate the self is not known. Nor is it known whether this may be a more general phenomenon among American first-grade girls in mixed-sex classes.
5. Individuals who continually strive for higher positions might also be expected to rank higher in their misperceptions of position.
6. A common fact of human existence is that groups will frequently be conquered by other groups. When confronted with a vastly superior "other group" rapid acquiring of the conqueror's behavioral patterns, social structures, and language can occur. In essence, this is an amalgamation with the conqueror that places yet to be conquered groups at a lower status level. The actual actions of the conqueror toward the conquered group can, of course, change this series of events.

Part IV

Dominance and the Individual

Chapter 14

Sexual Dimorphism and the Status Hierarchy

Daniel G. Freedman

The key to sex differences in behavior seems to involve the basic mammalian fact that males have far greater gametic production than females. One male can potentially account for thousands of offspring, 1 female for perhaps 15 to 30. If, as biologists and evolutionsists hold, the purpose of life is to maximize one's own genetic contribution to subsequent generations, males and females should develop different courtship strategies as a result of the difference in gametic production.

Mammalian males should tend toward promiscuity, while, contrariwise, mammalian females should tend to select and solicit males who give promise of help in raising their young. These basically different aims may be seen as forming a sort of polar dialectic that gives rise to a spectrum of possible compromises. Needless to say, many factors aside from gametic differences play a role in the male–female compromise. In birds, for example, females are always chromosomally XY and males XX (the opposite of mammals), which may be a factor in maintaining the 90 percent of monogamy in that phylum (Freedman, 1979). In pelagic fish, where neither parent cares for the fertilized eggs, the immensely high egg production eliminates gametic differential as an important factor in courtship strategies (Hediger, 1965).

Among mammals whether a species is preyed upon and precocial (on its feet soon after birth) or a predator and altricial is important in courtship strategies. Precocial mammals (e.g. deer, buffalo, zebra) are invariably herbivorous herd animals and almost always under some degree of pressure from predators.

Among them there is usually a rutting season in which males vie for central territories and within which territory a harem group is usually accumulated. Females tend to be attracted to victorious males with their more central territories (Wilson, 1975). In many precocial species (e.g., the impala) individuals are seen in either all-male bachelor troops or as one-male harem troops. Actual copulations amony prey tends to be very short in duration, taking some 45 seconds in the impala.

Mammalian predators are rather more variable in their social arrangements. Wolves, like packs of dogs, make no distinction between male and female in the dominance hierarchy, and this is reflected in the rather low physical dimorphism (females are but a few kilos lighter) and in hunting patterns (all save nursing females participate equally in tracking and attacking prey). Lions, on the other hand, usually tolerate no other adult males in the pride, and while a mature lioness is large and lethal, the male is substantially larger and looks far more imposing because of the mane. Other cats, e.g. tigers, hunt alone and are, therefore, rarely involved in dominance–submission encounters except at mating time. The duration of copulation among predators is substantially longer, reaching a figure of several hours in polar bears according to Hediger (1965).

Only when we get to the troop-living semiterrestrial macaques and baboons do we have a full-blown status hierarchy. Entire troops are usually organized along lines of dominance and submission, with males substantially larger, more aggressive, and more dominant than females. Because of some obvious parallels with human group behavior, it is tempting to place our own species in this category, terrestrial primate. Let us, however, examine the evidence for and against such a move.

In humans sexual dimorphism is moderate, some 8 to 12 percent in height on the Murdock world sample. Therefore, one might expect considerable variation in intersexual strategies in this most numerous of all primates. Let us start with an examination of sex differences in infancy.

Sex Differences in Human Infancy

In general, differences between the sexes are subtle at birth, and usually large numbers are needed before a trend is apparent. But differences are there. As soon as babies are born, boys cry more, respond less to vocal cajoling, and require more holding. Thus, within the first days, boys are exhibiting intimations of the temper tantrums they will exhibit at greater rates throughout later infancy and childhood.

Boys are rather more interested in inanimate, nonsocial objects than are girls. For example, infant boys are as attracted to a bull's eye as to a drawing of a face, whereas girls clearly prefer the face (Lewis, Kagan, and Kalafat, 1966). Boys also get into problem solving sooner, so that if a toy is put behind a barrier, 9-month-old boys will reach for it, while girls will often sit, watch and wait (Kagan, 1971). Later in the first year, when dolls are given children, boys tend more often to take theirs apart (to see how it works?). Even in children blind from birth, boys more quickly pick up how gadgets work and can, for example, easily replace a wheel on a wagon. Blind girls appear to be particularly helpless about such chores (Burlinghame and Robertson, 1957). Boys thus seem to evince a very early interest and ability in mechanical tinkering.

Bock has discussed the considerable evidence that such abilities (called spatial-visualization skills) are inherited via a major recessive gene located on the X chromosome (Bock, 1973). Without going into the theory of X-linked inheritance, in overly simplified terms this means that boys need but one dose of this gene to show the trait while girls would need two. This model accounts for the observed fact that only about one-quarter of the girls are as good as the average male on tests of spatial visualization, and that males are generally more successful mechanics and engineers (Bock and Kolokowski, 1973). In the same vein, Munroe and Munroe (1971) found that among East African children physical distance from home was related to performance on tests of spatial ability. Boys ranged further from home than girls; yet those few girls who were found further from home than typical for their sex outperformed their male counterparts on the measures of spatial ability.

Another long-term trend is heralded when girls are found to be more responsive to their mothers' voices in the first weeks of life. Throughout childhood they are more alert to vocal sounds. Further, they form words before boys do, and throughout their subsequent lives will do better at all verbal and linguistic tasks—whether these involve tests of spelling, grammer, verbal fluency, or learning new languages (Freedman, 1974).

Recently, a large number of studies have appeared which relate these differential skills to sex differences in brain organization, specifically to differences in hemispheric dominance. In brief, the left hemisphere is said to control verbal tasks; the right hemisphere, spatiovisual tasks. Kimura (1969), who is one of the most active workers in this field, has presented evidence that left hemispheric dominance for speech occurs earlier in girls than in boys and that by contrast boys eventually show right hemispheric superiority over girls to go with their better spatiovisual and arithmetic skills. Let us not, however, overstate the fatedness of these sex differences, for modifications can occur. Hoffman and Maier (1966), for example, were able to enhance female performance in problem solving by expanding the linguistic presentations of the problem; conversely, one of my students developed a reading test that requires constant problem solving so that it actually favors boys (reported in Freedman, 1979).

There are other sex differences to note. Girls start life more cuddly. In three separate studies of newborns, in three different laboratories, newborn girls were found to mold into the arms better, to kick less, and, in general, to offer less resistance to being cuddled (Freedman, 1974).

Older females apparently have the same tendency. Blanck observed students at the University of Chicago library and performed an analysis (see Study 6 in the Appendix, Freedman, 1979). She found that only females tucked their feet under their coats, only females tended to hold their heads to one side while rounding their shoulders—that is, even when older, girls tend to curl into compact spherical shapes. Males, by contrast, were usually squarely planted in their chairs, with shoulders level and head unbent. Both feet were often on the ground, with legs spread apart. (Wickler [1972] asserts that the legs-apart position typifies primate males, and he presents evidence that its function is to display the penis.)

Female newborns are also more sensitive tactually. They react to a fine camel's hair brush that hardly affects males, and they cry more readily when wet. That is, they react more immediately to any change of skin temperature or to being touched (reported in Freedman, 1974). Pair that with the fact that Kinsey found most women are more typically sexually aroused by touch than by vision (the male mode), and we have yet another possible tie-line from birth through maturity. Here, however, we must note dissenting arguments by authors who attribute Kinsey's finding to "culturally induced" repression (e.g., Kronhausen and Kronhausen, 1964).

The Smile

Perhaps the most interesting sex difference found at birth involves the "reflexive" smile, that is, smiling in the absence of any apparent sensory stimulation. Several studies have found that newborn girls smile more than do newborn boys. This early "reflex" smiling is clearly related to later social smiling, for the babies who do more of it turn out to be "smilier" youngsters (Freedman, 1974).

There are other indications that female infants are more socially oriented. Kagan (1971) and Lewis (1966) have found that infant girls, but not boys, can differentiate adult male and female voices; that girls prefer to watch faces, whereas boys are equally interested in nonsocial objects; and that, as early as 9 months, girls more than boys will babble back when spoken to. Baby girls will react differently to adult males and females long before boys make this distinction, and it is not surprising that the girls are also more worried by strangers (even as are rhesus monkey females; Mason, Green, and Posepank, 1959). We can only conclude that girls are more socially oriented right from the start.

Do females actually smile more than males beyond infancy? We have conducted ten studies with young persons and adults, and in every instance females smiled more readily and more often than the males in the study. Perhaps the "cleanest" study done along this line was one by Dixon McLean in which high school yearbooks from many schools and containing several ethnic groups were examined (reported in Freedman, 1979). In each yearbook there were more females smiling than males. It would thus appear that females are biologically predisposed to smile more readily than are males.

As for what this might mean, the best hints come from the primatologists. Their findings are that the "fear-grin" facial expression in monkeys, which calls off an attack by a dominant animal and is therefore called an "appeasement gesture," is more often exhibited by females (DeVore, 1965). The explanation has been that since females tend to occupy the lower half of the hierarchy, they have become more adept at warding off attack. But the human newborn data suggest that monkey and ape females, too, may be genetically programmed for lower thresholds to grinning.

It appears that women are prepared at birth to emit more readily this expression of "appeasement," and they do so the world around. In some cultures, for example the Navajo, smiles are harder to come by than, say, among the Australian Aboriginals, where, as a visitor, I experienced one wonderfully smiling face after another. As a corroboration of these impressions, the count of smiling faces in Navajo and Aboriginal yearbooks yielded the lowest percentage of smiling among Navajos (18.5 percent, sexes combined) and the highest among Australian Aboriginals (73 percent). In each culture, however, females smiled at significantly higher rates (Freedman, 1979).

There is one more thing to report about smiling. As we shall see, children of 6 years and older readily arrange themselves into hierarchies of "who is toughest." We found that when we paired children into tough and not-so-tough twosomes, the not-so-tough child consistently smiled and gazed more at the tough child. This was true of both boy and girl pairs, and it appeared to bear out the supposition that smiling is a gesture of appeasement. However, there is one exception. When a girl who was rated by her mates as tough was paired with a not-so-tough boy, she, nevertheless, smiled and gazed more at him rather than vice versa. In other words, although she was judged physically stronger and tougher, there was the tendency to act as if the reverse were true in the heterosexual pairing (Freedman, 1979).

In order to examine this tendency in detail, Cronin (present volume, Chapter 17) conducted a study in which 12-year-old girls and boys were independently assessed as to their prowess in dodgeball. In this game a child stands in the center and is encircled by the rest who try to hit him or her with the ball. A hit is counted as a tally for the thrower. Two teams were organized, one of the best girls and the other of the worst boys, and then matched against each other. Although the girls had the greater talent, they collapsed completely, allowing the less athletic boys to retrieve the ball and win the contest. They clearly did not try as hard as when playing a girls' team, and the boys, thus encouraged, tried harder than when playing a boys' team. Cronin repeated the study with a new group of children and then with a group of Hopi children on the Hopi reservation. In each case there was exactly the same outcome.

It thus appears that not only are girls poised on the threshold of a smile, but also that they exhibit at greater rates than males all of the behavior that goes with smiling: conciliation, appeasement of aggression, and avoidance of physical combat (especially vis-à-vis males). Recent results of the Harvard six-culture study bear out this conclusion (Whiting and Edwards, 1973).

Development of Status Hierarchies

We have already noted basic behavior characteristics. For example, boys are more prone to temper outbursts by the end of the first year, although there are such indications even in the earliest days of life, and by 3 years of age, where peers are available, boys are ready to run in groups and to play more aggressive games than do the less mobile girls. (Freedman, 1974; Omark and Edelman, 1976).

Probably as a result of basic temperamental differences, boys and girls, like monkey and baboon youngsters, tend to play with others of the same sex—from the first moment where such a choice is possible. In any sandbox and playground 2-year-old boys will be near boys and toddler girls will be near girls. By 2½ years the boys' groups have become more mobile, and by 4 years they are all over the playground, even as 4-year-old girls tend to play in the sandbox. The boys clump into rather larger groups as well, and by 6 years they already have the looks of what we eventually call a gang.

What about other cultures? Is it the same? We have observed Japanese children in both Tokyo and Kyoto, Chinese in Hong Kong, Balinese in a small village in Bali, Australian Aboriginals in and around Darwin and at Edward River, Ceylonese in Kandy; in India we made observations at the Auribindo ashram in Pondicherry and in several New Delhi schools (including a school for the blind), and in Kenya we observed Kikuyu and Masai youngsters. We also obtained drawings from these children to the translated instructions of "draw anything you wish." In the United States we have made similar observations and obtained drawings at Navajo and Hopi school settings in New Mexico and Arizona (Freedman, 1976).

In every setting, including the school for the blind, boys tended to be with boys, girls with girls; boys tended to cover more space in their play, to exhibit more rough physical contact; they were louder, more chaotic, more disorganized, and less neat. Girls tended to play in smaller groups and to engage in closer, quieter, more orderly games; they tended also to stay closer and to be more obedient to the teacher and to rely more on adults to settle disputes than the boys. Draper (1976) has found much the same trends among !Kung bushmen children, and this has since been reinforced in a report summarizing the entire Harvard six-culture study (Whiting and Edwards, 1977).

As for the drawings, boys everywhere were obviously more attracted to and intrigued by vehicles of transportation: trucks, airplanes, rockets, horses (among the Navajo). Even among the Kikuyu, where girls were rougher than any other girls in their play (found also in the six-culture study) and where they, too, drew many moving vehicles, boys, by contrast, were even tougher and drew even more vehicles.

Flowers also provided some interesting comparisons. While girls everywhere were more prone to draw flowers, *all* Balinese and most Ceylonese children had flowers in their drawings: Balinese boys, however, would intersperse airplanes and vehicles among their flowers while the girls never did so (see Freedman, 1976).

In conclusion, it is our experience that children are everywhere more alike than different. In some respects sexes differ far more than do cultures. While Balinese boys may appear softer and gentler than say, Kikuyu boys of Kenya, the contrast between boys and girls was far more consistent than were such cross-cultural contrasts.

Who Is the Toughest?

Omark and Edelman (1976) have shown that if you ask two 4-year-old boys "Which of you is toughest?", in 80 percent of such twosomes both will answer, 'Me!" By 6 years of age an important change occurs in the answers to this question. Boys now agree about 80 percent of the time on who is tougher; more than that, the class as a whole is in agreement. A hierarchy has been formed, known to all, and with the positions in it acknowledged by everyone. Such clear-cut hierarchies of toughness exist in every interacting group of children over 6 that we have looked at, and it can even be seen in incipient form among 4- and 5-year-olds. More often than not, teachers are completely unaware of it, and, in our experience, they usually do not believe it until they speak to the children and find it for themselves. The hierarchy is usually linear with boys at the top, girls at the bottom, with overlap in the middle.

It is an absorbing fact that within these groups everyone pays much attention to those high in the hierarchy and little to those at the bottom. Everyone can readily name the boys at the top and tend to agree more on their rank than on the lower ranking boys or girls. That is, there is simply greater accuracy in ranking the upper half of the hierarchy (Omark and Edelman, 1976). Chance (Chance and Larsen, 1976) has called this phenomenon "attention structure," inasmuch as a dominance–submission hierarchy may be ascertained by noting who pays attention to whom. Certainly if the leadership is constantly reasserting its dominance with random attacks, as among the macaques, a potential victim had best keep his attention upward. In humans something similar happens without physical attacks. Most of our attention, most of our talk and gossip is directed to and about dominants in our lives. Among graduate students, for example, the favorite topics of gossip are the professors. Among the professors it is the chairman or other percieved "bigwigs" in academia. One puts down those above one, complains about them, jokes about them, and tries to learn all about their personal foibles. In this way one can partially equalize the difference in status, but the net result is a lot of time spent on the topic of the head men and women.

We have tried to establish hierarchies other than toughness with our school children, such as who is smartest, best athlete, most handsome, and in each case we observed an interesting thing. While there may be overall agreement on hierarchical positions, the arrangement is kept in dynamic flux by one significant trait: the tendency for each child (especially boys) to think he is better than anyone else! We have called this *overrating*, and it is clearly a carryover from 4-year-old behavior when 80 percent of boys declare themselves invincible. Six-year-old-boys, possibly males of any age, tend to rate themselves several slots above where the group has them placed. It is as if a male achieves a preeminent concept of his invincibility early in life, and thereafter external circumstances must serve to modify this initial sense of limitlessness.

This sense of omnipotence seems to be part of an evolutionary heritage and comes packaged into hierarchically arranged species. It appears to be the crux of what motivates the hierarchy, the psychological basis for male vying with male. Without overestimation of self there can be no hierarchy, no challenge to the establishment. It is probably also the basis for sibling rivalry, for a son competing with his father, for the Oedipus complex, and for the substantial psychological literature accumulated supporting the existence of that complex.

If males exhibit an overblown sense of self-worth and competitiveness, does this not lead to greater disorganization in all-male groups? Savin-Williams (1977, and this volume, Chapter 12) has in fact demonstrated the opposite in 12- and 13-year-old summer camp cabin groups of boys and girls. While there were more challenges within male groups, males also arranged themselves almost immediately into followers and leaders. By contrast, the all-female groups often had the problem of "Who is in charge?"

What of the male hierarchies that started at 6 years of age? Is there much change over the years, and do more intellectual endeavors replace athletics as the major criteria? One might suspect that in an academically oriented high school, such as the University of Chicago Laboratory Schools, an academically oriented pecking order would take over. Weisfeld, Omark, and Cronin (this volume, Chapter 11) reexamined the boys first seen as 6-year-olds (Omark and Edelman, 1976) when the same group was in high school, and the order developed at 6 years of age, largely on the basis of athletic ability, held at 14. Since the majority of these children will doubtless become professionals as they enter adulthood, the subsequent history of this group may prove most interesting to document.

Same-Sex Groups versus Mixed-Sex Groups

Parker and Omark (this volume, Chapter 24), found that 6- to 9-year-old girls do take "toughness" seriously if in an all-girl class and, further, that they are more likely than coed girls to overrate their own toughness. In fact, they did so at higher rates than did boys in mixed classrooms. These young girls were more outspoken than were coed girls, engaged in more vigorous gymnastics, and in general were more spontaneous. Such effects seem not to be limited to these younger ages in that women educated in all-female colleges have been found to enter the professions with substantially greater frequency than do coeducated women. While carefully controlled studies have not yet been done, the implications are potentially important. It appears quite possible that females in the presence of males learn to inhibit their aggressive and competitive impulses (Cronin, this volume, Chapter 17), and there are further data that indicate that coeducation dampens male spontaneity as well (Yolles, 1967).

There are, then, practical "human engineering" implications here that take us into the much discussed current issues of women's equality with men. While we cannot do more than mention this issue here, it seems a reasonable note upon which to close this review. Human sex roles are not so fixed that different ecological situations cannot alter the phenotype; only further research and/or historical developments will help define the limits and possibilities of this malleability.

Chapter 15

Social Dominance and Human Motivation

Glenn E. Weisfeld

Intraspecific aggression is a means by which the organism can compete for scarce resources. Since aggression carries substantial risks and costs, we would expect it to be reserved especially for those occasions on which the resources are valuable. The available ethological evidence supports this generalization. For example, a prey animal will fight desperately against a predator, a territorial animal will defend its home region violently, a female mammal will defend her young fiercely, and a male hamadryas baboon will punish a straying member of his harem vehemently. This principle may also explain why social aggression is typically more frequent between males than between females. Males have more to gain or lose in the form of reproductive success than females since there is less variability for females in the number of offspring they will leave (Bateman, 1948; Trivers, 1972).

In social animals competition may occur over ranks as well as over tangible resources since both can greatly enhance individual fitness. For example, hierarchically organized primates do not compete solely for resources; they also vie for dominance ranks that permit preferential access to resources. As Washburn and Hamburg (1968) state:

> Being dominant appears to be its own reward—to be highly satisfying and to be sought, regardless of whether it is accompanied by

---

This work was supported by National Institute of Mental Health Predoctoral Fellowship No. 1 F31 MH05370-01.

> advantage in food, sex, or grooming. In the long run, position guarantees reward, but in the short run, position itself is the reward . . . (p. 473).

Man, too, seems to possess a desire, or "drive," to compete or to gain social recognition. A variety of terms have been employed to refer to this behavior: "social dominance," "rivalry," striving for "social approval," need for "positive evaluation," and, perhaps most generally, "achievement motivation." Regardless of the terminology, researchers seem to agree that this type of behavior unfolds at 3 or 4 years of age. For example, Heckhausen (1968) maintained that these activities require a mental age of 3½ years. This characteristic time of onset suggests that genes may underlie the behavior. Additional evidence for the biological basis of social striving includes cross-cultural similarities in the ways in which young children provoke each other to fight (Eibl-Eibesfeldt, 1974a, 1974b) as well as the virtually universal prevalence of hierarchization in human society. These signs that social striving may be rooted in human evolution indicate a need for a more refined, comparative analysis of both empirical findings and theoretical perspectives on social behavior. The present chapter attempts to provide such an analysis, using the comparative concept of dominance to integrate diverse information on human and nonhuman social behavior.

Formation of a Dominance Hierarchy: The Conditioning-to-Success Phenomenon

In many species certain controls have evolved to mitigate the dangers inherent in intraspecific aggression. For example, competition for dominance often is ritualized, or governed by rules of combat. In addition, animals seem to respect each other's hierarchical or territorial claims. Aggression is also minimized through experience in hierarchical encounters. The participants in a rank order are familiar with each other's fighting prowess, and so they can be judicious about where, when, and with whom they will fight. If the outcome is a foregone conclusion, there is no reason to risk a fight. Animals that are strangers can rely only on contemporaneous cues about each other's fighting ability, such as size, weaponry, vigor, and aggressiveness.

An animal's experience with a particular opponent affects it in various ways. First, the animal responds appropriately to the dominance or submission displays of its opponent; that is, the animal develops a response set to different social contexts. Second, the animal's own self-confidence and hence its self-assurance and aggressiveness are affected by its experience of positive or negative outcomes in the competitive arena. It is not known which of these effects is the more potent, but they probably complement each other. The net result is that "conditioning to success" occurs: an animal that has repeatedly experienced victory is more likely to initiate aggression, and hence more likely to elicit submission from future opponents, than is a habitual loser.

A similar phenomenon has been induced experimentally by subjecting animals to repeated defeats at the hands of very aggressive opponents or to repeated victories at the expense of very submissive conspecifics. The experimental animals become either submissive or aggressive, depending upon the type of social encounter that they experience. This effect has been demonstrated for mice (Ginsburg and Allee, 1942), monkeys (Warren and Maroney, 1958), bumblebees (Free, 1961), crickets (Alexander, 1961), and sunfish (McDonald, Heimstra, and Dankot, 1968). Children can also be conditioned to succeed or fail at competitive tasks (Sears, 1937). For example, a good performance at a task is more likely if the subject has had previous "successes" on tasks administered by the same teacher (Dweck and Repucci, 1973).

A third way in which experience in competitive situations affects future performance is that an individual who repeatedly fails tends to become anxious in such situations. While this anxiety is probably adaptive in prompting an animal to withdraw from dangerous encounters, it often interferes with performance of the competitive task. For example, low-ranking pigeons had more difficulty learning a discrimination task in the presence of a superior animal than in the presence of a subordinate (Diebschlag, 1940). Similarly, human subjects with a history of failure do relatively poorly under test conditions that increase anxiety, such as time pressure (Bartmann, 1963), competitiveness (Brown, 1963; Ryan and Lakie, 1965), and criticism (Johannesson, 1967; Lazarus and Ericksen, 1952; Thompson and Hunnicutt, 1944). Less self-confident subjects evidently have a greater fear of failure to begin with, and the experimentally induced anxiety aggravates this fear, severely detracting from performance. Habitually successful subjects, on the other hand, are not as debilitated by experimentally induced anxiety.

The conditioning-to-success phenomenon can account for additional observations about competitive individuals. High-ranking monkeys persist longer at problem solving (Richards,

1972) and participate more socially (Crook, 1970a) than low-ranking ones. Similarly, successful human subjects seek out competitive or social encounters (Bales, 1953; Coombs, 1969; Coopersmith, 1968; Savin-Williams, 1977a; Sherif, 1956), participate actively (Anderson and Brandt, 1939; Bass, 1949; Hilgard and Sait, 1941), and tend to persist longer at a problem (Coopersmith, 1960; Gewirtz, 1959; Ollendick, 1974). Successful human subjects also prefer to take high risks, i.e., to put their ability on the line (Morris, 1966; Sears, 1940), such as by undertaking tasks at which they can fail as easily as succeed (McClelland, 1958a). Unsuccessful subjects tend to withdraw socially and to give up easily.

Studies on humans suggest further that success induces individuals to overestimate their abilities and thus to issue or accept challenges more readily (Coopersmith, 1968; Harvey, 1953; McClelland, 1958b). Other people also overestimate these subjects' future performances (Harvey, 1953; Sherif, 1956); so both self-confidence and self-display seem to be operating as status-enhancing forces for these individuals.

In summary, we can say that conditioning to success works in these ways. Success promotes self-confidence, the use of intimidating dominance displays, and aggressiveness. These effects, along with the tangible and psychological rewards of success, engender future success. Failure brings about the opposite effects; the less formidable combatant withdraws discreetly and avoids future conflict. The net result of these processes is that the rank order within the social unit becomes increasingly differentiated; that is, hierarchical linearity increases over time. Thus, conditioning to success may constitute the means by which individuals learn to accept and to behave according to their hierarchical ranks.

Intraspecific Competition and Uncertainty

If one assumes that animals usually fight only when their comparative fighting abilities are in doubt, it follows that aggression should be relatively common before a stable hierarchy forms, when a new animal joins the group, when a hierarchy is breaking up, between strangers, near territorial boundaries, and between individuals close in rank. These effects, of course, have been documented in ethological studies of many species. Among animals familiar with each other's fighting prowess, the easiest targets will be preferentially attacked. This has been referred to in chickens as the "peck right," the opportunity to abuse subordinates. Bullying of this sort, however, cannot lead to an improvement in one's own status. The result seems to be that the drive for dominance reaches a compromise with the dictates of prudence, such that an animal often challenges its immediate superior, the most vulnerable target that it can displace in the hierarchy. By following this strategy, animals can gradually build up their self-confidence and experience at fighting and thereby rise as high in the rank order as their native ability allows. Thus, a working hypothesis might be that animals that are close in rank fight frequently not only because their ranks are relatively uncertain, but also because challenges are often issued against an animal's immediate superior.

This pattern of issuing challenges is also consistent with the results of social psychological research on level of aspiration. Lewin, Dembo, Festinger, and Sears (1944) and Child and Whiting (1949) established that people raise their expectations for themselves somewhat following success and lower them after failure. Similarly, the theory of achievement motivation (Atkinson, 1958) holds that when deciding to compete, people take into account their estimated probability of success and the value of the payoff. They try to maximize both, which means that they usually choose a goal that is slightly above the level of their previous performance.

An interest in determining one's place in relation to others and the influences of these evaluations on individual performance have been emphasized in social comparison theory. Festinger (1950) posits a drive to evaluate and improve one's social standing. Some of the studies cited as evidence for this theory can be interpreted within the dominance paradigm as well. Subjects have been shown to select as a competitor someone close to them in ability (Hoffman, Festinger, and Lawrence, 1954; Whittemore, 1925), to exhibit their greatest interest in those individuals who are slightly superior to them (Latané, 1966), and to choose to compare themselves with persons of slightly higher rank (Thornton and Arrowood, 1966; Wheeler, 1966). All of these tendencies have been discussed in the preceding chapters as characterizing children's reaction to and participation in the group dominance hierarchy.

Cooperative Aspects of Intraspecific Competition

In most species hierarchical rank is determined by fighting prowess. Nevertheless, various controls, mentioned above, mitigate the intensity of competition. Contests for dominance are often restricted by the ritualization of combat. Once relative dominance is established, the outcome of future encounters can be predicted with some accuracy by both participants, thereby reducing the necessity for actual combat. Finally, hierarchically organized animals need not fight over individual resources. When the prerogatives of rank are recognized, resources need not be contested piecemeal.

These controls or conventions are interrelated. Competition for dominance can be ritualized and resorted to judiciously because hierarchical prerogatives are respected. Instead of fighting to the finish the combatants can negotiate terms of surrender. In essence, the victor spares the vanquished, who, in turn, concedes the spoils.

Beyond the contestants' inevitable disparities in physical strength and, consequently, in access to resources, a sort of justice prevails. Dominant and subordinate animals benefit equally from the order and control over aggression that hierarchization confers. Wynne-Edwards' (1962) concept of "conventional competition for conventional rewards" expresses this mutuality of benefit to each party. Peace prevails not as a result of unmitigated coercion but through realistic capitulation and giving of quarter.

This same idea of justice among unequals was put forth by the social psychologist Homans (1974). In his view people who are engaging in a group task soon rank themselves according to ability. Ideally, the individual with the greatest skill is elevated to a position of leadership where his efforts can be maximally useful to the group. He thus gains certain prerogatives from his subordinates in exchange for his special contributions. Contention is then minimized, and all members benefit from agreeing on the allocation of roles, obligations, and prerogatives. In human societies the rules governing competition and cooperation are extremely complex and variable. In nonhuman societies the conventions appear to be simpler and more species-specific. Nevertheless, parallels can be demonstrated between the way the rules operate in human and nonhuman societies.

Kummer's (1971) studies of baboon social systems provide a revealing illustration of this point. He found that hamadryas males herd females into harems but that these males do not attempt to herd females already integrated into a one-male unit. By contrast, male gelada baboons, although possessing harems, do not herd their females. These latter males fail to exhibit the hamadryas males' respect for the *droit du seigneur*. By way of compensation female geladas assume a greater responsibility for stability of the one-male unit. They assist the male in repelling attempts by rival harem leaders to usurp his females. These differences suggest that the distribution of prerogatives in a social group is controlled by species-typical capacities to learn the rules of that particular social structure. As Kummer expressed it:

> It appears, then, that hamadryas baboons, in evolving a system of one-male units in adaptation to their environment, also evolved a mechanism to cope with the concomitant problem of highly possessive males; possessiveness was complemented by a "respect" for possession. The inhibition against encroaching on another's social property is at least analogous to certain rules of human behavior, and the fact that such inhibitions occur in nonhuman primates is of great interest in the search for the roots of human social behavior (1971, p. 105).

Social Feelings that Stabilize Social Conventions

Moralistic Anger

Any agreement or convention is susceptible to violations. An animal may transgress the conventions of ritualized combat. A subordinate animal may try to usurp a superior's prerogative. A dominant animal may abuse a subordinate. When such violations of the conventions of a hierarchy occur, the victim usually reasserts its rights vigorously. A dominant primate will often discipline a subordinate severely for what seems a harmless peccadillo, such as preceding it onto a branch; an otherwise meek subordinate may turn violent if abused by a dominant animal (Bernstein and Gordon, 1974; Maynard Smith and Ridpath, 1972). Similarly, violations of other types of convention about intraspecific aggression, such as territorial infringements, prompt punitive retaliation "on principle."

The reason for the vigor of these "rage reactions" is that much is at stake. Beyond any specific resource itself, the animal must defend its investment in the social order, that is, the advantages that conventional competition confers. The entire agreement about relative hierarchical position and prerogatives is thrown into question by such transgressions. In the absence of an enforcement mechanism the social order would break down.

Interestingly, both the social psychologist Homans (1974) and the sociobiologist Trivers (1971) explained the utility of the corresponding phenomenon in man. Homans said that anger arises in people who feel that their output for the group outweighs their returns, that is, in people who feel they are being exploited. Trivers maintained that "moralistic anger" is the force behind the aggression necessary to punish violators of an agreement to return favors, i.e., to practice "reciprocal altruism."[1] A wealth of social psychological studies support these interpretations that human anger serves to discourage exploitation by others. In a number of studies aggression was evoked when the other person violated social norms or received undue bounty (Epstein and Taylor, 1967; Geen and Berkowitz, 1967; Patchen, 1961). Aggression has also been elicited when the subject merely observed but was not himself victimized by the transgression (Weisfeld, 1973). Similarly, in some primate species a "control animal" often disciplines erring members of the troop, such as juveniles that play too roughly, even though the control animal itself has not been victimized. These last studies suggest that moralistic aggression is not confined to dyadic social exchange but rather that perceiving the violation of a social convention, even without actually being harmed or frustrated, may suffice to prompt intervention.

If a victim of a transgression is not in a position to exact punishment on the spot, he may bear the transgressor a grudge and punish him later. This process is analogous to what Homans describes as symbolic compensation for a group member's accomplishments when they are too great to be repaid immediately. In the latter case accomplishments are rewarded through elevation to a higher status, which can be redeemed for prerogatives in the future. In the case of a transgressor who cannot be punished satisfactorily at the time, status may be lowered, followed by denial of some future prerogatives.

Gratitude

In his analysis of reciprocal altruism, Trivers (1971) also discussed the emotion of gratitude. This affect is the complement of moralistic anger: it serves to reward others' acts of altruism or conformity to conventions. The recipient of an altruistic act is motivated by gratitude to reciprocate, thereby encouraging cooperation in the future.

Gratitude is, indeed, an unpleasant feeling that is expiated only by fulfilling one's obligations. It arises when we feel that the other person's contributions outweigh his returns. It has been found empirically that this type of imbalance leads to compensation. The beneficiary rewards the altruist through either tangible rewards or symbolic forms of tribute: praise, thanks, or promises (Heider, 1958; Lerner and Lichtman, 1968; Macaulay and Berkowitz, 1970; Rosen and Bielefeld, 1967). The unpleasant aspect of social gratitude has been emphasized in a study by Greenberg and Shapiro (1971). These researchers showed that subjects frequently declined favors that could not be repaid. Accepting "charity" seemed to entail a burden to reciprocate.

Guilt or Shame

In addition to their agreement on the beneficial role of anger and gratitude in a system of social conventions, Trivers and Homans both maintained that the emotion of guilt helps to enforce equitable exchange. To Homans, guilt arises in the individual when his contributions are perceived to fall short of his dividends, that is, when he senses that he is exploiting others. Similarly, Trivers maintained that guilt is an unpleasant emotion that serves to discourage one from exploiting others and thereby incurring their antipathy.

Guilt, or shame, arises when a person's achievement does not meet societal standards. When a person fails in this way, he typically acts guilty or embarrassed. This emotional display, together with any verbal apology or expression of modesty, constitutes acknowledgment of error, or inferiority. In effect, the person temporarily assumes a lower social rank where he has fewer prerogatives than before. Accordingly, the bodily expression of shame in man resembles the submission displays of mammals; both are characterized by inconspicuous demeanor and furtive glances (Darwin, 1872; McDougall, 1921). Furthermore, like guilty people, subordinate mammals often behave in ways that suggest obeisance and the concession of prerogatives. For example, a subordinate primate that offends a superior may hasten to groom or present sexually to the latter. These displays appease others and preempt retaliatory aggression.

Subjects in social psychological experiments who are made to feel embarrassment, shame, guilt, or failure typically make some kind of reparation (Apsler, 1975; Aronfreed, 1968; Isen, Horn, and Rosenhan, 1973; Leventhal, Weiss, and Long, 1969; Macaulay and Berkowitz, 1970; Regan, 1971; Wallace and Sadalla, 1966). Once the individual's ratio of accomplishments to rewards is again equal to that of his peers, the acute distress of shame apparently dissipates. Anyone, regardless of social rank, can feel ashamed on occasion and must acknowledge his failure in order to get relief. The emotion of shame may be said both to maintain the individual in an equitable relationship with his peers and to spur him away from low status and its privations.

Pride

Man seems also to possess a complementary affect that encourages people to fulfill societal expectations. If shame is the stick, pride is the carrot serving to encourage social success, motivating the individual to seek high status and hence to gain access to the material rewards associated with social position. Pride is the opposite of shame in several other respects also. The affect of pride is accompanied by conspicuous posture and demeanor in man and other animals, in contrast with shame or guilt postures. In addition, success leads to self-confidence and extraversion, while the opposite is true of failure. Finally, pride, or high self-esteem, causes one to arrogate privileges to himself, but shame disposes the individual to relinquish social privileges and rewards.

Neither Homans nor Trivers discussed the positive affect of pride. Empirical studies suggest, however, that the successful person entertains a high opinion of himself and behaves accordingly. If the individual's contributions outweigh his returns or if his self-esteem is raised, he demands compensation (McMillen, 1971; Mischel, Coates, and Raskoff, 1968; Moore, Underwood, and Rosenhan, 1973; Pepitone, 1964). As in the case of anger and shame, third parties can be involved. Furthermore, the individual's accomplishment can be tendered to a group or even to the whole society rather than being a favor bestowed on a single person. Likewise, reciprocation can come from a group or a single beneficiary.

The emotions just discussed can be seen to serve an important role in the establishment and maintenance of social conventions necessary for a stable social unit. Depending on one's conceptual scheme, these affects may be said to (1) enforce equitable social exchange, (2) promote reciprocal altruism, or (3) maintain the appropriate distribution of prerogatives in a dominance hierarchy.

Social Dominance and Conventional Altruism

The parallels between human conventional social systems and animal dominance hierarchies suggest that the former evolved from the latter. Perhaps the main transitional event was the evolution of gratitude, which seems to be a recent innovation in vertebrates. In species in which individuals act either hostilely or neutrally toward each other but have no means of active assistance, there simply is nothing for which to be grateful. In these species gratitude does not exist. However, as Trivers (1971) argued, gratitude facilitates the operation of reciprocal altruism, so that this emotion probably evolved concomitantly with the development of a potential for actively assisting others. Many of the examples of reciprocal altruism in the higher social mammals appear to derive from elements of parental care. For instance, food sharing among nonkin has been observed in African hunting dogs (Kühme, 1965; van Lawick-Goodall and van Lawick, 1971), gibbons (Berkson and Schusterman, 1964), orangutans (Cook, 1975), and chimpanzees (Teleki, 1973); chimps also signal the location of food to each other (Menzel, 1971; Reynolds and Reynolds, 1965; Sugiyama, 1969). Similarly, the males of certain Japanese monkey troops "adopt" juvenile females (Itani, 1959).

It is important to distinguish an act of altruism from a simple concession of a hierarchical prerogative. In sociobiological terms, an altruistic act is conferred not out of coercion but because of an expectation of reciprocation later. The altruist typically receives "credit" for his deed in the form of a rise in status vis-à-vis the beneficiary. Like any other rise in rank this promotion might be presumed to entail certain privileges—in this case, the right to demand repayment of the favor in the future. Human beneficiaries typically express their obligation to their benefactor explicitly with praise or with promises to reciprocate. Animal recipients in some cases seem similarly to signal their inclination to concede prerogatives to the donor by means of submission gestures. For example, canids beg food by nuzzling the donor's muzzle, a generalized submission gesture (Schenkel, 1967; van Lawick-Goodall and van Lawick, 1971). Thus, an altruistic act seems to entail a temporary change in the relative dominance of the two parties (although not necessarily a complete reversal of their ranks), whereas simple concession of a hierarchical prerogative reflects and even reinforces the existing dominance relationship. This change in relative dominance following an act of altruism would ensure that effective reciprocation occurred later.

Detailed studies of meat sharing in chimps seem to establish that this behavior is truly altruistic (Teleki, 1975). Even the alpha male requests food by means of standard sounds or (submission ?) gestures rather than demanding it by threat of force. Seldom does any conflict occur even if the subordinate refuses indefinitely to share the meat. As Teleki (1975) comments:

> Qualitatively, meat exchange among chimpanzees seems more similar to the regulated sharing seen among human gatherer-hunters than to the competition commonly seen among baboon collector-predators. . . . This . . . is very reminiscent of meat distribution patterns among human gatherer-hunters . . . (p. 151f).

Even more convincingly, Nissen and Crawford (1936) reported that almost all of the food they had observed being shared by chimps proved to have been exchanged on a reciprocal basis when considered over time. In their words, the animals traded food on a "credit" or deferred basis (p. 415). If two chimps occupied the same cage, the food was divided up fairly evenly.

The utility of cooperation in hunting prey may have expedited the evolution of altruism in canids and primates (Campbell, 1966; Kawabe, 1966). Tiger and Fox (1971), on the other hand, proposed that solitary hunting by a well-armed hominid was sufficient for the evolution of altruism:

> Accumulation for a hunter is, after a certain point, useless: he cannot consume more meat than he needs or use more weapons than he can handle. The successful hunter, therefore, has to convert surplus into prestige (p. 124).

"Prestige" here means the same thing as "credit," as used above. As a result of his contribution, the altruist rises in status and can then invoke privileges when the need arises. He thus holds an insurance policy in the form of the good will or gratitude of his fellows.

Tiger and Fox suggest here that altruism and gratitude not only permitted exchanges to take place over time, but also provided a way of trading a resource of little value to one person for something of greater value to him. This, of course, constituted commercial trade and eventually led to the great specialization of roles that characterizes human society.

The ontogeny of competitive behavior in man seems to resemble the evolution of dominance behavior as envisioned here. Between 2 and 5 years of age children often use "instrumental aggression" to gain control of desired objects (Dawe, 1934). This type of aggression appears to be merely a means to an immediate end. This stage may correspond to the evolutionary period before the occurrence of dominance hierarchies. At 3 or 4 years, rivalry, or dominance, begins. Fights become more personal and may not involve any object. Shouting occurs; the term "hostile aggression" is indeed fitting. Hartup (1974) found that subsequently the total number of fights steadily decreases, through the time (6 or 7 years of age) when the dominance hierarchy becomes clear-cut and well understood by the members. This is the stage of "toughness" (Omark and Edelman, 1973) that may correspond to the "brute force" period of evolution. In adolescent boys dominance seems to be determined by athletic prowess (this volume, Chapters 11 and 12). This may correspond to the advent of hunting and warfare in hominid evolution since skill at running, throwing, etc. is required in all these pursuits.

In adults the criteria of social status seem to reflect the ecological needs of the particular culture (von Fürer-Haimendorf, 1967). These needs, and hence these criteria, are usually complicated. In most societies a high degree of specialization occurs, so that different people excel at different activities and a person's status shifts as he moves from one setting to another. Each subgroup within the society has its own dominance relationships, roles, and network of sympathies and antipathies. Each member of the society gravitates toward those groups where he can gain recognition or credit. Conversely, other members of the society, through their praise and criticism, channel his efforts along desirable lines.

Conclusion

The concept of the dominance hierarchy has provided a single coherent framework for the preliminary integration of diverse data and theories from the behavioral sciences. Ideas such as achievement motivation, exchange theory, conditioning to success, level of aspiration, altruism, and aggression can be integrated sensibly and parsimoniously by invoking this sort of comparative analysis. Perhaps alternative explanations of these phenomena will prove more compelling than those offered here. Hopefully, in any event, biologically oriented studies of humans—adults as well as children—will lead to clearer understanding of these and other social processes.

Note

1. In this chapter "altruism" is used in the sociobiological sense of aiding another organism for the sake of some ultimate biological dividend to oneself. In the case of "kin altruism" one aids a close rela-

tive, say, a sister. If the sister's survival and reproductive fitness are thereby enhanced, the altruist ultimately benefits to the extent that he shares genes with her offspring. Thus his own "inclusive fitness" has been promoted by his altruism. In "reciprocal altruism" the altruist and beneficiary are not related. The altruist benefits biologically only if the beneficiary returns the favor later. To ensure reciprocation and to avoid being exploited, the altruist apparently must interact with the beneficiary in certain ways, some of which are explained in this chapter (see also Trivers, 1971).

Chapter 16

The Representation of Social Dominance in Children's Drawings

Janet Strayer
F. F. Strayer

The main questions that led to the present research project grew out of our earlier assessments of dominance relations among preschool children (Strayer and Strayer, 1976). We were interested in knowing whether the children themselves were aware of the social dominance structure operating within their group. One way of assessing this is to ask them directly, though this task does have its own set of problems. The present study proposes another route for investigating children's perception of their social relations—requesting children to draw a picture of themselves and a friend. The question explored was whether children are aware on a representational level of social relationships evident from an ethological analysis of their spontaneous behavior. Assuming that children might be implicitly aware of such relationships, we wanted to know if children would express these relationships in their drawings and if independent judges unacquainted with the children could correctly assess relative dominance relations from drawings produced by young children. Finally, assuming such judgments were possible, we wanted to know *how* such social information is conveyed in the drawings.

The use of drawings as a psychological tool has had an erratic history, often divergent from mainstreams of empirical investigation. Children's drawings, in particular, have been of theoretical interest from a variety of viewpoints. An argument for progressive changes in representational and symbolic development shown in children's drawing has been made by Piaget (1962) and Werner (1961). Basic principles of perceptual organization are thought to be evident in children's art (Arnheim, 1965). Children's drawings have also served as measures of intelligence (Harris, 1963), as measures of maturation (Kellogg and O'Dell, 1967), and as projective techniques for assessing creativity, social development, personality, and psychopathology (Bettleheim, 1959; Lebo, 1962; Lowenfeld, 1970; Machover, 1964). Notwithstanding the diverse theoretical interest children's drawings have provoked, their usefulness as a psychological technique has been criticized on grounds that drawings have not been developed as a tool that would give reliable differentiation (Anastasi, 1958). The difficulties of using drawings as measures of individual differences seem better understood upon consideration of at least three potential sources of confusion that hinder the reliable interpretation of such indirect expressions.

The first problem is to establish, independetly of the drawing, the ecological validity of the criterion which the drawing is presumed to index: the problem of defining the criterion in the "real" world. A second problem is how to measure this in the drawing. Finally, the most critical problem has been given the least attention: the demonstration of reliable relationships among drawing measures and the behavioral measures of the criterion. The most obvious difficulties arise when psychological criteria are extracted entirely from differences among drawings themselves. In contrast, the present study gave priority to the demands of providing an ecologically valid criterion and then to investigating children's drawings in relation to it.

Behavioral Observations

Daily observations were made of 37 children ranging in age from 9 to 11 years during 40- minute free-play periods at their school. Dyadic interactions were scored over a five-week period. A total of 16 hours of observational data were collected. Dominance assessments were based on conflict categories found to be reliable indices of social dominance in an earlier study (Strayer and Strayer, 1976).

Figure 1 is a summary of the social dominance rankings in the group. The circles represent girls, and the squares represent boys. The filled units represent children who drew a friend also within the observed group. Clusters along the same row indicate tied

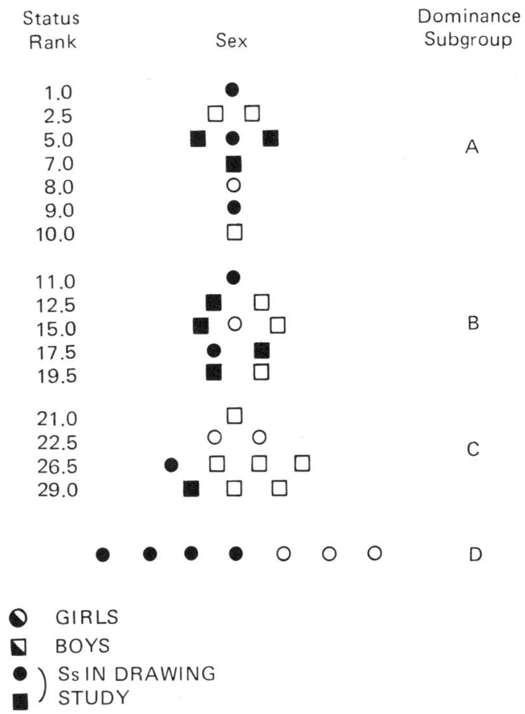

Figure 1. Social dominance structure with status rank and dominance subgroups.

status. A, B, and C indicate high-, medium-, low-status subgroups. The bottom row D represents girls who could not be assigned a status ranking since they did not engage in any agonistic interactions. Figure 2 shows how these various dominance subgroups interacted.

Figure 2 is summary of conflict among the high-, medium-,

Dominance Subgroup	Wins from: A	B	C	D	Total Wins
A	22	24	24	0	70
B	3	28	12	0	43
C	0	0	11	0	11
D	0	0	0	0	0
Total Losses	25	52	47	0	124

Figure 2. Summary of social conflict episodes among dominance subgroups.

low-status and no-agonism subgroups. Members of Group A win interactions from all other subgroups. Group B wins predominantly from Groups B and C but seldom from Group A. Members of Group C win only over individuals from the same subgroup.

Drawing Measures

Following the five weeks of observation, all of the children were asked during an art period to draw a picture of themselves and a friend. They were given as much time as they needed to complete the picture. Twelve children drew someone outside the observed social group, and their drawings were not considered in this analysis. The 18 children (8 girls and 10 boys) who drew someone within the social group ranged in age from 109 to 133 months (mean: 122.6; SD: 9.18).

Figure 3 shows the choices each child made when asked to

Drawer		Classmate		Status Difference
1	●	17.5	●	+ 16.5
5	●	1	●	−4
5	■	2.5	□	−2.5
5	■	7	■	+ 2
7	■	5	■	− 2
9	●	D	●	0
11	●	D	○	0
12.5	■	15	□	+ 2.5
15	■	7	■	− 8
17.5	■	21	□	+ 3.5
17.5	●	1	●	− 16.5
19.5	■	10	□	− 9.5
26.5	●	1	●	− 25.5
29	■	17.5	■	− 11.5
D	●	D	●	0
D	●	D	●	0
D	●	D	○	0
D	●	D	●	0

Figure 3. Dyadic choices and status differences in drawings.

draw a friend. All children drew a friend of the same sex. In general, most of the children selected as friends were in the two higher subgroups, except for the girls in Group D, who chose each other. The status difference among dyads is indicated in the last column. Since members of Group D did not engage in conflict and could not be placed within the status structure, drawings in which they appeared were assigned a zero dominance-difference score.

All 18 drawings were analyzed initially for detail of each figure by two judges according to instructions in the Goodenough-Harris Manual (1963). The judges did not know the observed dominance status of either figure. Differences in scoring occurred on only three occasions; for these items the judges consulted to provide an agreed-upon score. Each drawing was also scored on each of the following graphic dimensions: closest distance between the two figures, farthest distance between figures, figure orientation (degree to which figures tilted toward or away from each other), relative height (ratio of own portrayed height to that of the other figure), and relative percent of page occupied by each figure.

Finally, eight different judges (four males and four females) were asked independently to rate the set of 18 drawings on the following dimensions: maturity of style, friendliness, amount of interaction, hostility, and social dominance. Judges were instructed to examine and score all drawings on one dimension before proceeding to the next. The first four dimensions were rated on a 7-point scale. Dominance ratings required judges to indicate if one of the figures appeared more dominant than the other and to indicate the extent of perceived dominance difference by placing a mark for each figure at any point they thought appropriate along a 15-mm line. This line was labeled at one end "most submissive" and at the other "most dominant."

Intercorrelations were computed among the six graphic dimensions and observed dominance differences as well as among the five judgment dimensions and observed dominance differences. Both of these correlation matrices were analyzed using a principal components solution with normal varimax rotations.

Results

High interjudge agreement on each social dimension permitted a pooled consideration of ratings for each dimension. Pearson product-moment correlations ranges from $r = .75$ to $r = .90$ (all, p's $< .01$). The main question of whether behavioral dominance can be assessed correctly from drawings receives affirmative support. Results provide evidence that adult judges *do perceive* social dominance differences in children's drawings of themselves and a classmate, that interjudge assessments are reliable, and, most importantly, that judgments are *correct* with respect to observed dominance differences.

The 18 pictures drawn could be grouped into three *post hoc* categories: 8 pictures portrayed a classmate higher in status than the drawer, 4 portrayed a classmate lower in status, and 6 portrayed a classmate tied in status with the drawer. For the 12 drawings for which there was an observed difference between members of the dyads drawn, mean ratings of dominance were 75 percent correct. For those drawings depicting dyads for whom tied dominance status was observed, the distribution of direction judgments was not significantly different from chance. Mean ratings of direction of dominance were 100 percent correct for the 4 drawings in which the drawer was higher in observed status than the other child drawn, and 63 percent correct for the 8 drawings in which the drawer was lower in status.

Similar trends were evident in more detailed analyses of individual judgments. Table 1 shows the distribution of judgments for the three *post hoc* categories. Judges showed the greatest agreement on dominance differences when the drawer was higher in status than the friend drawn. They were less correct about direction of difference when the drawer was of lower status than the child drawn. Finally, not only *direction,* but also extent of judged dominance difference between the two figures drawn correlated significantly with observed differences in group dominance status ($r = .55$, $p < .01$, $df = 16$). Figures 4 and 5 are reproductions of actual drawings which illustrate this finding. In both figures JD

Table 1. Individual Judgments of Direction of Dominance (In Percentage)

	Relative Status of Drawer		
Judgment of status difference	*Higher Status*	*Equal Status*	*Lower Status*
Judged higher	63	13	31
Judged equal	31	55	31
Judged lower	6	31	38
Total judgments	N = 32	N = 48	N = 64

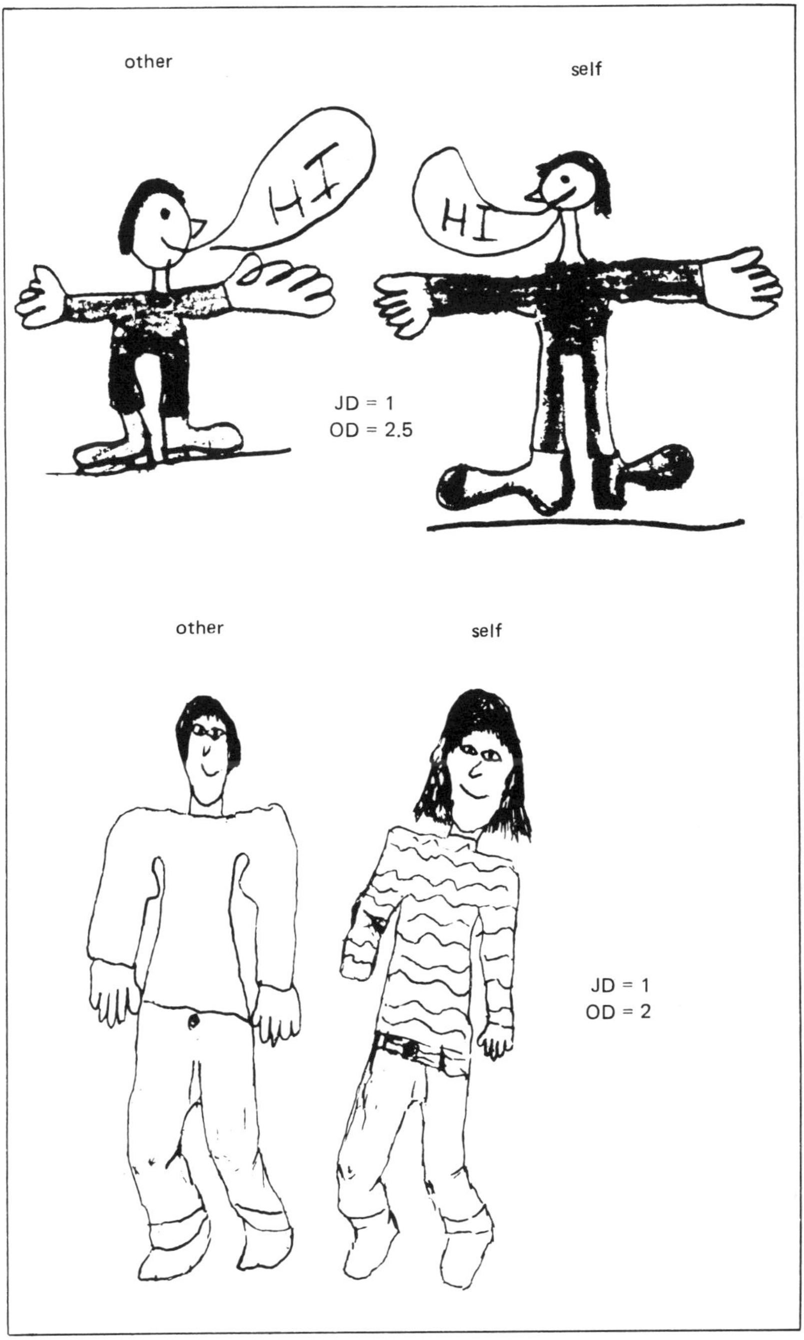

Figure 4. Drawings representing nearly equal dominance status.

Figure 5. Drawing representing highly discrepant dominance status.

indicates the mean judged dominance difference while OD refers to observed status difference. In Figure 4 both drawers and classmates are close in observed dominance status, and in Figure 5, there are marked differences.

Other interests concerned the relationship of observed dominance to other social dimensions rated and to the graphic variables scored. A principal components analysis of the correlations among social dimensions was computed, and four orthogonal factors were indicated: friendliness–hostility, dominance, amount of interaction, and maturity. Both judged and observed dominance differences loaded on the second factor. This analysis underscores the finding that while judged dominance differences are related to observed differences in dominance status, they are independent of the other social dimensions rated.

Our final questions concerned the relationship of both judged dominance differences and observed dominance differences to the graphic dimensions used in scoring the drawings. The correlations of dominance judgments and observations with graphic measures are shown in Table 2.

Judgments of dominance correlated most highly with relative percent of page occupied ($r = .58$, $p < .01$, $df = 16$), and relative height ($r = .59$, $p < .01$, $df = 16$). The notion that certain graphic dimensions might be most important for the portrayal of actual observed differences in dominance was also supported. Observed dominance differences correlated significantly with the same graphic dimensions used for judgments of dominance differences (for relative percent of page, $r = .56$, $p < .01$, $df = 16$;

Table 2. Correlations of Judged and Observed Status Differences with Graphic Measures

	Judged Dominance Difference	*Observed Dominance Difference*
Judged dominance	—	.56*
Observed dominance	.56*	—
Closest point	−.21	−.02
Farthest point	−.18	−.06
Figure orientation	−.10	.06
Relative height	.59*	.54**
Relative % of page	.58*	.56*
Detail difference	.39	.22

*$p < .01$
**$p < .05$

for relative height, $r = .54$, $p < .01$). A principal components analysis of intercorrelations between both observed and judged dominance differences, as well as graphic measures, revealed two significant factors. Both dominance measures loaded with relative height, relative percent of page, and detail differences on one factor, while distance measures such as closest point, farthest point, and figure orientation comprised the second factor.

In summary, these findings indicate (1) that judges can perceive differences in dominance status portrayed in children's drawings of themselves with a classmate and (2) that there is a high percentage of correct judgment with respect to direction and extent of observed dominance differences. The graphic dimensions which were most related to judgments of dominance differences were relative percent of page occupied, relative height, and detail differences. Observed differences in dominance status among children were also significantly correlated with these same graphic dimensions.

Discussion

Whatever the limitations may and should be on making projective inferences from children's drawings, it does appear that they have some ecological validity when examined in relation to behavioral observations. Elementary school children portray social dominance when drawing themselves with a classmate, and these portrayals are apparent to adult judges who have no experience with the children. In addition to substantiating the usefulness of drawings as an indirect observational method (McGrew, 1972), our present findings may have some useful clinical applications. That children should portray a social dimension, such as dominance, that is important in their peer organization is not surprising. We now have some indications of *how* this information is encoded in a drawing. It was noted that when the drawer was higher in status than the classmate drawn, mean judgments of dyadic dominance were 100 percent accurate. It is possible that differences in the graphic medium are more clearly accentuated by these children. In general, the particular graphic measures that convey information about dominance status are height, relative percent of page occupied, and detail differences. All of these measures were positively related to both judged and observed dominance differences. A more refined analysis of graphic representations, however, would probably be more informative. It seems likely, for example, that differences in dominance status may be portrayed by a combination of drawing variables which are differently weighted but which together are effective as a basis for judgments that correlated with observed dominance status. Like most forms of artistic expression, the effect produced by such differential weightings may vary systematically between individuals.

The use of drawings as a tool for investigating the expression of independently observed social variables seems a valuable and interesting channel connecting indirect and direct methods of observation. Further investigations of why some children drew classmates higher in dominance status and others drew classmates equal to or below themselves may be revealing of personality variables, such as self-esteem, interacting with or emerging from early peer-group social structures. It would, of course, also be interesting to see whether the representation of other social relations, such as leadership, affiliation, and cooperation, are amenable to this analytic approach. Finally, it seems equally interesting to investigate whether children, as well as adult judges, perceive particular graphic dimensions in peer drawings as representations of social relationships.

Acknowledgments

A shorter version of this paper was presented at the annual meeting of the American Psychological Association, Chicago, 1975. The authors gratefully acknowledge the assistance of Barry Harmon in collecting and analyzing portions of the behavioral data. In addition, we would like to express our appreciation to the eight members of the Psychology Department at Simon Fraser University who served as independent judges in the analysis of the children's drawings.

Chapter 17

Dominance Relations and Females

Carol L. Cronin

Females are not very prominent in discussions of dominance hierarchies for one obvious reason. In a primate society females usually occupy unstable subdominant positions in the hierarchy. Sexual access to females, in fact, is often one of the rewards merited by a dominant male. These circumstances raise some questions about the nature of female participation in the dominance hierarchy. As a starting point, the following four questions may be asked:

1. Do females end up at the bottom of the hierarchy by competing for rank as males do?
2. Is subdominance the same for a female as it is for a male?
3. Are females even capable of structuring and functioning within a dominance hierarchy, in the traditional sense of that term?
4. If left to themselves, what sort of social structure do females build?

In an attempt to answer at least the first question, this chapter will review some of the data gathered on females and dominance, and it will report on two naturalistic experiments involving boys and girls in competition.

When investigating competition in human children, developmental psychologists have usually asked simply which sex

The work for this chapter was supported by grants from the U.S. Public Health Service and the Harry Frank Guggenheim Foundation to Daniel G. Freedman.

is more competitive. As Maccoby and Jacklin (1974, pp. 247–254) explained in their review of the literature on cooperation and competition, competitive and cooperative behaviors must sometimes be combined in real life. For example, players on a basketball team cooperate to compete against players on an opposing team. Many studies of competition in children involve both cooperation and competition as dependent variables. Some of these studies utilize the Madsen marble-pull game, which is played with a marble holder that can be pulled back and forth across a table between two players. The game actually requires cooperation between the players, who wind up empty handed if they simply compete against each other. With nursery school childen Szal (1972) obtained the following results: In boy–boy pairs neither player obtained many marbles, because the boys frequently pulled the marble holder apart (by simply competing); girl–girl pairs obtained the greatest number of marbles because they arrived quickly at a turn-taking strategy; girl–boy pairs scored between the same-sex pairs, with the girls becoming more competitive and the boys more cooperative as the sessions progressed.

One could conclude from these results that the boys were more competitive and the girls more cooperative. One could assert instead that the boys were foolish competitors and the girls were smart competitors since the girls did gain the objective stated in the experiment, many marbles for both players. One could also argue convincingly that since cooperation was the unstated objective in this experiment, the only conclusion to be drawn is that the girls were good cooperators and the boys were poor cooperators. Another alternative explanation is suggested here—that in this study girls were better at a cooperative-competitive strategy than boys were. The boys behaved in a simply competitive way, being unable to introduce cooperative behaviors into this game strategy. Perhaps the boys were able to operate on only one track with male opponents, the track of competition. This suggests that the competitive drive between two male opponents is stronger than that between any other two opponents, for the girls were able to keep the cooperative option open when opposing other girls. Finally, the cooperative option opened for the boys when they competed against girls.

In their "box score" summarizing the findings on sex differences in competitiveness, Maccoby and Jacklin (1974) listed 24 studies, 12 percent of which found females to be more competitive and 25 percent of which found males to be more competitive (pp. 250–251). In her correction of the box score Block (1976) added two more studies, changing the figures to 11 percent for females and 31 percent for males. The remaining studies found no significant sex differences, over a wide variety of age levels. It is important to remember that, like the Szal study, most of these studies contrasted competition with cooperation. This makes interpretation difficult and generalization almost impossible.

Four studies that did not confound cooperation and competition to the same degree yielded data which are more easily interpreted. Jacklin and Maccoby (1978) observed 33-month-old children playing in same-sex and mixed-sex dyads. The subjects showed higher levels of social behavior in same-sex pairs. The girls showed passivity and withdrawal in the presence of boy partners, but not girl partners. Kaplan (1970) found that in cooperative tasks even girls who were judged high in dominance acted submissively toward boys, no matter how the boys were ranked. Parker (1976) and Parker and Omark (this volume, Chapter 24) investigated whether girls would act more competitively toward one another if they were in an all-female group. Parker observed children in an all-boys school and an all-girls school. Although he did not find evidence of a dominance hierarchy in the all-female group, he did find evidence of more competitive feeling among the girls, as expressed in overrating themselves on the criterion of toughness. It may be the case that girls understand the structure of the dominance hierarchy (Omark, 1972; Edelman, 1973) but are not willing to compete for position in the hierarchy if boys are participating. Megargee (1969) measured dominance and leadership in same-sex and opposite-sex pairs of adults. After taking a paper-and-pencil test assessing dominance, the subjects were placed in pairs, with one member high and the other low in dominance. In same-sex pairs, the high dominant person became the leader 60 to 75 percent of the time. In dominant male–less dominant female pairs, the male became leader 90 percent of the time. In dominant female–less dominant male pairs, the male still became leader 80 percent of the time. Perhaps the most interesting result in this study is the fact that in the last group it was usually the dominant female who decided that the male should lead. It suggests a strong female collaboration in the assumption of power by males.

The situation may actually be very complex in adults. Peplau (1976) had dating couples compete under two conditions: against each other and as a team against another couple. They were presented a verbal skill task. Women with "traditional" attitudes about women's roles performed less well when competing against their boyfriends than when joining them to compete against others. All of the men and those women with more "liberated" attitudes performed equally well against all competitors. Apparently, many women feel inhibited about competing against men with whom they are emotionally and / or sexually involved.

Using open-ended stories which deal with success, Horner (1972b) found a fear of success or "motive to avoid success" in teenage and adult female subjects. Horner's results gained empirical support when Peplau's women who performed poorly against their boyfriends also demonstrated the "motive to avoid success" in finishing Horner's open-ended stories.

Information from primatology may prove to be very helpful to psychologists who are attempting to construct some comprehensible picture of competition in humans. Primatologists interested in the social bond in monkeys have asserted that the main force keeping monkey societies together is the bond between the rank-ordered males, that is, the bond between males in the dominance hierarchy (Chance, 1963; Chance and Jolly, 1970; DeVore, 1965). The dominance hierarchy, in turn, is largely determined because of and by means of competition over valued resources. The dominance hierarchy, once stabilized, functions to keep competition at a low level. Competition is no longer necessary, since higher ranking males have won priority of access to resources. Rhesus and baboon societies are similar to each other in that they maintain their rank-order structures agonistically, that is, by means of aggressive and submissive gestures used whenever an individual needs to be reminded of his rank. In most species animals rarely actually fight one another over resources, once a stable hierarchy is formed. When aggression is resorted to, it is probably being used to maintain the social order (Bernstein and Gordon, 1974).

In contrast to monkey societies, chimpanzee and gorilla groups are marked by more relaxed, friendly exchanges between individuals. Encounters between dominant and subordinate animals are often accompanied by touching and other invitations to join in grooming, play, sexual, or mothering behavior with the one who began the attention-demanding display. Chance and Jolly (1970) concluded that the rank order established by attention-demanding displays is directly comparable to other forms of rank order in that it also determines preferential access to food.

The important point is that it is *male* competition that must be brought under control to keep the social group functioning in an orderly way. Besides competing for food, resting places, and grooming or other forms of attention, males also compete for sexual access to females (females do not compete for similar exclusive rights to males). While virtually every female is assured of leaving offspring, there is probably a great range in the number of offspring a given male might leave, one hypothesis being that some low-ranking males leave none at all.

Female groups, especially in rhesus and baboon societies, tend to be much noisier and more disorganized than male groups (Chance, 1963). When females of these species form separate assemblies for any length of time, rank-order agonistic behavior is occasionally apparent, but it is more subtle than male agonistic behavior. In at least one troop of captive baboons, however, females were able to form a stable dominance hierarchy after their male leader died (Berger, Haynes, and Stettner, 1979).

In a variety of nonhuman primate species, then, competition among males is more crucial (since leaving offspring is at stake), more fierce, and more direct than is competition among females. Females of these species even attempt to get high-ranking males to do their competing for them (Kummer, 1968). There is a very complex sex difference in competitive behavior in these various primate groups.

The question then arises whether or not male primates (including humans) have some biological preadaptation for competition. The biological evidence is particularly compelling on this point. Androgens (especially testosterone present during early prenatal life in humans) seem to have a unique role in promoting some types of competitive behavior. Genetic females who, through some malfunction or mistreatment, are fetally androgenized, later display malelike rough-and-tumble play and often prefer to be with boys (Money and Ehrhardt, 1972). Normal males and androgenized females become more aggressive if their testosterone levels are increased later in life, but normal females show little or no increased aggressive reaction to exogenous testosterone (Joslyn, 1973).

Testosterone injections have been shown to raise hierarchical rank in nonhuman male primates (Clark and Birch, 1945); and androgen therapy raised the self-esteem of a group of timid human adolescent boys (Sands and Chamberlain, 1952). Hutt (1972) pointed out that androgens also facilitate the focusing of attention and perseverance at a task, both of which are essential in competition. Hutt speculated that androgens may promote not only physical aggression but competitive drive in general. Males, then, would be expected to exhibit competitive behavior more frequently, and at a more intense level, than females do. This difference is influenced by testosterone, particularly in prenatal life.

If males are more competitive toward other males than females are toward other females, the question remains how the two sexes behave toward each other. It is quite likely that, in both nonhuman primate and human societies, the sexes are predisposed *not* to compete with each other. That is, it is quite likely that males are specialized for structuring society through competition with other males, while females are specialized in such a way that they do not challenge males except in unusual circumstances, as when females have been segregated for some time (see Bernstein and Gordon, 1974). Hoyenga and Hoyenga (1979) present evidence that in some species the sexes have special adaptations which reduce the likelihood of intersex competition. For example, there is a sex difference in the auditory sensitivity of tree frogs such that the female does not even perceive the aggressive call of the male (Narins and Capronica, 1976).

Callan (1970) explained that in a species such as ours, in which the sexes must cooperate in pairs for long periods to insure the survival of the young, open competition between males and females is "wasteful and inefficient" (p. 131). After a male and female begin a family, the male usually continues to compete for resources outside the home. If his mate wants him to succeed in that competition, her best strategy is probably to let him begin by ranking at least above her. Meanwhile, the female may turn her attention to child rearing, which includes frequent nursing, giving an infant total care, and providing the close personal bond that a human infant requires (Bowlby, 1969; Freedman, 1974). It is likely that, in the course of our history as a species, men have evolved being specialized for competition while women have evolved with less competitive behavior patterns in general and submissive behavior patterns that they display toward men.

This is not to say that every man seeks the most submissive woman he can find. Such a strategy would be of small benefit to him. As Guthrie (1976) observed, the status of one's partner affects one's own status. The optimal plan is for a man to seek the highest ranking female he can find who still ranks below him. This plan allows for the operation of sort of a "sexual syllogism" which might read as follows:

> My woman is high ranking.
> I rank higher than my woman.
> Therefore, I am *very* high ranking.

Obviously, the "sexual syllogism" has a favorable implication for the female, if her mate is very high ranking and she ranks just below him. In terms of competition for resources in the public realm, however, the male point of view is more important. The ideal is for a male to find a female with the perfect balance, one who is high ranking and yet submissive to him.

Callan (1970) suggested another adaptation that may be found in women. It may be that women are capable of more direct competition than they are willing to demonstrate with men and, furthermore, that females demonstrate this competitiveness only in the presence of other females. Generalizing from the research on nonhuman female groups (especially in domesticated animals), Callan suggested that females might have the ability to compete for dominance all along, so that they might be able to take over should the need arise. Sex segregation might be one of the conditions under which the need for a female hierarchy would arise.

The situation is probably even more complex. Young (prepubertal) girls are sometimes quite similar to boys in their competitiveness (Aldis, 1976; Strayer and Strayer, 1975), and older women in some societies have as much formally recognized authority as men (Evans-Pritchard, 1965). It may be that there are different options open to a female at different points in her life cycle. Gutmann (1973) suggested that the submissiveness of women of childbearing age is under the influence of a "parental imperative." Perhaps there are many roles possible for a female during preadolescence, including submissive and competitive roles; menarche and the onset of the childbearing years may eliminate the competitive option for those females who wish to bear children; the climacteric may once more broaden a female's options to include competition with males. The role of female hormones may be essential in facilitating these alternative strategies, submission or competition. Although not much is known about the behavioral effects of estrogen in humans, it has been found that estrogen has an inhibitory effect on aggression in male mice (Suchowsky, Pegrassi, and Bonsignori, 1969). Whether or not hormones play a role in the inhibition of competitiveness in human females, it seems that females are less competitive toward males during their childbearing years than at any other time. It is obvious that a women is better off being noncompetitive toward males if she wants a competitive male to help her during those years of potential parenthood.

Four studies just mentioned showed that girls were more willing to compete with one another than with boys. They raised the question of whether or not observable differences could be found in an experimental competitive game situation. The task became one of observing girls in some type of game, first competing against other girls and then competing against boys, to see if the girls behaved differently under these different conditions. The first experiment involved an athletic competition, in which the boys, being generally better players, had an advantage over the girls. In an attempt to test the two variables "sex" and "skill level" separately, a second experiment was conducted, exactly like the first except that it involved an academic competition in which the sexes were equal in ability. In both studies the players were classified according to skill, in an attempt to keep this variable somewhat controlled.

It was hypothesized that in both types of competition girls would be inhibited when competing against boys. Inhibition would be demonstrated if all girls, even those who performed well against other girls, performed poorly against boys. Second, it was hypothesized that boys would not be inhibited when competing against girls in either competition. Boys were expected to show consistent performance with male and female opponents.

In the context of this chapter, the experiments were designed to provide an answer to the first question posed at the beginning, which asked whether females end up at the bottom of the hierarchy by competing for rank. Since trophies and ribbons were awarded in these competitions, the participants were actually competing for objects symbolizing rank.

Experiment I: An Athletic Competition

Methods

The setting was an elementary school playground in Chicago. The school was a small (280 enrollment) private school, whose members were middle-class Afro-Americans from 6 to 14 years of age. The students spent about 40 minutes of their lunch period every day on the playground adjacent to the school. The playground was a rectangular paved area about one-fourth the size of a city block, surrounded by a 6-foot-high cyclone fence. There was no equipment on the playground except what the children brought (balls, bats, jump ropes, and so forth). The only adult present every day was an overseer, an elderly man who never supervised any activities.

The subgroup chosen for the study was the intermediate unit

in the school, which included 9-, 10-, and 11-year-old children. The main investigator went into their classrooms one day and announced that there would be a dodgeball tournament for their unit, out on the playground, at lunchtime during the following two weeks. It was explained that anyone who wished to participate could sign up and that four of the best players would get trophies while several others would get ribbons at the end of the tournament.

Of the 90 eligible children, 54 signed up to play, 24 girls and 30 boys. Of these 54, 4 girls and 1 boy never showed up to play; several others missed some of their scheduled games due to absence, forgetfulness, or lack of interest. The average age of the girls who participated was 10.4 years; the average age of the boys was 10.6 years.

The equipment used for the game included a leather soccer ball and an 80-foot length of clothesline rope tied at the ends to form a circle which marked the playing area. The rope was always set on the same spot near the fence on one side of the playground. Posters showing children's individual scores were mounted on the fence nearby during the entire tournament. The posters reminded the children of their scheduled games as well as kept their enthusiasm high. The observers used clipboards to make written records, a watch to time the games, and a whistle to signal the beginning and end of each game.

The game of dodgeball itself is very simple. It was chosen for the study because, according to the children's gym teacher, both boys and girls had learned the game in (sex-segregated) gym classes, and neither sex had more experience at it. The game begins with one person chosen to be in the middle through some random process. Everyone else stands around the person in a circle (here along the rope which lay in a circle). One person out on the circle has the ball and throws it into the circle, trying to hit the person in the center. If the ball misses the one in the center, that target person, who has just "dodged" the ball, gets a point and the ball bounces out to the circle, where someone else will retrieve it and throw it at the target person again. The target person gets a point for every time he or she dodges a throw. If a thrower hits the target person, the latter goes back to his or her place on the circle, and the thrower becomes the target person. Again, whoever retrieves the ball when it is loose gets to throw it. There were two additional rules in this game: no player could step across the rope and into the circle for any purpose; and although a player could run away from the circle to retrieve the ball, he or she had to return to the assigned place on returning. This latter rule, as it will become clear, was established to maintain some controls in the game.

The entire tournament involved eight games, played on separate days over a two-week period. The first games played were same-sex games. A girls' baseline game enabled us to break the girls up into two groups, according to skill. The girls were ranked by the number of points (dodges) each one made in the baseline game. The girls in the top half were designated high skill; the girls in the bottom half were regarded as low skill. In the next game the high-skill and low-skill girls played the game from assigned places. They were arranged high skill and low skill, alternately, around the circle, and instructed to keep these positions during the course of the game. With this arrangement, we reasoned, high-skill and low-skill players had (theoretically) equal probabilities of retrieving the ball. High-skill and low-skill players would, in fact, have to compete with each other for possession of a loose ball. This gave us an opportunity to see whether or not one group tended to win competitions for the loose ball.

The next game was the boys' baseline game, after which the boys were similarly divided into two groups, high and low skill. These two boys' groups played, alternating by skill, in the fourth game.

Then came the mixed-sex games, in this order: high-skill girls against low-skill boys, high-skill girls against high-skill boys, low-skill girls against low-skill boys, and low-skill girls against high-skill boys. (Our impression from informed observations was that the average boy player was a little better at the game than the average girl player, so we gave the girls the advantage in scheduling the mixed-sex games in that their best players played first.) In these games the children were arranged boy and girl, alternately, to give each sex an equal probability of retrieving the ball. Our reasoning was the same as described for the same-sex games. From the children's point of view they were playing, not in groups, but in leagues, in which every player was trying to score his or her own points for dodges. We, of course, were interested in the total statistics for each group in each game.

Each game lasted 15 minutes. The statistics shown here are only for the first 6 and last 6 minutes of each game. The middle 3 minutes of each game were used as a breathing period for the observers, although one observer continued to record points for dodges during the 3 minutes, for the players' sake.

There were two observers present for each game. Both observers learned the names of the children before the tournament began, since all notes were to be made on children by name. The first observer recorded, for each loose ball, the name of every child who moved some part of his or her body (other than the head) toward the ball. This was later scored as a competition, and we found that, for most loose balls, one, two, or three children competed for the ball. The name of the person who actually retrieved the ball was circled. The second observer recorded who was in the middle (the target person), how many dodges he or she made (one point for each dodge), and who hit him or her (that person became the next target person).

Results and Discussion

The most important statistics for each group in each game are shown in Table 1. The games are listed in the order in which they were played (the results for baseline games are not shown because the children were identified according to skill by means of those games). In game 1 the high-skill girls made significantly more points than the low-skill girls did (86–8, $p < .02$). Although the differences in hits and retrievals were not significant, the high-skill girls tended to get more of these.

In game 2 the high-skill boys did not score significantly more points than the low-skill boys did, nor did they score significantly

Table 1. Children's Skill in Same-sex and Mixed-sex Dodgeball Games

Game	*Players Skill, Sex* (N)	*Dodges*	*Hits*	*Retrievals*
1	High girls (9)	86*	8	59
	Low girls (8)	8	4	41
2	High boys (11)	37	15	63*
	Low boys (10)	26	9	30
3	High girls (10)	4*	3*	36
	Low boys (10)	67	12	51
4	High girls (10)	6*	4*	23*
	High boys (10)	58	17	66
5	Low girls (7)	0*	4	26*
	Low boys (6)	71	6	54
6	Low girls (8)	0*	0*	4*
	High boys (11)	38	19	51

*χ^2 test significant ($p \leq .02$)

more hits than the low-skill boys. We discovered by examining individual scores that one of the low-skill boys had scored 19 of the 26 points made by that group. This boy was probably misplaced since he had not participated in the baseline game and had been assigned to the low-skill group because of his small stature. If we were to throw out his scores for the entire tournament, it would leave a significant difference in this game's score without eliminating significant differences in the other scores. So, with that explanation, his scores remain in the totals. High-skill boys retrieved the ball significantly more often than low-skill boys (63–30).

The scores shown for the next four games form the central evidence for the first experiment. They are quite similar in their onesidedness (67–4, 58–6, 71–0, and 38–0); all show differences well above a level of significance ($p < .02$). No matter what the skill levels were, boys always scored significantly more points than the girls did. In three of the four games they made significantly more hits; in another three games they made significantly more retrievals. Even the *best* girl players were soundly beaten by the *poorest* boy players.

There are a couple of ways in which one might attempt to explain these data. It is obvious that the skill difference between girls' groups observed in the first game was wiped out when the girls played boys. High-skill girls did just as poorly against boys as low-skill girls did. One could argue that all of the girls were much poorer players than all of the boys. This would not be likely since the girls and boys had had equal exposure to the game of dodgeball before the study was done. Some overlap in skill level would be expected; it was our impression that there was considerable overlap in skill between girls and boys.

Another explanation of the findings, and a more likely one, is that in the presence of boys the girls were simply not displaying the kind of skill of which they were capable. In other words, while they actively competed in the all-girl game, they "held back" when they played the boys. Inhibition seems to be the more reasonable explanation of why the girls' skill difference disappeared when the girls played the boys.

When the data were examined on who competed for the ball, a difference between girls and boys was discovered in the way they competed for the ball against same-sex players. On any given loose ball, more boys went after the ball than girls did in the same-sex games. In the boys' game there were seven occasions on which more than seven boys went after the ball at the same time.

The largest number observed in the girls' game going after the ball at one time was four. It seems that even though only one person can end up with the ball, any boy who thinks he has the slightest chance takes off after it. It may be that the girls, in contrast, compete with one another only when their chances of winning are very high (perhaps they are being smart competitors). Another possibility is that girls simply are not as eager to compete as boys are, regardless of the odds. At any rate, the girls in this tournament competed with other girls less readily than the boys competed with other boys.

When the girls faced the boys, they competed even less frequently than they did against one another. Table 2 shows the data for each same-sex game as well as the total number of times a boy competed for the ball and total number of times a girl competed for the ball in each mixed-sex game. Whereas the better players tended to compete more frequently than poorer players in the same-sex games, girls competed less frequently than boys in all four mixed-sex games, regardless of skill levels involved.

Table 2. Competitiveness in Same-sex and Mixed-sex Dodgeball Games

Game	*Players Skill, Sex (N)*	*Total Competitions**	*Solo Retrievals*	*Retrievals in Cross-Skill or Cross-Sex Competitions*
1	High girls (9)	87	29	18
	Low girls (8)	77	22	10
2	High boys (11)	130	20	27
	Low boys (10)	105	12	14
3	High girls (10)	85	13	18
	Low boys (10)	114	17	24
4	High girls (10)	73	9	10
	High boys (10)	116	24	31
5	Low girls (7)	60	6	12
	Low boys (6)	79	22	23
6	Low girls (8)	25	1	2
	High boys (11)	82	21	15

*On some loose balls five or more children competed for the ball. As identification was impossible when this occurred, heads were counted and the competitions were divided evenly between the two sides. Such competitions were numbered, by game, 5, 68, 43, 79, 10, and 13. Thus the differences between groups, especially in games 2 through 6, are probably even greater than the totals indicate.

The second column in Table 2 shows solo retrievals; these were loose balls on which one person retrieved the ball without being challenged by anyone else. It is clear from the totals that boys' retrievals went unchallenged by girls (who, of course, were standing right next to them) far more often than the reverse. The third column shows that when children of different skill levels competed for the ball in same-sex games, the better player tended to retrieve the ball. But in mixed-sex games boys kept the ball more frequently than girls did, again regardless of skill levels.

It may be concluded, first, that the girls in this study competed less readily against same-sex opponents than the boys did; second, that the girls were even less competitive against the boys than they were against one another.

The question arises whether or not the girls were conscious of holding back when they played against the boys. In interviews held after the tournament the two best girl players were asked, "Were you trying as hard when you played against the boys as you were when you played against the girls?" Both girls answered simply, "Yes." These two girls and two of the best boy players were then asked, "Why did the girls do so poorly when they played the boys?" The answers follow:

Nina: "See, boys are tougher than girls. They throw hard and they throw whichever way you move." (She explained that boys fake, watch the person move, and then throw in that direction. She said that she had tried to do that too.)

Wendy: "The boys are strong. They're rougher than we are."

Victor: "Boys are stronger, and boys have better aim."

Gary: "When the girls throw, they miss the person. Boys throw better."

It seems that the girls (and one of the boys) were aware of the boys' superior strength. The girls maintained that they tried against the boys, but they blamed their poor performance on the roughness and toughness of the boys. It is very likely that the girls held back against the boys out of fear of being overpowered or hurt.

These results cannot be accounted for by asserting that these girls have been taught by their parents that competition is not suitable behavior for girls to display against boys. These girls were not even conscious of playing submissively against the boys. It is more likely that the girls have learned from experience that the boys are stronger and more competitive. The boys' greater strength and desire to compete, of course, are greatly influenced by biological factors, as pointed out in the introduction to this paper.

Because a question was raised about the effect of the context of this experiment (an athletic competition) on the girls, an academic competition was the next event planned.

Experiment II: An Academic Competition

Methods

The setting was the same school. The spelling tournament was held indoors during school time. The children were released from art or creative writing classes in order to participate in the spelling contest. This occurred about seven months after the dodgeball tournament.

As in the dodgeball tournament, the children chosen for the study were from the intermediate unit in the school. The students were informed that all could volunteer to participate in the spelling tournament. Of the 87 eligible children, 69 signed up, 34 boys and 35 girls. About half of these volunteers, 16 boys and 16 girls, were selected to participate in the spelling tournament described here. Of the children participating in the spelling tournament, 22 percent had also participated in the dodgeball tournament. In the spelling contest the average age of the girls was 10.6 years; the average age of the boys was 10.7.

A spelling contest was chosen for the academic competition for several reasons. First, girls and boys in the school spent equal amounts of time in (sex-integrated) spelling classes. Second, since most of these children had never participated in an old-fashioned spelling bee, it was expected that they would respond enthusiastically. Again trophies and ribbons would be awarded.

The written pretest, based on the list prepared by Gates (1937), allowed the division of the children into these two sets of matched groups: high-skill boys and girls, and low-skill boys and girls. The scores of the high-skill children ranged from 19 to 30 correct out of 36. Each boy was matched with a girl who had scored within one point of him. The scores of the low-skill children ranged from 5 to 15, and boys and girls were matched in the same way. One low-skill girl moved away before the contest was held, and she was replaced by a girl with a higher score.

The words for the spelling contest were taken from *The New Iowa Spelling Scale* (Greene, 1954). This list includes several thousand English words, showing for each word the percentage of children in each elementary school grade who spelled that word correctly in a written survey. For example, the word "universal" was spelled correctly by nearly 50 percent of the eighth graders surveyed, while the word "face" was spelled correctly by a similar percentage of third graders. To create a contest word list, 52 words were chosen which had been spelled correctly by 45 to 55 percent of third graders. Then 52 words were chosen which had been spelled correctly by 45 to 55 percent of fourth graders. The process continued through the eighth-grade level. Contractions and capitalized words were not used, and no word was used twice. By means of a random number table (Fisher and Yates, in Kirk, 1968) these 312 words were rearranged to create a contest list composed of words covering a wide range of difficulty. (The precise details of the randomization process, as well as the list itself, are available from the author.)

The contest began with oral spelling quizzes organized somewhat differently from traditional "spelling bees" so that children would be required to volunteer to spell words. At any given time the two competing teams stood behind a table which faced the two experimenters. On the table were two light bulbs, which were attached to batteries and simple doorbell buttons. One child from each team sat down with a button in front of him or her. When one of the experimenters pronounced a word from the contest list, the two children who were sitting could volunteer to spell the word by pushing the button, which lit up the corresponding bulb. Whoever lit up his or her bulb first was given the chance to spell the word correctly and score a point. If the word was spelled correctly, the volunteer was given one point, and *both* children got up and went to the end of their respective lines. If the word was misspelled, the other child was given a chance to spell the word correctly and score a point, after which both children went to the end of their lines. If two misspellings were given, no point was awarded, the children went to the end of their lines, and the word was discarded. Ties in volunteering were given first to one team and then to the other. After each turn the next two children sat down and were given a new word. Each child on each team had three turns in each game, which meant that each team of eight spellers had 24 turns during a game. Although prizes were awarded according to personal scores, strong team solidarity developed during the contest, with children cheering and clapping for their teammates. No coaching was permitted, but those not participating were permitted to sit in the room and watch.

On the first day the high-skill and low-skill girls competed against each other, and the high-skill and low-skill boys did the same. The second day saw the high-skill boys compete against the low-skill girls and then high-skill girls. Finally, the low-skill boys competed against the low-skill girls and then high-skill girls. In this tournament, as is evident, the game arrangement was altered to give the boys the best chance of having good early experiences since boys are poorer spellers on the average than girls are.

Duties were split between the two experimenters. The first experimenter presented the word and used it in a sentence. The second experimenter watched the light bulbs in order to decide which child volunteered first. Then the first experimenter judged the spelling of the word and noted who volunteered first and who won a point. These games were also videotaped for further analysis (Warren, 1979), and the second experimenter also took *ad lib* notes.

Results and Discussion

Totals for volunteering and scoring points for each team in each game are shown in Table 3. The totals for points scored are pretty much what one would have expected. In both same-sex games the high-skill players scored significantly more points than the low-skill players did ($p < .05$). High-skill boys scored significantly more points than low-skill girls. For high-skill girls and low-skill boys the trend was in the same direction, although at a level below significance. Games involving spellers of equal ability were split, with the high-skill boys and the low-skill girls winning by narrower margins.

The totals for volunteering are more interesting. It is clear that high-skill girls out-volunteered low-skill girls (20–3, $p < .01$). When their matching boys' teams competed, they nearly split the volunteering opportunities, with low-skill boys actually volunteering more than high-skill boys, 11 times to 9. In all four mixed-sex games the boys out-volunteered the girls, with the difference reaching a level of significance in game 3 and narrowly missing in game 6. A test for determining the significance of combined probabilities (Jones and Fiske, 1953) shows that, in all of the mixed-sex games combined, boys volunteered first significantly more times than girls did ($p < .005$).

Apparently the girls and boys had different approaches to this

Table 3. Children's Skill in Same-sex and Mixed-sex Spelling Games

Game	*Players Skill, Sex (N)*	*Ties in Volunteering*	*Times Volunteered First*	*Total Points Scored (24 possible)*
1	High girls (8)	1	20*	16*
	Low girls (8)		3	1
2	High boys (8)	4	9	17*
	Low boys (8)		11	2
3	High boys (8)	4	19*	14*
	Low girls (8)		1	2
4	High boys (8)	5	11	11
	High girls (8)		8	8
5	Low boys (8)	8	10	4
	Low girls (8)		6	8
6	Low boys (8)	6	13	6
	High girls (8)		5	12

*χ^2 test significant ($p \leq .05$)

competitive situation. The high- and low-skill groups were quite different from each other in terms of ability. The girls seemed to recognize this difference, in that their superior spellers volunteered while their inferior spellers deferred. With the boys, in contrast, vastly inferior spellers tried to out-volunteer the superior spellers among the boys. On six occasions in the boys' game a low-skill boy volunteered and misspelled the word, after which his opponent spelled the word correctly to score a point. These occurrences did not become less frequent as the game progressed. The only change that was seen was that more ties occurred toward the end. Even after many failures the low-skill boys continued to volunteer first. Not even occasional laughter from the audience (at ludicrous combinations of letters) discouraged these would-be spelling champs.

In the mixed-sex games, no matter which teams were involved, the boys always out-volunteered the girls. This was expected when high-skill boys opposed low-skill girls simply because of the skill difference. It was not expected, predicting from skill, that low-skill boys would out-volunteer high-skill girls 13 to 5. The explanation has been offered that the girls may have been unaccustomed to the mechanical arrangement or even intimidated by it. The videotapes showed, however, that more often it was the high-skill girl who had her hand on the button, ready to go, before the low-skill boy did. The girl simply did not *push* the button first to volunteer first (Warren, 1979). Although the high-skill girls did beat out low-skill girls, they did not beat out boys of precisely the same spelling ability as those low-skill girls.

Discussion

The pivotal issue in the dodgeball contest was whether the high- and low-skill girls' groups, which performed at different levels against each other, would perform at different levels against the boys. In fact, the high- and low-skill girls looked alike when performing against boys. They were indistinguishable, both groups performing very poorly.

In the spelling contest skill was more carefully controlled, and performance was separated from desire to perform (volunteering). While both girls and boys performed fairly consistently, it was clear that girls were not eager to demonstrate their ability against boy opponents. Instead girls deferred to boys, even if the boys were poor spellers. One wonders what would have happened if girls had not been "handed" the opportunity to spell after their boy opponents volunteered and misspelled the word.

An attempt may now be made to answer the first two questions that were posed at the beginning of this chapter. The girls in this study did not compete for prizes (trophies signifying status) in the same way that the boys did. Regardless of who their opponents were, the boys played to win. Girls played to win (though less intensely than the boys) when their opponents were girls. Girls performed poorly (in dodgeball) and did not take advantage of opportunities (in spelling) when their opponents were boys.

If this is generally the case, that girls do not compete in the same way that boys do, then rank in the dominance hierarchy is not the same for girls as it is for boys. Perhaps some other type of social structure, or some modified view of the dominance hierarchy, would more adequately describe female participation in the social order. There is evidence that females in sex-segregated groups form hierarchies which are less rigid than males' hierarchies (see Savin-Williams, this volume, Chapter 12), but something altogether different happens when males and females interact. This may seem obvious to those who are familiar with the structure and function of dominance hierarchies, but those who approach social relations from another point of view often assume that competition is unidimensional. Competition, for females, involves at least two strategies, one for female opponents and another for male opponents. Obviously, these differences will not be erased by requiring coeducational gym classes and athletic teams. It may be that these differences are adaptive for our species and in fact should *not* be erased.

There are other variables that should be examined if this line of research on inhibition is pursued. First, it would be quite helpful to discover whether one's stage in the life cycle makes a difference, that is, whether women in their childbearing years are more inhibited when competing against males than other females are. Second, the opponent's rank in the hierarchy may determine the degree of a female's inhibition, allowing for a type of sexual selection. Third, in mixed-sex competition the ratio of males to females may be important. When females outnumber males in physical competition, it may be that some females are no longer inhibited because they do not expect to attract any male anyway.

The relationship between the sexes in every species is quite complex. Better answers to the questions raised above may help us gain some small insight into the ways that human males and females relate to one another.

Acknowledgments

Gratitude is expressed to Drs. Daniel G. Freedman and Susan S. Stodolsky for their guidance and helpful advice and to colleagues Glenn Weisfeld, John Callaghan, Vicki Ruzicka, and Ronald Warren for their insights and generous assistance.

Chapter 18

Prestige and Self-esteem: A Biosocial Interpretation

Jerome H. Barkow

The social sciences rest on simple psychologies. If economics is often accused of viewing man as rational but insatiably greedy, anthropology does something similar. Anthropological man hungers not so much for money and material goods as for prestige. Whenever anthropologists discuss competition and jealousy, speak of potlatches or of counting coups, describe displays of ritualized boasting and ostentation or even the emergence of social stratification, they are making a basic assumption: men seek prestige and status, deference and respect. Such an assumption calls not for challenge (there is no reason to suspect its validity) but for analysis. This chapter seeks to explore the idea that men,

---

This chapter is based on the much longer article, "Prestige and Culture: A Biosocial Interpretation," which appeared in *Current Anthropology* 16:553–572, 1975. Copyright 1975 by The Wenner-Gren Foundation for Anthropological Research. Published by the University of Chicago Press. Portions are reprinted here with the permission of that journal. In the original, comments appeared from A. A. Akiwowo, T. K. Barua, M. R. A. Chance, E. D. Chapple, G. P. Chattopadhyay, D. G. Freedman, W. R. Geddes, B. B. Goswami, P. A. C. Isichei, M. S. Knudson, C. F. Parker, J. A. Price, H. B. Sarles, S. M. Manson, and G. A. Schwirtz. Omitted from the present chapter are discussions of the phylogenetic background of self-esteem; the relationship between the latter, cultural evolution, and the maintenance of stability in traditional societies; and the inadequacies of the concept of *n-ach*. Space limitations have caused the editors to remove the discussion of self-esteem and inclusive fitness theory originally included in the revision; readers interested in how the ideas here presented are related to inclusive fitness theory (or "sociobiology") should consult Barkow (1978b).

everywhere, require prestige to maintain self-esteem and that their efforts form part of our pan-primate heritage.

We must begin with a pair of warnings: At no point is the assumption being made that we and other primates are single-mindedly obsessed with social rank and prestige. If this chapter occasionally gives that impression, it is because to focus on any one topic is necessarily to neglect others. Evolution can deal with all selection pressures simultaneously but exposition is limited to one topic at a time. The second warning is against easy analogizing between ourselves and other species. To stress the continuity between human and nonhuman primate social dominance is not to claim that our status systems are to be understood solely on the basis of research on chimpanzees—or, for that matter, that studies of first graders will provide much insight into such products of sociocultural evolution as the caste system of India. We are not our ancestors, nor is child behavior adult behavior.

Though the earliest hominids doubtless shared the general primate tendency to form dominance hierarchies, in our own species natural selection has considerably modified this tendency. We apparently evolved under considerable selective pressure for the ability to modify rapidly and drastically troop size and social organization in order to adapt to differing and changing ecological settings without speciation (Barkow, 1978a). For our species this pressure has resulted in the ability to form social hierarchies based on abstract principles and cognitive evaluations. Much of this chapter will deal with how our distinctively human hierarchies may have evolved from the general primate trait of social dominance.

Do we really know, though, that contemporary human social hierarchy and primate social dominance do indeed share a common origin, that is, are they homologous? Since our ancestors were nonhuman primates and primates exhibit social ranking, our ancestors presumably were socially ranked. We evolved from these ancestors while organized in cooperative bands of hunter-gatherers (DeVore and Washburn, 1963; Laughlin, 1968; Pfeiffer, 1972; Washburn, 1961; Washburn and Avis, 1958; Washburn and Lancaster, 1968) which may have continually fought against one another (Bigelow, 1973). Social organization implies at least some ranking; so it would seem that at no point during our phylogenesis did we lack social hierarchy. If human and nonhuman social dominance have apparently been continuous, their bases are likely to be homologous. Cultural systems of social rank are, therefore, based on, rather than opposed to, our primate heritage.

The concepts of self and self-esteem form a bridge between theories of biological evolution and the actual social organization of both human and nonhuman primates (as discussed in the other chapters of this volume). To construct this bridge, it will be necessary to introduce a simple vocabulary, adapted from Miller, Galanter, and Pribram (1960), and then to move on to a brief reconstruction of some aspects of hominid behavioral evolution.

Vocabulary

Human behavior may be conceived of as consisting of a series of hierarchically ordered plans (subplans, subsubplans, and so on); some or all of these are end-tested, that is, their outcomes are compared with a goal-state and either judged successful, in which case the next subplan begins operation, or unsuccessful, in which case the subplan may be reexecuted or a different one with the same goal put into operation. All plans are in the service of a limited number of goals, though any given piece of behavior is likely to be involved in more than one plan or goal. Behavior is, in general, multipurposive. The vocabulary of goals replaces those of "motives," "behavioral dispositions," "instincts," and the like. The origins of the goals remain rather obscure, but those we will be concerned with here are assumed (following Bowlby, 1969) to be products of biological evolution. A goal is not necessarily a discrete end point—it can be the maintenance of a continuing state. Self-esteem is one such state. Different general plans for achieving a goal may be referred to as "strategies," each strategy encompassing a hierarchical array of plans and subplans. Alternative plans (or subplans) on the same level of a given hierarchical array may be referred to as "tactics."

The Self

It was Hallowell (1955, 1960) who first discussed the *self* as a product of biological evolution.[1] Evolving man was selected for the ability to conceptualize the universe and its laws in what may be termed a cognitive map. As he developed the capacity for this internal representation of the universe, he automatically included in it a representation of himself. As the world became "objectified" (Hallowell's term), he became an object to himself. The representation of himself in his own cognitive map may be termed his *self*. The other higher primates, too, probably share with man at least a rudimentary sense of self—or at least, the work of Gallup (1975) and Griffin (1976), as well as the language-teaching experiments of the Premacks (1971), the Gardners (1971), and the LANA project (Rumbaugh, 1977) would so indicate.

The self functions in part as an evaluation device, monitoring both the internal state of the organism and its position in the physical and social environments. It evaluates the state of the organism in terms of the goals and goal-states built into it by biological evolution. These goals and goal-states are probably at least somewhat modifiable by experiences, while socialization (and culture) undoubtedly determines the strategies and tactics the individual will follow in achieving them. Though the attainment of a goal-state must once have maximized inclusive fitness, it does not necessarily do so today, any more than our sense of taste is today a reliable guide to good nutrition.

Our sense of taste and nutrition *were* once closely related—for our ancestors. In their environment preferring the sweet to the bitter was selected for because it meant a ripe rather than unripe piece of fruit (and not a tooth-destroying candy bar, as it often does for us). In the same way the maintenance goal-states became linked phylogenetically to pleasure because their maintenance enhanced the inclusive fitness of our ancestors (and regardless of their effect on *our* inclusive fitness). The *seeking* to maintain goal-states can also be pleasurable: when the self evaluates its plans and the strategies and tactics in their service as likely to be successful, positive emotion results. The individual is happy.

A full theory of the self would include a detailing of all goals and goal-states of the human organism, an explanation of the selective pressures responsible for their existence, an analysis of the strategies commonly used to attain them with particular attention to their cross-cultural variability, and an account of how ongoing behavior can be switched from serving one set of goal-states to serving another. Such a task would obviously be far beyond the scope of this chapter. The relevant goal-states here are limited to those having to do with the individual's social position and security. Grouping them together permits us to speak, heuristically, of a single goal-state, that of *self-esteem*.

Individuals strive to achieve and maintain self-esteem. Though the strategies they follow in their striving vary enormously, by far the most common is that of seeking social prestige—the approbation, respect, and even deference of those important to one. There is no reason to believe that human beings strive for "dominance" unless they happen to have had personal socialization experiences such that their major self-esteem strategy involves physical or political power over others (cf. Tiger and Fox, 1971).

Whether a concept of self-esteem is actually helpful in interpreting nonhuman primate behavior is moot, however; it would certainly not lend itself to operationalization. But let us assume that our protohominid ancestors, at least, did indeed strive to maintain self-esteem and that their chief strategy in doing so involved seeking high social rank. Let us further assume that *relative standing in protohominid bands was evaluated by the self in terms of the attention and deference received from other members of the group. For protohominids, then, self-esteem was based on attention and deference.* The protohominids, however, were developing into a symbol-using species under strong selective pressure to incorporate social norms into their cognitive maps.[2] While the self-evaluation process which gives rise to self-esteem or its lack continued to rely in part on received attention and deference (and still does, even today), new possibilities now arose, possibilities based on the new, shared norms.

The existence of norms meant that the self could be evaluated in terms of them. Hallowell argues for an intimate connection between the development of the self and the development of a normative order, suggesting that the very concept of the self implies evaluation in terms of norms. Initially, the self-evaluations and, therefore, self-esteem of our early ancestors would have been based on actual behavior of others—on the sum total of attention, threats, and deference received from the rest of the band. With the development of a more differentiated sense of self and a set of norms in terms of which it could be evaluated, it became possible to judge the self as of higher rank, not solely on the basis of observable interaction but as a result of evaluation in terms of abstract principles. Judging the self as of higher rank than others cognitively, in terms of norms, is one of the chief self-esteem strategies of contemporary adult human beings..

Perhaps an early norm involved food sharing, originally arising as an appeasement gesture to a higher ranking individual, or as an attention-getting device, or a courting ploy. Socially transmitted by a group with the cognitive capacity for memory and reciprocity, the behavior would have become a norm (see note 2). An individual might now have begun to evaluate himself not primarily on the basis of agonistic encounters, for example, but also in terms of his success in getting sufficient food to share, the value of generosity had been born. As one evaluates the self, one simultaneously judges others. It would have been a small step to move from evaluating oneself in terms of food sharing to evaluating others in the same terms, and so granting them deference and attention to the extent they were ready to share food (cf., this volume, Chapters 7 and 13).

If we assume this process to have continued, progressively, self-evaluation would have come to involve the application of abstract principles (norms). The maintenance of self-esteem would have become a cognitive process. Cognition created the new possibility of distortions of preception and evaluation: the individual could now ignore some data and emphasize others. Thus, an unpopular but successful hunter might maintain his self-esteem in terms of the generosity norm. As both his culture and his cultural capacity grew more complex, he could even choose which sets of norms to use in evaluating himself and others, again so as to maximize his perceived relative standing. As we will see shortly, such self-esteem tactics are today quite common.

Why should selection have favored a self-esteem maintained by complex cognitive processes rather than nothing but success in, say, attracting mates or winning agonistic encounters? Perhaps the low-ranking individual who might otherwise have left the group remained because he overevaluated himself cognitively, thereby increasing his inclusive fitness. His presence might have aided the defense of those with whom he shared genes; or perhaps, the enormous behavioral flexibility permitted by a cognitive type of self-esteem enhanced fitness by permitting individuals to occupy different social positions at different times, depending on the task at hand. Certainly, if females began to accept cognitive criteria of relative standing and to choose mates in terms of them, there would have been sexual selection for the capacity to maintain self-esteem cognitively.

It is probably an error, however, to seek a particular adaptive value for the capacity to evaluate relative standing in cognitive terms. We were being selected for the Gestalt of characteristics summed up in Spuhler's (1959) term "capacity for culture" and not for isolated psychological traits. If Hallowell is right, the essence of culture is normative structure, and a normative structure requires by its very nature a sense of self and an ability to evaluate that self in terms of the norms. Self-esteem, therefore, becomes one product among many of this poorly understood selection for cultural capacity.

Self-esteem Theory

Were there a single, generally accepted theory of self-esteem, this section would be straightforward. Unfortunately, the psychological literature offers many theories concerning the nature of the self or ego and the manner in which it strives to maintain its esteem (strength, integrity). For purposes of the present discussion, "self-esteem" is defined as a goal-state maintained as long as the self can evaluate itself as higher than at least some others. This definition is intended to emphasize the continuity between human self-esteem and nonhuman primate social dominance.

LeVine and Campbell (1972, pp. 140–141) examine the work of a number of personality theorists (including Karen Horney, Harry Stack Sullivan, Carl Rogers, Erich Fromm, and Daniel Miller) for ideas of self-esteem. Among their findings are the following tenets:

1. Self-esteem is a prerequisite for mental health, in general, and for appropriate esteem of others in particular.
2. A low level of self-esteem leads to mental pathology, in general, and to negative attitudes (hostility, distrust, contempt) toward others in particular.
3. Low levels of unconscious self-esteem lead to the use of one or more distortions of information concerning the self that maintain the minimum of conscious self-esteem required for nonpsychotic functioning in most interpersonal environments.

To these three additional (and more problematic) assumptions are suggested:

4. Socialization experiences interact with the individual's genetic given to establish the minimum level of self-esteem necessary for the individual to function without pathology.
5. The major strategy for the maintenance of self-esteem is the pursuit of prestige.
6. Socialization determines the major strategies the individual will utilize in seeking prestige.

The first three assumptions are widely accepted and in little need of discussion here. Since our definition of self-esteem involves an evaluation of relative standing, it seems reasonable to assume (assumption 4) that some of us are satisfied with a lower rank than are others and that it is the interaction of particular genetic constitutions with particular experiences which determine at which rank we cease striving to go higher. The Macaque data, (Sade, 1968), in which status of mother strongly influenced status of offspring, are consistent with this assumption. Given that both genotypes and socialization techniques probably have modal patterns within particular cultures, it may be that the modal self-esteem level of some societies is higher than that of others. It is perhaps even more likely that, within a single society, some social strata will require higher average levels than others.

Assumptions 5 and 6 are both the most vital and the most problematic. That we maintain self-esteem primarily by seeking prestige is a surprising assertion when we consider that (to the author's knowledge) none of the personality theorists who have treated this topic has come to this conclusion. In part, this is because none has thought to use "prestige" in so broad a manner. For our present purposes the approval and respect of one's neighbors, one's kinsmen, one's parents and ancestors, one's gods and spirits are all tantamount to prestige. Such approbation and respect permit the self to evaluate itself as being of higher standing than at least some others, thereby maintaining self-esteem.

Assumption 6, that major prestige strategies for the maintenance of self-esteem are determined by socialization experiences, is not meant to rule out the possibility of endogenous factors playing a role in self-esteem. One is tempted to speculate that sex-differences in dominance behavior (see this volume, Chapters 11, 12, and 17) and prestige strategy in part result from a propensity to favor some criteria for the evaluation of relative standing over others; e.g., males might have a greater tendency than do females to utilize criteria having to do with physical prowess or endurance. Presumably, sociocultural evolution would tend to amplify slight endogenous factors so that socialization practices would exaggerate the differences between the way the sexes evaluate relative standing. It might even be that various primate species differ in dominance behavior chiefly because of differences in the nature of the evaluation criteria which determine whether the self-esteem goal-state has been reached. Leaving these biosocial speculations aside, we can move on to a discussion of the types of prestige strategies available to members of our own species.

Prestige Strategies

The usefulness of so broad a conception of prestige will perhaps be apparent if we divide the universe of possible prestige strategies into four categories in terms of their orientation (individual or group) and the physical presence or absence of the prestige-givers.

Individual-oriented Strategies with the Prestige-givers Present

Striving for the respect and approval of those with whom one is associated is the simplest and perhaps the most powerful of all prestige strategies. We see it most clearly in children and adolescents, but throughout adulthood the opinion of those with whom we daily converse has a powerful effect on our self-esteem.

Individual-oriented Strategies with Prestige-givers Absent

Here the individual is striving for the respect and approbation of others physically distant from himself, or dead, or wholly mythological. Such striving is possible because our cognitive map is to a considerable extent a product of what we have been taught. If we have been taught that the ancestors watch us daily, we may be striving to win their respect rather than that of those physically about us. Apparently, some socialization techniques make us permanently sensitive to internalized representations of signifi-

Table 1. Sources of Prestige

	Location of Prestige-giver	
Orientation	*Present*	*Absent*
Individual	Onlookers	Introjected "others" Gods or spirits Reference groups Ancestors Leaders
Group	Other groups	Gods or spirits Ancestors Leaders Other groups

cant others, at times almost to the point of withdrawal of attention from those with whom we interact. It may be that the old distinction between shame cultures and guilt cultures (Piers and Singer, 1953) was actually a distinction between societies in which individuals tend to maintain self-esteem by seeking the approbation and respect of their immediate fellows and those in which individuals rely more on the evaluation of internalized representations of others, who may or may not have once existed.

Many individual-oriented strategies involve groups, though the individual does not necessarily strive to enhance the relative standing of the group as a whole. Rather, the individual perceives membership in the group or category as prestigious in itself and, therefore, follows a tactic of ostentatiously communicating his (claimed) membership. Thus, the new citizen may ostentatiously display the flag, the adolescent female may wear ostentatiously provocative clothing to communicate to others that she now belongs in the category of "woman" rather than that of "girl," and the *nouveau riche* may purchase an ostentatiously large mansion so that the world may know they belong in the "wealthy" category.

Even individuals more secure in their membership in a high-status group must constantly evaluate themselves (and be evaluated by others) in terms of normative criteria having to do with their fitness for membership. Such criteria have to do not so much with the individual's relative standing as with whether his comportment matches that expected of group members. This type of evaluation criterion has to do with honor. "Honor" always involves membership in a prestigious group or category. One has "honor" as a Frenchman or "honor" as a master chef. Campbell (1966) and Pitt-Rivers (1966) use "honor" in a manner involving the right to compete for prestige, a right accorded only to group members. In their case the group in question is that of men, a prestigious group indeed in the Mediterranean societies with which they are concerned.

Group-oriented Strategies

Here one receives prestige not as an individual but as part of a group. The prestige-givers may be physically present, and if so, they are most likely to be other groups. They may be physically absent; in this case they may also be gods or spirits, ancestors, or distant leaders. Hsu (1971) argues that while North Americans work for themselves as individuals, the Chinese and Japanese strive to increase the relative standing of their group (initially, a kin group) vis-à-vis other groups. We seem to be dealing here with a fundamental difference in the boundaries around the self. Apparently, for Chinese and Japanese the self includes the group, so that the group's success means self-esteem for the individual. Of course, presumably an individual who helps raise the standing of the group as a whole simultaneously earns the respect of his fellow group members, so that in a sense he is following an individual-oriented strategy even while he utilizes a group-oriented one. For Westerners the self appears to be more tightly drawn, so that much less self-esteem can be derived from membership in a group. Nevertheless, the differences between East and West in reliance on one type of strategy rather than the other are probably only a matter of degree.

Cognitive-distortion and Reference-group Tactics

When the prestige strategies just described fail, we make use of more psychological, covert tactics. One of these is fantasy. Fantasy may be viewed as a sort of homeostatic mechanism which serves to return low self-esteem to the level required for functioning. All of the distortion tactics to be discussed in this section partake of this homeostatic quality.

Cognitive-distortion tactics have to do with distortions of information concerning the self (cf. Omark, this volume, Chapter 13). Either the evaluation criteria associated with a particular strategy are manipulated or else the incoming information is distorted. In short, the self cheats. Most of these tactics are essentially the same as the Freudian "defense mechanisms." Anna Freud (1966, pp. 43–44) discusses ten of these mechanisms: regression, repression, reaction formation, isolation, undoing, projection, introjection, turning against the self, reversal, and sublimation. The most important, for present purposes, are rationalization and introjection.

Rationalization is perhaps the most easily recognized: most of us are highly skilled in discounting the import of an encounter or event damaging to our self-esteem. Students fail examinations because the professor is "unfair"; men fail to be promoted as a result of "nepotism"; the comedian fails to get his laughs because the preceding performer did not "warm up" the audience; and so on.

We constantly but subtly distort incoming information in our own favor, thus preserving self-esteem in the event of failure to meet an evaluation criterion associated with a strategy for prestige.

Introjection seems related to what, on the nonhuman primate level, is usually described as a capacity to accept subordinate social rank (Eibl-Eibesfeldt 1970, p. 446). Introjection (or identification) is, in psychoanalytic theory, the process whereby the child internalizes representations of his parents. Faced with an evaluation of another as far higher than ourselves, we, in a sense, become him. Thus, Bettelheim (1943) reported that some prisoners in the concentration camps aped their guards, while Solzhenitsyn (1974) describes a few Soviet political prisoners as imitating the professional criminals who so tyrannized them in the prisons and labor camps.

Reference-group tactics involve shifting from one individual-oriented group-membership strategy to another. Particular groups tend to have particular sets of evaluation criteria associated with them: under certain circumstances individuals may discover that their relative standing is higher in terms of one group's set of evaluation criteria than another's and so shift reference groups to their own advantage. For example, a pious Moslem Hausa in Maradi, Niger Republic, judges himself not in terms of his Western sophistication, but in terms of his knowledge of Islam; this allows him to evaluate himself as of higher rank than the powerful government functionaries, who, in turn, rank themselves as superior on the basis of their European-style knowledge (Barkow, 1975a). The opportunity to make such shifts in reference group and even social identity is especially common during times of rapid social change.

Cognitive-distortion tactics are far less effective in maintaining self-esteem than is the actual achievement of prestige. Were this not so, what would be the advantage of remaining in contact with reality? Further, the excessive use of cognitive-distortion tactics is associated with psychopathology. By distorting incoming information, these defenses make it less likely that the individual will be successful in adapting to his or her social and even physical environment. If one conceives of individuals acting according to plans with goals, then any process which distorts incoming data makes it more difficult to evaluate accurately progress toward a goal. Thus, excessive reliance on cognitive-distortion tactics in the maintenance of self-esteem may drastically interfere with the individual's functioning by damaging the self-correcting nature of ordinary plan behavior. Those who suffer from this kind of failure are often labeled "neurotic" or "psychotic."

Conclusions

This chapter represents a brief effort to link nonhuman primate social ranking to human symbolically mediated, culturally patterned systems of status and prestige. It has used the concept of self-esteem as a bridge, but much has been left out. Self-esteem is only one goal-state among many. Most plans and strategies are in the simultaneous service of a number of goals. For an adequate description of human motivation a larger theory is needed, one involving many goals and a set of rules for switching from one plan to another.

Similarly, to discuss self-esteem, it is necessary to consider the self, but it would be more accurate to speak of the "self-system" (Maddi, 1972): we have many selves, related in part to our group memberships. What we have been calling reference-group tactics involve a decision as to which self to emphasize in a given situation. At the same time, the assertion that prestige is the main strategy we use in seeking self-esteem pays no attention to its other possible sources. Surely, the feeling of competence, at least—what White (1959) has termed "effectance motivation"—enters into self-esteem. If so, to what extent can this nonsocial source of self-esteem replace prestige?

What are the sex differences in self-esteem and prestige striving? Are they consistent with sociobiology's "parental investment theory" (Trivers, 1972), in which sexual dimorphism is considered a product of the extent to which fertilization differentially commits the total reproductive potential of the two sexes? What is the relationship between sex differences in culturally provided prestige-strategies and the boy–girl differences in dominance behavior described by other contributors to this volume?

Finally, what of a developmental theory of social dominance, self-esteem, and prestige? One gets the impression from other contributions in this volume that the agonistic aspect of social ranking in young children is remarkably similar to that in nonhuman primates. As the child matures, he increasingly comes to supplement the primate agonistic hierarchy with one based on culturally constituted, symbolic criteria for the evaluation of relative standing. In many cultures, at least, symbolic criteria and cognitive distortion tactics will almost entirely replace, some time during childhood or adolescence, the old agonistic criteria. But there does seem to be a tendency for agonism to lurk in the background; physical size affects our behavior toward others throughout our lives, and in times of stress hierarchies based on agonistic behavior may reimpose themselves. It may well be, too, that nonverbal communication of dominance and submission, regardless of the criteria of evaluation upon which they are based, contain many transcultural elements derived from our common primate heritage.

A human ethology must grapple with the language and symbol-using nature of our species and not limit itself to mere motor behavior. The present discussion has tried to link primate social dominance and human symbolically mediated behavior. The results suggest that a human ethology must move toward a closer integration with the other human social and behavioral sciences.

Notes

1. See Barkow (1978a) for a discussion of Hallowell as a pioneer in the study of the evolution of human behavior. See Shafton (1976) for a theory of the phylogenesis of the self and its relationship to social hierarchy.
2. See Barkow (1976, 1977, 1978a) for a discussion of selection for the ability to internalize norms and the interactions between this selection pressure and primate attention structure. See Chance and Jolly (1970) for a treatment of the concept of attention structure as a way of understanding primate social organization.

Part V

Dominance and Social Ecology

Chapter 19

Cross-cultural Notes on Status Hierarchies

Daniel G. Freedman

No one has yet described an intact cultural group without some form of hierarchical power arrangement (Osgood, Miron, and Archer, 1963). In general, the status hierarchy is emphasized in patrilineal societies and tends to be relatively muted in matrilineal groups. This makes logical sense in that males are demonstrably more competitive, and given official sanction to do so, they will develop in that direction. Contrariwise, a culture that traces descent through the female tends to be matrilocal and to place more power in female hands (Schneider and Gough, 1961). This dichotomization seems to hold even in polygynous and Moslem societies (there are matrilineal Moslem groups), and matrilineal females are, on average, more independent, more self-serving, and less obviously subservient to males than are women of patrilineal groups.

In a Moslem patriline, such as the Hausa of Nigeria, both religious and familial forces conspire toward the hierarchical arrangement so that between any two Hausa boys the older is automatically more dominant. In a pilot study of the "toughness" hierarchy among Hausa boys the oldest was always deemed toughest, even if it were a matter of only being older by a few days.[1] Doubtless an informal, less age-based hierarchy coexists with the formalized one, but it is subtle and cannot be elicited easily at the verbal level. And so it is with Hausa society as a whole—highly hierarchized, each with his age, family and preestablished caste rather firmly aiding and limiting his claim to status.[2]

In contrast to the Hausa, American Indian groups have been notorious for their distrust of status-seeking individuals, often to the chagrin of white government employees who wished to simplify their duties through agreements with an overall "leader." The so-called nonadherence of the Navajo to treaties in the last century was largely a matter of the scattered leadership they insisted on. A treaty with one subgroup meant little to another subgroup. Even today, the Navajo tend to grumble over the centralization of their leadership, made necessary by the federal government. It is quite simple: no one who actively seeks power is to be trusted. Leaders arise out of example and emulation. If someone is successful at growing corn, he is emulated and to that extent is a leader. If someone knows many verses to a curing chant, he is respected for that accomplishment and his status as a "singer" is considerable. Politicking, handshaking, the ready smile have no place in traditional Navajo society. Such behavior is probably perceived as a disease that some of their numbers have contracted from the whites. (Witherspoon, 1977, is a good general source on Navajo ways.)

Knowing all of this, we should not be surprised that the Navajo are mainly matrilineal, that mothers and grandmothers generally wield considerable power within families, and that women are often as vocal as men in chapter-house political discussions. That is, as in many other matrilines, male power is reduced relative to patrilines, with women taking up the slack. Other correlative events involve greater sexual independence of matrilineal women, less emphasis on ascertaining paternity, a lowered male fear of cuckoldry, easier divorce, and the frequent rearing of young by the mother's mother (Schneider and Gough, 1961; Freedman, 1979).

With regard to competitiveness versus noncompetitiveness in American Indians, a certain amount of myth still persists from Ruth Benedict's having tagged pueblo culture as Apollonian (an emphasis on self-control) in contrast with the Plains Indians who were tagged Dionysian (giving in to one's emotions). While it is true to this day that the pueblo-dwelling Hopi and Zuni are not as prone to get into a bar fight in Winslow, Arizona, as are the Apache-related Navajo, the notion that the Zuni or Hopi are nonaggressive or anticompetitive is not based in fact. Among the Hopi, for example, aggressive gossip is rampant, and their basketball teams are as competitive as that of any other group.

It is, however, apparently true of American Indians in general that in a "fun" game it is a hidden rule that no one should emerge as a clear victor. This is what Benedict emphasized when she reported that in foot races among Zuni boys the apparent winner slows up at the finish line. However, the nonpueblo Navajo are similarly inclined, and the author has witnessed a most interesting clash of culture along these lines.

Two young Caucasian Mormons, on a preaching mission at Navajo Mountain, Utah, joined a group of Navajo young men in "fun" basketball. It soon became apparent that two sets of rules were in operation, the all-out, "play it by the book" rules of the Mormons and the relaxed, laugh-a-lot rules of the Navajo. The more the two Mormon lads scored, the more they were elbowed and bumped—subtly at first, then not so subtly. One grew increasingly irritated and quit, and the other soon followed him to the showers; the game then continued on its sloppy, merry way. The author has participated in and seen a good deal of Hopi and Navajo basketball, and, for both groups, playing well despite being drunk is the mark of success. Clearly, the Appolonian-Dionesian dichotomy is not congruent with these observations.

The Hopi, like the Navajo, are also matrilineal. They are exceptional in that some of their political leaders have been women. The current leader of the conservative faction at Old Oraibi is a woman of considerable influence, although the officially elected political leadership in New Oraibi are and always have been males. Respected elder "sages" tend to be men, and it is fair to say that traditional division proposed by Parsons (1955) holds for the Hopi. That is, for the most part, men wield political power, whereas women tend to be spokespersons for the family.

Even at the level of family, however, female power among the Hopi can be somewhat misleading, as my colleague and I found when trying to arrange the filming of an early morning hair-washing ceremony performed two weeks after a baby's birth. We knew the baby's mother well, obtained her consent, and felt we were all set to film, since she appeared to speak for the family. Her mother-in-law, who was to wash her hair in Yucca suds, also consented, albeit more reluctantly. The real objections, however, were registered by the men in the house, whose sleeping bodies we had to step over at 4 in the morning! They had not been consulted at all and, rightfully, threw a collective tantrum. Our friend reluctantly had to call the whole thing off.

The reverse situation can be seen among the patriarchal Hausa, where a wife appears powerless. Her anger, however, can and often does influence male decisions.

Universality of Male–Female Status Seeking

What all this adds up to is that despite substantial cultural and kinship variation, the male–female dichotomy in status seeking seems much the same everywhere. Our own data, based on films taken in playgrounds over a wide sample of cultures, bear out this generality. For example, 5- through 13-year-old Navajo boys and girls at Crown Point, New Mexico, are largely indistinguishable from their patrilineal counterparts, e.g., the Kikuyu of Tigoni, Kenya. In both locales boys engage in much more rough-and-tumble play, tend to gauge one another on relative toughness and strength, and apparently develop peer hierarchies almost exactly as do white Caucasian males (e.g., Savin-Williams, this volume, Chapter 12). In all three groups, Navajo, Kikuyu, and Caucasian-American, girls tend to play more sedentary and less aggressive games. They tend to be more rhythmic and to sing songs in their games. They tend also to gather near the teachers, and they tend to be first in line on returning to class after recess. In their free drawings ("Draw anything you want") 5- to 7-year-old boys in all seven cultures sampled, were more prone to draw moving machinery or animals (planes, cars, trucks, horses) whereas similarly aged girls are everywhere more prone to draw flowers (Freedman, 1976, 1979). Patrilineality versus matrilineality appears to have precious little effect on peer behavior over these preadolescent years.

Our interim conclusion, then, is that while differing kinship systems tend to correlate with certain differences in status systems, there is, at the same time, considerable continuity over the various systems. At present potential universals are most easily described in terms of sex differences: males are apparently everywhere the more aggressive, the more competitive, and the more status-oriented sex. They are everywhere more attracted to aggressive display by others (Maccoby and Jacklin, 1974), and they apparently exhibit a greater sense of their own omnipotence than do females. They thus tend to engage in more disputes over status and in overt competition. This is not to say that females are disinterested in such disputes, only that they tend to be less direct. Among the Yonomamo, for example, women may incite male relatives to fight when there is a challenge to their subgroup's honor. In Chagnon and Asch's film *Ax Fight* (1976), the village elders are hard put to quiet several angry women, who maintain male tempers at a high pitch with their incessant verbal abuse.

The Harvard six-culture socialization project, now in the summing-up stage (Whiting and Edwards, 1973), is perhaps the best source of objective evidence currently available on cross-cultural consistency of sex differences; but see also D'Andrade's (1966) excellent review. While it is clear that much more cross-cultural documentation is needed, if only to enrich our understanding of the limits and malleability of the human tendency toward social hierarchization, an objective inspection of these currently available materials would strongly suggest the universality of the sex role differences just discussed. This in turn strongly suggests that sex differences are best considered within a non-dualistic (neither hereditarian nor environmentalist) biosocial framework.

Notes

1. J. H. Barkow, personal communication, 1969.
2. We had hoped to do a comparative study with Ibo and Hausa children, but our visit to Nigeria in 1969 coincided with Ibo-Hausa warfare. According to LeVine (1966), individual achievement and ability is the primary path to status among the Ibo. He accounts for their great economic success in Nigeria largely on this "individualistic" tradition. Their success led, in turn, to Hausa animosity and, finally, to extremely bloody anti-Ibo "pogroms." Here is a lesson, too, for Westerners interested in training tribal groups for "success" (e.g., McClelland, 1953).

Chapter 20

Playground as Laboratory: Naturalistic Studies of Appeasement, Altruism, and the Omega Child

Harvey J. Ginsburg

> Sit down before a fact as a little child. Be prepared to give up every preconceived notion; follow humbly wherever nature leads, or you will learn nothing.
>
> Thomas Huxley

A consideration of the sage advice given by Huxley seems an appropriate way to begin a discourse on naturalistic child research. In one sense, the ability to drift with the stream of behavior, rather than to struggle against it, is the most important skill for a naturalist to learn.

When it was initially decided to use the playground as a laboratory in which to study behavioral interactions between children, there was no particular investigation in mind, no strategy for obtaining data, nor any set procedure to follow. Indeed, the first several days of observation were spent simply looking at, not for, behavior.

During the initial period of time, one pattern of interaction began to stand apart from the rest. When the children being observed engaged in fights, there seemed to be a common element across the skirmishes by which one child (the child being attacked) indicated to the other that the fighting should stop. Some signal seemed to be given which urged, "All right, enough is enough, you win!" As this became the focal point of our observation, the behavior in question became all the more intriguing. Although it was still unclear what that sense of commonality was during the fights, now there was something to look for as well as at.

Appeasement Gestures in a Natural Context

Numerous ethological studies of infrahuman organisms have provided evidence that species-specific behavioral displays exist that effectively inhibit further attack during episodes of intraspecific aggressive activity. For example, wolves have been noted to lie passively on their backs (hind legs wide apart, abdomen exposed) as an indicator of defeat by a rival member of the pack (Schenkel, 1967). When such behavior occurs, the agonistic interaction is halted; attack by the dominant animal ceases.

Eibl-Eibesfeldt (1970, 1972) observed that among nonhuman organisms intraspecific aggression rarely leads to the actual death of one of the combatants and suggests that such "appeasement displays" serve a highly adaptive function. For instance, if indeed members of same species regularly killed their rivals in battle, the preservation of the species obviously would be jeopardized. Eibl-Eibesfeldt noted that across many human cultures types of ritualized greeting systems are apparent (bowing, kneeling, prostration, and so forth) when subordinates defer to higher ranking (more powerful) members of the social order. Such appeasement postures suggest that Darwin's (1872) notion of "making oneself smaller" (via body posture) is perhaps an effective inhibitor of human intraspecific aggression.

The initial study attempted, within this framework, to classify the types of behaviors that most effectively inhibited acts of aggression between children during unstructured playground activity. More specifically, the intent was to test the hypothesis that displays of human submission involving diminution of body stature have a higher probability of terminating attack by a aggressor than other behaviors exhibited by a child under attack during an aggressive encounter.

Methods

Over a six-week period of time 16 hours of videotape recordings were made of the behavior of the children (third- through fifth-grade male elementary children) during periods of unstructured (and, for the most part, unsupervised) playground activity. The number of children on the grounds at any one time remained highly constant, between 28 and 34. The recording was made from a Volkswagen van parked on a hillside street which provided a panorama of the playground.

The videotape unit was activated whenever the initiation of agonistic interaction between children was observed.[1] Thus, while the nature of the behavior to be videotaped was preselected (aggression), it was not possible for the observer to predict in advance the outcome of the fight or the type of behaviors which preceded the cessation of the aggressive encounter.

There were 72 instances of aggressive interaction evident in the videotape records. Aggressive behavior was defined in terms of five specific modes of attack: *hitting*, where one child beats another with clenched fists; *jumping upon*, where the aggressor attempts to force another child to the ground by wrapping his arms around the trunk of the other after gaining momentum from a running start; *pushing*, where the aggressor causes a child to lose balance by rapidly extending his arms and with an open palm making contact with the body of the other child; *pulling*, where the aggressor grasps clothing or body parts of another child and forces the child to move in the direction of the pull; and *kicking*, where the aggressor first moves his leg back and then forward so the foot makes contact with a body part of the child under attack. As might be expected, these behaviors usually appeared in combination with one another.

We kept the camera fixed upon the child under attack for a two-minute period after the aggressive interaction halted. "Cessation of aggression" was thus defined as no physical contact by the aggressor upon the child under attack during this period.

After collecting the field data, analysis of the videotape record focused upon the behavior of the child under attack which just preceded the cessation of aggression. Although descriptive categorization of behavior presented little problem, deciding upon an objective criterion of "making oneself smaller" presented some difficulty. It was finally decided to obtain a baseline measure of the child's stature by measuring the child under attack (from ground to top of head) at the beginning of the fight sequence. This was accomplished by superimposing a transparent ruler (in centimeters) over a stop-action picture of the child. Using an identical procedure, we made a comparison measure of the child under attack during termination of the aggressive encounter (just prior to the cessation of attack by the aggressor). This allowed an objective determination of change in body stature of the child under attack during the terminal sequence of an agonistic episode.

The 72 samples of agonistic activity were thus analyzed in terms of the ongoing behavioral interaction which occurred between children during the course of an aggressive encounter. This analysis focused upon behaviors exhibited by the child under attack that preceded the cessation of agonistic activity by the aggressor.

Results

A sign test (Siegel, 1956) was used to determine whether a general decrease in body stature on the part of the child under attack preceded the cessation of aggression (compared to baseline measures of stature obtained during initial sequences of the fight). The data indicate that such a change in body posture does occur with high probability just prior to termination of attack by the aggressor ($p <$ .001, onetailed, $z = 5.07$). There were 58 instances of diminution of body posture prior to cessation of aggression out of the 72 videotaped episodes.

Further classification of behavior that preceded termination of aggression was performed by providing a descriptive analysis and a frequency count of the events which occurred prior to cessation of attack by the aggressor. These results are shown in Table 1. As can be seen from this table, threat display, backing off, and face-to-face verbal interaction were the categories not consistently associated with decreased body stature. In addition, it was also observed that eye aversion frequently accompanied a number of the appeasement displays listed in Table 1. Lack of eye contact on the part of the child under attack usually occurred in combination with diminution of body stature. Of the 58 episodes, 56 involved eye aversion.

One of the more interesting observed behaviors was shoe tying. This apparently symbolic appeasement gesture (in which the child being aggressed upon assumed a no direct eye contact, head down, kneeling posture) effectively inhibited further acts of aggression. Eight instances of shoe tying (all of which preceded the cessation of aggression) were observed during agonistic interactions. In contrast, a shoe-tying baseline rate of zero was observed when no aggressive behavior between children was occurring on the playground. The observation of one child who became intent upon "tying" his loafer (a shoe without laces) is indicative of the fact that shoe tying during a fight is more a case of appeasement than a problem of loose laces. It also conceivably demonstrates to peers that, "if only my shoes stayed tied, I really could win this fight!" Thus, not only does this behavior suggest submission to the opponent, it additionally serves as a face-saving device for onlookers.

It appears that Darwin's notion of "making oneself smaller" (via body posture) is an effective inhibitor of human intraspecific aggression. Contrary to the speculation of Lorenz (1966), natural inhibitors of aggressivity may have evolved in parallel with human aggressive behavior. Such inhibitors may be largely ineffective, however, since man's ability to kill rapidly with weapons does not provide sufficient time for the target of an attack to offer a gesture of appeasement. However, in hand-to-hand agonistic behavior displays by children one can observe a number of behaviors (kneeling, bowing, waxy passivity, and so on) which effectively diminish the probability of continued aggression.

Brannigan and Humphries (1972) have noted the great value of being able to classify the function of behavioral signals, not in terms of their supposed emotional causation but in terms of the response relationships which under specified conditions typically follow that signal. They suggest that a behavioral signal has an appeasement function if it reduces the probability or intensity of attack in a given situation. This study demonstrates that behaviors found to precede the reduced probability of continued aggression typically involved a diminution of body stature in combination with lack of direct eye contact. Thus, to the extent that children's displays of aggressive behavior in a playground situation represent at least one mode of human agonistic interaction, these results suggest that a number of human behaviors do exist which serve an appeasement function.

Altruistic Behaviors during Episodes of Social Conflict

In our study of appeasement displays a relationship had been noted between the intervention in a fight by other children and the assumption of a submissive posture on the part of the child under attack. In all but one instance (see Table 1) aid giving was observed while the child displayed a behavior resulting in di-

Table 1. A Description of Nonverbal Inhibitors of Aggressive Behavior

Behavior	*Description*	*Frequency Prior To Cessation of Aggression*	*% of Time Accompanied by Diminished Body Posture*	*% of Time Accompanied by Eye Aversion*
Head bow	Child stands upright but lowers head to a position nearly horizontal to ground.	4	100	100
Shoulder slump	Child stands upright, lowers shoulders, and bends forward.	6	100	100
Head bow/ shoulder slump	The co-occurrence of head bow and shoulder slump.	9	100	100
Kneel	Child positions himself with one or both hands on ground; head bow always accompanies kneel.	18	100	94
Shoe tying	Child kneels on ground, averts eyes, and ties shoelaces.	8	100	100
Lying motionless on ground	Child lies with back on the ground and limbs extended.	2	100	50
Waxy manipulation of limbs	Child allows aggressor to manipulate limbs without any muscular resistance.	2	50	100

Table 1 (Continued).

Behavior	*Description*	*Frequency Prior To Cessation of Aggression*	*% of Time Accompanied by Diminished Body Posture*	*% of Time Accompanied by Eye Aversion*
Threat display	Counterattack stops antagonism by aggressor; fists are clenched; child under attack becomes aggressor.	3	0	0
Back pedal	Child extends open palms, slowly backs away from aggressor.	3	0	0
Face off	Children place hands on hips while face-to-face verbal interaction occurs.	2	0	0
Aggressor distracted by ongoing events	Aggression ceases when another event in playground catches attention of aggressor.	4	50	50
Intervention by adult	During adult intervention child under attack has assumed a diminished body posture and lack of eye contact toward aggressor.	2	100	100
Intervention by other children	Other children intervene, typically on occasions when child under attack assumes no eye contact, diminished body posture toward aggressor.	7	86	86

minished body stature. With this suggestive evidence at hand, it was decided to embark upon a more systematic study of altruistic behavior in children. Most definitions of altruism (Heider, 1958; Leeds, 1963) have typically suggested three criteria to be met if a given behavior is to be considered altruistic: (1) the behavior must be emitted voluntarily, (2) the recipient in some way gains from the action of his benefactor, and (3) some degree of self-sacrifice for the benefit of another must occur.

Although many experimental variables associated with increases or decreases in self-sacrificial helping behavior have been identified (Krebs, 1970), little systematic evidence has been gathered which outlines the course of a particular altruistic act as a naturally occurring phenomenon. Most naturalistic accounts of altruism have involved anecdotal reports of isolated incidents, after-the-fact interviews of individuals who have aided during crisis situations (Form and Nosow, 1958), and the highly debated accounts of infrahuman helping behavior (Hebb, 1971; Krebs, 1971). While a number of studies have been quasi-naturalistic in design (usually involving some staged accident during which time the behavior of naive observers is monitored), unstaged, noncontrived naturalistic documentation of altruistic phenomena is lacking.

The reasons for the dearth of unobtrusive, naturalistic studies of helping behavior are obvious. First, for most circumstances, it is difficult to predict beforehand when a situation necessitating aid giving will occur. Assuming such an event could be predicted in advance, one would presumably be hard pressed in an ethical sense to allow an aid-giving situation to run its course without becoming the agent responsible for intervention. Also, in order to study altruism in a natural setting, the behavior must occur consistently and with sufficient frequency; it should be easily quantifiable and meet the definitional requirements of an altruistic act. Fortunately, the activity that was chosen for study circumvented these problems. During an aggressive interaction between two children the intervention of another child on behalf of the individual under attack may frequently be noted.

Methods

As in the first study, the subjects of this study were third- through fifth-grade male elementary school children ($N = 102$). The children were unobrusively observed over a nine-week period during their respective hourly recess periods (N varied between 23 and

34 per period; the playground activity was unstructured). Videotape monitoring of activity was similar to that of our initial study of appeasement displays. The videorecorder was activated whenever the initial occurrence of a fight between two children was observed. Thus, whether another child eventually intervened could not be determined beforehand.

The analysis of altruism focused upon (1) the physical size characteristics of the aggressor, benefactor and recipient, (2) the mode of intervention, (3) the consequences of aid giving for the recipient as well as benefactor, and (4) the behavior of the child under attack which immediately preceded intervention by the benefactor. This was measured by occluding from view a portion of the playback monitor so that a naive observer could see the intervening child but could not see the two children engaged in the fight. The observer was asked to indicate at what point the aid-giver initiated helping behavior. A stop-action frame was provided of the point at which aid giving was initiated, and a second naive observer was independently asked to state what activity the child under attack was performing (kneeling, striking back, and so forth).

Results

The videotape record extending across approximately 45 hours of observation provided 36 instances of this form of aid-giving behavior. "Aid giving" was defined as intervention during a fight on behalf of the child being attacked. On a single occasion 22 children intervened; 4 children intervened on two separate occasions, while 2 children intervened on three separate occasions.

We initially hypothesized that the aggressor in a fight would be physically larger than the child under attack and that the intervening child would be larger than the primary aggressor. Only the former portion of this hypothesis was substantiated. Nearly 80 percent of the aggressors were larger than the child under attack ($p < .0008$, one-tailed, binomial test). In contrast, the aid-giving child was larger than the primary aggressor in less than 40 percent of the episodes ($p < .12$, two-tailed, binomial test).

A consistent mode of intervention was also observed. There were 26 aid-giving interactions that involved an "attack from the rear." The benefactor typically jumped upon the back of the aggressor ($p < .012$, two-tailed, binomial test). Of the remaining episodes, 5 involved direct frontal confrontation, while 5 simply involved the intervening child stepping between the combatants.

Further, a consistent form of interaction between the aggressor and child under attack was exhibited just prior to the intervention of the aid-giver. In 27 instances intervention was preceded by a failure of the aggressor to heed an appeasement signal exhibited by the child under attack.

It should be noted that the appeasement signal–inhibition of aggressive phenomena does not necessarily occur in rapid fashion. A child exhibiting a submissive gesture may do so for a somewhat extended period of time before the attack subsides. All that is suggested is that such behaviors have a high probability of occurrence just prior to the cessation of aggressive activity. If a child continues to exhibit an appeasement display and the fighting persists, it is during this time that aid giving is likely to occur. That is, if the child being aggressed upon exhibited a diminution of body posture (kneeling, shoulder slumping, head bow, and the like) and the aggressor continued his attack, then intervention was most likely to occur ($p < .005$, two-tailed, binomial test).[2]

Finally, while the child under attack invariably gained from the intervention of the benefactor with 29 out of 36 instances eventuating in successful escape following intervention ($p = .0002$, one-tailed, binomial test), the intervention frequently involved a significant cost to the benefactor. On 25 occasions aggression was redirected toward the third child by the primary aggressor ($p < .03$, two-tailed, binomial test). It is of interest to note that on just 7 of those 25 occasions did the initial child under attack return to give aid to his aid-giver. For the remaining 18 episodes, when the intervening child and the primary aggressor fought, the recipient of aid fled the scene.

The results summarized suggest that the behavior under consideration does qualify as altruistic. It meets the definitional criteria: the intervention occurs voluntarily, the recipient gains from the behavior (a successful escape is usually assured), and the behavior involves some degree of self-sacrifice (frequently the intervening child himself becomes the target of attack).

The data of this study call into question an issue that has been referred to as the "altruistic paradox" (Cohen, 1972). It has been argued that a perplexing paradox of human nature is that people may at one and the same time be altruistic and caring and also be aggressive and cruel. This paradox apparently stems from a penchant for viewing prosocial (altruistic) and asocial (aggressive) behaviors as mutually exclusive events. In terms of the results of this study, viewing many behaviors in this dichotomous manner may be unwarranted. A variety of expressions of aid giving may involve activities that are simultaneously altruistic and aggressive. Indeed, the data of this study suggest that an act of aggression may be under certain conditions the optimal vehicle for an altruistic, aid-giving motive.

The data of this study are also relevant to current sociobiological theory (Wilson, 1975). A frequently stated sociobiological argument is that reciprocity played an important role in the evolution of human altruism (Trivers, 1971). In terms of immediate reciprocation on the playground, however, very few instances were actually observed. Sociobiologists would no doubt argue that in viewing only playground behavior a very limited spectrum of potential reciprocal aid giving has been tapped. For instance, the recipient of physical aid may offer assistance to his benefactor in some scholastic endeavor (mathematics, grammer, and so forth). While this certainly appears to be a reasonable possibility, an empirical demonstration has yet to be provided by sociobiological theorists.

We are currently in the midst of a study concerning the issue of reciprocal altruism on the playground.[3] Our early data appear to tentatively suggest that dominance more than simple reciprocation is a key factor in this form of aid giving. Individuals at the upper strata of the play hierarchy continually and consistently provide agonistic aid to lower echelon children. Once again, little reciprocity has been noted in this more elaborated study of agonistic intervention.

The Omega Child in a Competitive Hierarchy

The following school year children on the same playground were observed in hopes of obtaining more information about the interactions of children during unstructured play. Much to our consternation, the rules had changed. Instead of free-for-all activity, teachers were now supervising the children in organized team sports. Thus, the stream of behavior had shifted course. A decision was made to follow it. Again, the first few days of observation were spent without any real focus upon specific patterns of play. As in the first study, however, a rather compelling form of interaction began to emerge. It was noted that the last youngster chosen for the kickball team (the omega child in a competitive hierarchy) seemed to interact in an altogether different manner from other children in the kickball hierarchy.

One of the more intriguing phenomenon among species

which organize themselves into social hierarchies is how the lowest ranking (omega) organism relates to the rest of the group. We commonly expect that the omega member receives the brunt of negative social interaction from all higher ranking conspecifics. He is viewed as the individual constantly being picked upon, taunted, and physically intimidated by others in the group.

Studies of infrahuman organisms and children, however, suggest this picture may be somewhat distorted. Schjelderup-Ebbe (1922) observed that in the pecking hierarchy of chickens the lowest ranking hen was generally left alone by higher ranking ones. Similarly, Imanishi (1957) noted that in the Japanese monkey (*Macaca fuscata*) the omega member generally occupied a position at the periphery of the group and rarely engaged in any form of social interaction with other members of the troop. Savin-Williams (1975) found that the omega youth in a group of boys at summer camp was least involved in dominance interactions and rarely displayed agonistic behavior toward others. Similarly, data obtained by Strayer and Strayer (1975) indicated that omega children in a preschool hierarchy were both the most infrequent initiators and targets of aggression. Such repeated evidence about the unique role of lowest ranking members in the social hierarchy provided an impetus for the present study, which was designed to examine more general aspects of social relationship between omega children and other members of the play group. More specifically, we focused upon a comparison between social activities of omega children and higher ranking members during organized athletic periods at school. Our specific measures included (1) average interpersonal distance during a game and the location of omega children in relationship to the entire group, (2) the amount of physical contact with others, and (3) frequency of episodes of aggression during the play period.

Method

The children were 156 third- to sixth-grade elementary school males observed during structured and supervised playground activity. The boys were organized into football teams (approximately 12 per team) during each play period. The time of the play periods averaged 40 minutes. The selection of team members was left up to the children. The general procedure involved each team leader selecting an initial child for his team. These two then selected a third team member, and so on, until all children were placed on a team. An omega child was, therefore, defined as the one chosen last by the two teams on the field. A total of 12 omega children were studied (2 each for six play periods). The teams were reorganized at the beginning of every week. Even though reorganization occurred, children chosen last remained consistent throughout the course of observation. Continuous videorecordings of all events that occurred during each play period were provided from our concealed VW van. The focal length of the camera was kept constant throughout the study so equal comparison data could be gathered. The midarea of the play group (defined as the place of greatest congregation of children) was used as a center point for the viewfinder of the camera. Approximately 24 hours of team activity was videotaped (six 40-minute observations for six different play periods).

A stop-action videotechnique was used to analyze the tapes. At every 40 counts of the digital counter built into the playback unit, action was frozen. A transparent ruler was placed over the videomonitor and the distance was measured (in millimeters) between the child under consideration and the group member nearest him. These proximity measurements were obtained for omega children and a comparison child selected at random prior to stop-action analysis. Additionally, each child under consideration was scored in terms of whether any form of physical contact occurred with other group members in the stop-action frame. If physical contact occurred, the proximity score was zero.

A quantifiable index of the location of the omega child (or randomly selected comparison child) in relation to the rest of the group was obtained by superimposing a transparency of two concentric circles (12 and 27 cm in diameter) over the videomonitor. The center point of the transparency matched the center point of group aggregation for each stop-action frame. A frequency count was made of the number of times the child under consideration was located in the inner or peripheral circle.

Agonistic behavior was also measured during the stop-action frame. This was defined as the child hitting, kicking, pushing, pulling, or jumping upon another play-group member, or conversely, being the recipient of such activity.

For each child 25 stop-action segments were obtained. Thus, a total of 300 segments were measured for omega children and an equal number for the play group members randomly selected for comparison. It should be noted that a blind procedure was used for all measurements (scoring was performed by a naive observer unaware of the type of child under consideration).

Results

Table 2 summarizes the results of our observation. Significant differences were obtained when measures of interpersonal distance were compared between omega children and randomly selected comparison members of the play group. The average distance observed on the videomonitor between omega children and the child in nearest proximity was 24.75 mm; in contrast, the average for nonomega comparison members was 11.67 mm ($p < .001$, one-tailed; Mann-Whitney U-test, $U = 15.5$). Given the spatial distance maintained by omega children and their general lack of

Table 2. A Comparison of Social Scores between Omega and Nonomega Children

	Child	Proximity Score (Avg. in mm)	Periphery Score	Physical Contact with Another Child	Agonistic Interaction
Omega children	1	27	22	0	0
	2	24	18	0	0
	3	25	18	0	0
	4	31	21	1	0
	5	38	23	0	0
	6	22	19	0	0
	7	17	11	2	0
	8	34	23	0	1
	9	20	18	1	0
	10	26	17	0	0
	11	15	12	2	0
	12	17	14	1	0
Nonomega children	1	8	12	5	1
	2	13	14	4	1
	3	6	1	11	3
	4	5	4	9	2
	5	11	6	7	0
	6	23	10	3	0
	7	9	9	7	2
	8	18	13	2	0
	9	7	3	11	1
	10	5	11	6	2
	11	26	18	0	0
	12	9	6	8	3

physical contact, it was not surprising to find that only one of the 300 samples obtained included an agonistic interaction. In contrast, 15 of the 300 segments for nonomega children involved some form of agonism. In terms of the general location of omega children relative to the entire play group, our data showed they were typically represented at the periphery of the group (defined by position in the outer circle of the location transparency). Omega children were observed in the periphery on 216 occasions, compared to 107 instances for nonomega comparison children ($p < .001$m one-tailed; Mann-Whitney U-test, $U = 17.5$).

Omega children also engaged in significantly fewer physical contacts than nonomega children. Of the 300 total stop-action frames analyzed, 7 instances of physical contact were noted for omega children. In marked contrast nonomega comparison children were observed in contact with other group members in 73 of the stop-action frames.

The results of this study suggest that the social behavior of omega children differs in many ways from that of a random group of grade-school children. Omegas spend most of their time at the group periphery, maintain a significantly greater interpersonal distance from other children, and are involved in physical interaction less frequently than other members of the group. Rather than being the target of aggression by higher ranking children, however, omega children are generally left to themselves. They rarely engage in agonistic behavior with members of the group. This finding is highly compatible with previous investigations of social hierarchies in groups of children (Savin-Williams, 1975; Strayer and Strayer, 1975).

A striking form of nonquantifiable interaction was also observed. It frequently appeared that the stream of ongoing activity completely bypassed omega children. When these lowest ranking members of the hierarchy did move into closer proximity with the group as a whole, they often remained motionless while the entire entourage of children ran past. On such occasions the physical presence of the omega child was typically ignored by other playgroup members.

Discussion

As noted by Blurton Jones (1972a), unless we can state beforehand the norms from which variants depart, the study of behavioral variations will not prove very informative. Our first study of appeasement–submission offers a statement of some norms with respect to the types of behaviors which precede the cessation of aggression in male elementary school children. In the context of studying the norms of aggression and inhibitors of aggression it would be of great interest to study children who consistently fail to exhibit appropriate inhibitory gestures; secondly, it may be highly meaningful to study children who, in the role of the aggressor, consistently fail to heed the type of signals described by this study.

The study of altruism evolved from the first set of observations on inhibitors of aggression. A persistent issue about aid giving concerns the adaptive context in which to place various forms of helping phenomena. It may be that longitudinal naturalistic research might be of some value in providing that context. The observation that the recipient of aid typically fled the immediate scene without repaying his benefactor does not necessarily preclude the possibility of benefactor–recipient role reversal at a later date. Also, it would be interesting to determine if the notion of some relationship between kinship genetics and altruism is viable. If brothers or cousins were frequent benefactors, future observations of this nature would certainly strengthen the tentative tie between kinship and reciprocal altruism (Trivers, 1971).

The stream of behavior took us far afield in our study of omega children, literally to the periphery of the play group. One cannot help but wonder the extent to which being an omega child on the playground influences future personality development. Smith (1974) has suggested that although the long-term effects of being a child who does not join in rough social play are unknown, being an outsider (and presumably timid) may continue to be a handicap at school. He suggests that the pacing of social interaction may be helpful, for example, placing such children in a younger play group. However, in doing so, by virtue of the nature of dominance hierarchies, the burden of "omeganess" would be shifted to a different child, so this seems to offer no broad solution. Of course, in offering a solution, one presumes there is a problem. There may not be a problem; it is not known to what extent "omeganess" generalizes to other situations and social contexts. Indeed, one would probably be very hard pressed to find a total omega person, an individual at the bottom rung for all social endeavor.

In conclusion, we hope that this series of studies has not only provided a good measure of quantitative data, but that it will also serve as a source of stimulation for future investigations of child–child interaction. Additionally, we hope that some of the flavor of naturalistic research with children has been captured. Following where nature leads is not only profitable as a format for data collection, it is a genuinely pleasurable experience as well.

Notes

1. At the initiation of a fight the observer fixed the telephoto lens and kept the focus constant throughout the measurement period. Thus the size of the image was consistent during the taping and provided equivalent samples for comparison. On a few occasions, however, the children moved substantially during the course of the fight (to the foreground or background of the play area.) When this occurred, the baseline measure of body stature was the standing size of the child after the fight, in the area where fighting had terminated.
2. A friend of mine, William Braud, made the comment that perhaps the best intervention would be none at all since if appeasement gestures are all that effective, the fight should eventually stop of its own accord. But at least intervention speeds the process.
3. This research is currently being supported by NIMH grant 1 RO3 MH30406-01 (Biosocial aspects of altruistic behavior in children).

Acknowledgments

The author wishes to thank the students who made this research possible: Vicki Pollman, Mitzi Wauston, and Michael Easley. Well over one hundred hours of observation went into this effort. They were responsible for most of the data collection; the ‘we’ referred to is a collective one.

Chapter 21

Social Organization among Preschool Peers: Dominance, Attention, and Sociometric Correlates

Brian E. Vaughn
Everett Waters

The concept of "dominance" and questions as to the role of dominance hierarchies in social organization have been debated in ethological and psychological literature since the concepts were first introduced. The extensive literature on dominance relationships among nonhuman primates and the resurgence of interest in comparative analysis of human social behavior (e.g., Bowlby, 1969; Eibl-Eibesfeldt, 1970; Hinde, 1974) have combined to introduce the concept into the literature on human behavior. The translation was initially quite direct, and the term "pecking order" has even been used in a few descriptions of human social groups (Esser, 1968). Unfortunately (perhaps), even the most imaginative observers (e.g., Morris, 1967; Tiger and Fox, 1971) have been hard put to make a convincing case that the social organization of human adults can be outlined from an analysis of aggressive interactions. It is too easy to think of examples of the little man who wields great influence, to think of human social organization without also thinking of the importance of other determinants of status (property, family, intelligence, the ability to make coali-

---

The research reported in this chapter was supported in part by a Program Project Grant (NICHHD 5 POI HD 05027) to the Institute of Child Development. Authorship was determined by the flip of a presumably fair coin provided by the first author.

tions, and so on). The social organization of adult groups is replete with concepts and dimensions that cannot be appreciated within the framework of familiar dominance theories. This has also proved to be the case in recent analyses of social organization in nonhuman primates (see Crook and Goss-Custard, 1972; Wilson, 1975).

The study of dominance relations among adults has never progressed so far as to suggest important developmental questions. Instead, the concept of dominance has been applied to the analysis of social organization among children, on the (unwarranted) assumption that earlier must be simpler. It seems plausible, however, that if we could determine the *origins* of human social organization (assuming these are to be found in the childhood play group), we might be able to make a new and more productive approach to adult behavior. The question is, What kinds of constructs will best capture social organization among young children?

At present, the bulk of the behavioral data on dominance relationships among humans is based on observation of small groups whose members average 7 years of age or less. Abramovitch (1976), McGrew (1972), and Strayer and Strayer (1976), among others, have reported observations of numerous pairwise aggressive interactions between young children and have been able to determine whether or not linear hierarchies exist in such groups by analyzing tables reflecting the outcomes of such interactions (cf. Sade, 1967). In most cases it has been possible to derive hierarchies of varying degrees of linearity from such data (see Cronin, 1976; Missakian, 1976; Slukin, 1975). Most of these studies have drawn heavily from primate research (e.g., Bernstein, 1970; Sade, 1967, 1972), and in several cases (most notably Missakian, 1976) studies have pointed out the similarities between dominance interactions observed in nonhuman adults and those seen in human juveniles. Even with these encouraging findings, a number of investigators have not been entirely satisfied with the dominance concept as a model of social organization in groups of young children.[1]

Dissatisfaction with the unidimensional conceptualization of social organization stems primarily from the fact that "dominance" rankings based on the outcomes of aggressive encounters have not proved very revealing with respect to other aspects of behavior. They also tell us little about the roles individuals play vis-à-vis the group as a whole.

An alternative approach to the description of social organization has been suggested by Chance (1967) and elaborated by Chance and Jolly (1970). The proposal is that the *structure of attention* within a group, rather than aggressive interactions, priority access to food, objects, mates, and so forth, be used to draw the outlines of social organization. Chance (1967) argued that even in the absence of aggressive contact "dominant" animals are often kept visual track of by subordinates; it is the direction and amount (distribution) of attention that should be recorded. Chance and Jolly (1970) cite numerous examples of the use of attention as an index of dominance and have proposed a classification of primate societies based on patterns of attention characteristic of their social groups. More recently, Chance and Larson (1976) have edited a volume of papers specifically dealing with attention structures in monkeys, apes, and humans (children). These papers provide ample evidence that some individuals receive much more visual attention than others and that these individuals fill social roles conveniently labeled "dominant."

There is some evidence that human adults in groups distribute attention toward "dominant" or high-status individuals and that lower ranking individuals are more likely to look at dominants than vice versa (Exline, 1971; Mehrabian, 1968), though the assessment of dominance here was not based on agonistic interactions. In addition, Abramovitch (1976) has reported that dominant children (assessed by outcomes of object struggles) are more often looked at by lower ranking peers than would be expected if looking behavior were essentially random. Abramovitch found that her class could be divided into what she called "transitive thirds" with respect to dominance. Children from the top third of the class always won object struggles with children in the two lower thirds. Within the thirds, however, there was not a linear arrangement.

Hold (1975) has assessed attention structure, aggressive encounters, and a variety of other interactive behaviors in several classes of preschool children (ages 4 to 7). The correlates of the designation "center of attention" included receiving a variety of prosocial initiations from peers (such as offers of toys, being imitated) and indications of social poise or competence (being obeyed, breaking up fights, initiating constructive interaction). On the basis of these data she concludes that neither the number of aggressive encounters nor their outcomes should be the sole criterion in assessing dominance among preschool children. Abramovitch (1976) reached a similar conclusion, noting several instances in which a subject who dominated a particular peer might, nonetheless, have won fewer total fights than the lower ranking peer.

With respect to children of nursery school age, Chance's "attention structure" model has fared well as a descriptive model. It is not yet clear *why* high-ranking children in an attention structure receive more visual attention. Neither is it clear that it is "dominance" in the usual sense that is being assessed by attention-structure measures. Researchers at present lack data that go beneath the structure of social groups and reflect the dynamics maintaining social structure. In the preschool peer group these structures seem to be largely the dynamics of *social play*. If play is practice (Bruner, Jolly, and Sylva, 1976), then competence, especially in the social realm, should be the long-term goal (see Anderson and Messick, 1974). To the extent that dominance relationships, as they have been conceptualized in recent research—i.e., as initiated agonism (Strayer and Strayer, 1976) or as the outcomes of agonistic interaction (Abramovitch, 1976)—and the dominance hierarchy are indices of social competence, they will serve to reflect the dynamics supporting social structure. However, the present authors suspect that, just as in adult social groups, there will be other dimensions either equally or perhaps more important than dominance relationships maintaining the social structure of children's groups.

The data reported here were collected in a study designed to determine the correlates of attention structures in a preschool play group. The authors were interested in the relationship between attention structure, dominance, and each child's understanding of the social organization within the class. Thus behavioral and nonbehavioral measures have been combined in an effort to uncover the roots of attention structures.

Observations and Analysis

Subjects

An entire class in a laboratory nursery school at a large midwestern university was observed in this study. The class was in session from late September to early June and observations were made throughout the session. Originally the class consisted of 19 children (8 girls and 11 boys) ranging in age from 4.8 to 5.2 years. Two children retired from the class at the Christmas break and were replaced by two other children after the break; the sex ratio was unchanged. During the spring quarter one of the girls missed 75 percent of the class sessions and 80 percent of the behavioral observations. Her data were not included in the analyses presented here. All children, with the exception noted, were present in class for more than half of the class sessions and/or the observational periods. The socioeconomic status of the families from which the subjects were drawn was heterogeneous.

Setting

Observations were made in two settings at the laboratory school of the Institute of Child Development—inside the classroom and outside on the nursery school playground. The classroom was organized into a variety of play areas (block area, painting easel corner, reading area, dramatic play corner, and so on). The outside play area contained large-muscle play equipment (swings, slide, jungle gym) as well as a large sandbox and a lawn area over 150 feet square. Rough-and-tumble play (see Blurton Jones, 1967; McGrew, 1972) was more easily maintained on the playground than indoors. A variety of both riding toys and sand toys was available at the teacher's discretion for use on the playground. There were always more toys than children indoors, though not always more preferred toys.

Children were observed during the "free-play" time, which varied from 45 minutes to 2 hours depending on the weather, the inclination of the teacher, and the proposed schedule of nursery school events (field trips and the like). Whether the children played outside or inside during free play was almost completely dependent on the weather. During wet or excessively cold weather the children played indoors (especially from December to late March).

The Behavioral Data

AGGRESSION AND DOMINANCE. During the last ten weeks of the nursery school term, over 100 30-second observations of the free-play behavior of each child were made. Among the data collected were the occurrences, participants, and outcomes of aggressive or agonistic encounters. As with the attention data, the order of observation was random, with the restriction that no child was observed twice by one observer before all children present that day had been observed once. If a child was not in view of the observer, the next child on the list was observed, and the missed child was observed during the next interval in which he or she was visible.

The length of the observation period was tailored to the categories of behavior being observed; that is, it was long enough to allow both reliable frequency estimates of the behaviors and assessment of the outcomes of a variety of interpersonal interactions. To determine outcomes of aggressive interactions, it was occasionally necessary to watch longer than 30 seconds. In that event the only information recorded from the excess time was the outcome of the interaction.

Since these observations were made during the spring term, the bulk of the data are from free play on the outdoor playground (about 75 percent). Four observers collected the dominance data. All had participated in the development and refinement of the categories of interactive behavior and were in agreement as to category definitions. At the outset of these observations mean interobserver agreement as to the presence or absence of a given category of aggressive behavior was over 95 percent. Most of these agreements, however, were agreements that no such interactions had occurred. (There were 222 aggressive interactions, as defined below, in over 2,100 observations of 30-second duration.) Approximately 25 percent of the observations were made by observers working in pairs. This "conferencing" of observations by two raters was done to insure that individual raters did not "drift" into idiosyncratic definitions of the behaviors during the ten weeks of observation. Odd–even correlations were used to determine the adequacy of the number of observations for each behavior scored.

Two different classes of interactions were scored with respect to physical domination of peers. The first class of dominance interactions included two categories labeled hazing (H) and specific hostility (SH).[2] Hazing consisted of unprovoked taunting or teasing which seemed to have the intention of "getting a rise" out of the recipient. This teasing could be either verbal (e.g., calling another child a derogatory name) or physical (e.g., touching another child without tacit permission or physically interfering with a child's ongoing play with no constructive purpose). The SH category consisted chiefly of out-and-out fights where the intent seemed to be the infliction of injury. The fights were not over objects or places, and the *reaction* of the recipient did not seem to control the behavior. (e.g., crying on the part of the recipient often modulated H behavior or stopped it entirely, whereas SH episodes often continued in spite of protest or crying by the recipient). Quite often the SH was provoked by the recipient's immediately preceding behaviors (e.g., accidentally knocking over a block structure built by another child).

Since hazing and specific hostility are similar in that they generally have harm as the observed outcome, these categories were combined in the present analysis. A total of 158 instances of H and SH were observed, mostly H. The odd–even reliability for this category was .84 ($p < .01$).

Second, object struggles (OS) were recorded (i.e., struggles over the possession and/or use of a preferred object or space). A total of 64 such struggles were observed. Not included in this assessment were fights over objects or space that were resolved by a teacher and struggles with unknown children from other classes, even though the latter occurred with some regularity on the playground. This category of interactions had an odd–even reliability coefficient of .48 ($p < .05$), which indicates a moderately stable assessment of individual scores in this category. More observations would have been desirable.

Tables 1 and 2 are dominance matrices constructed from the *outcomes* (not the number of occurrences) of H and SH and OS interactions. A child was said to dominate another child if the *ratio of wins to losses* in the object struggle category was 2 to 1 or greater, or the *ratio of initiations to receipts (without retaliation)* in the hazing and specific hostility category was 2 to 1 or greater. This criterion for the assessment of dominance is somewhat more conservative than the criteria used in other reports (e.g., Strayer and Strayer, 1976), but it was considered to be necessary given our choice of focal animal, rather than event, sampling methods. Since cases of agonistic interactions were not being sampled specifically, there was a reasonably large number of such interactions, occurring between children not under observation at a given moment, that were missed. Consequently, a child who was the receiver in three of five observed dominance interactions may well have evened the score in the sixth. Therefore, the stricter criterion was chosen, with the exception that when only one interaction was observed, the initiator (in H and SH) or winner (in OS) was considered dominant. The matrices in Tables 3 and 4 were constructed to maximize the transitivity of the dominance relationships.

In the H and SH matrix (Table 1) only 9 of 70 of the table entries indicate low-ranked children dominating higher ranked children (i.e., most of the entries are in the upper half of the matrix, above the diagonal). When two instances in which the recipient (M3) did not notice the initiation of hostility are removed (see

Table 1. Dominance Matrix Derived to Maximize Transitive Relationships Based on Hazing and Specific Hostility

	M1	M2	M3	M4	M5	F1	M6	M7	M8	M9	F2	M10	F3	F4	F5	F6	F7	M11
M1		X	X	X	X	X				X						X		X
M2			X	X	X	X		X	X			X		X	X		X	X
M3					X	X	X		X		X	X	X	X	X			X
M4					X	X					X		X	X	X			
M5						X	X								X			
F1							X			X			X	X		X		
M6		X						X		X					X	X		
M7					X				X									
M8																		
M9											X	X						
F2	(X)													X	X		X	
M10																	X	X
F3														X				
F4																	X	
F5																	X	X
F6			[a](X)														X	
F7			[a](X)				(X)											X
M11					X						(X)					(X)		

X—column dominates row (2 to 1).
Blank cells indicate either no H and SH interactions or outcomes not meeting the 2 to 1 criterion.
(X)—cross-sex rank inversion.
[a]—recipient did not notice or counterattacked and subsequently displaced initiator.

Table 2. Dominance Matrix Derived to Maximize Transitive Relationships Based on Object Struggles

	M3	M11	M6	F1	M8	M4	F5	F2	M2	F4	M1	F7	M5	M9	F6	F3	M10	M7
M3			X								X							
M11			X		X						X							X
M6					X									X		X		X
F1								X			X	X						
M8																		X
M4							X										X	
F5								X				X	X			X	X	
F2									X		X		X	X		X		
M2											X		X	X			X	
F4											X							
M1							(X)											
F7						(X)							X					X
M5		X	X			X												X
M9																		
F6																		
F3																		
M10																		
M7																		

X—column dominates row (2 to 1).
Blank cells indicate either no OS interactions or outcomes not meeting the 2 to 1 criterion.
(X)—cross-sex rank inversion.

Strayer and Strayer, 1976), this changes to 7 of 68 entries or only 10 percent that do not conform to a hypothesis of a linear dominance hierarchy. When the formulas for linearity and rigidity proposed by Strayer and Strayer (1976) are used, the corrected matrix is 89.8 percent linear and 76.7 percent rigid. For the object-struggle matrix (Table 2), an ordering was found for which only 5 (of 39) relational reversals occurred. This matrix is 87 percent linear and 75 percent rigid. The figures show that, even though there are few relational reversals in the dominance rankings, a relatively large proportion (about 25 percent for both H and SH and for OS) of the aggressive encounters were initiated by lower ranking children toward higher ranked individuals. Thus, both H and SH and OS data can yield the kind of ordering of individuals approximating linearity that is central to most dominance research. However, the ranking based on the hazing and specific hostility category is not significantly similar to the ranking derived from the object struggle data (r_s = .20).[3]

ATTENTION STRUCTURE. Observations of the visual regard of peers by each child were made from late October until late March in order to determine how much visual attention each child *received* from peers. Each child was observed for ten seconds at a time, and the name of each child looked at or glanced at was recorded. A "look" was defined as an orientation of the face and eyes toward another child for two seconds or more. A "glance" was defined as a similar orientation of face and eyes toward another child for less than two seconds. Care was taken to notice whether looks and glances were truly at another individual, not at a toy or other object held or looked at by the other child. Ambiguous instances were not included in the data reported below. The order in which the children were observed was randomly determined, with the restriction that each child was observed once before any child was observed a second time by the same observer.

A total of 125 observations of each child was made between October and March. Since it was expected that any "attention structure" would take time to emerge, only the last 50 observations (500 seconds) on each child were used in the present analysis. Since all of these were collected in the winter term, all are observations during free play indoors.

Independent observers agreed as to which child was looked at by a target child (the child *we* were looking at) more than 80 percent of the time. When one observer reported a look or glance, the second observer also reported it more than 80 percent of the time.

To determine whether 50 observations of each child were adequate to give *reliable* estimates of the number of looks or glances received, the observations of each child were divided into sets of odd and even observations, on the basis of the order in which the observations were made. The correlation between odd and even data for the looking data was .79, which is quite high enough to satisfy our present purposes.

The attention structure of the class was determined by ranking the children in terms of the absolute number of looks and glances received by each child. Separate rankings based on looks alone and glances alone were significantly correlated ($r = .60$), and thus the larger data base of combined looks and glances was selected.

The Picture Sociometric Data

The techniques and correlates of picture sociometric assessment of peer acceptance are well documented in developmental literature (see Hartup, 1970; Moore, 1967). Most standard sociometric tasks are easy to administer and yield data which are usually more stable over time than accompanying teacher ratings or behavioral assessments. In the present research two different sociometric methods were employed—a paired-comparisons task and a free-choice task. Both tasks were administered at the end of spring term, eight weeks *after* the attention data had been collected.

For the paired-comparisons task a black-and-white photograph (from the front, head and shoulders, relaxed face) was made of each child, and 20 copies (3 in. x 3 in.) were printed. The photographs were mounted in pairs on cards (4 in. x 7 in.) and covered with transparent plastic film for protection. There are 171 possible pairs of children in a class of 19 [$N \times (N - 1)/2$]. Each subject appeared an equal number of times in the left and in the right position on the cards, and the deck was arranged so that neither child on a given card reappeared until all other children had been seen. This was done to prevent a child's judgments from being influenced by position or previous choices.

Each child in the class was tested individually in 15- to 20-minute sessions. The child was asked to indicate for each pair of photographs, "Which of these children do you especially like?"[4] Every child completed all 171 comparisons in two or three sessions. A child's sociometric rank was determined from the total number of times (s)he was the preferred child in the choices of his/her peers. The reliability of sociometric data lies in part in the large number of data points collected on each child with such procedures. The major limitations of the paired-comparison method are the difficulties involved in preparing the stimuli and the large number of responses required when the number of subjects in greater than 15 to 18.

The free-choice picture sociometric technique used in this study was developed by Moore (1973). Each child was presented with an array of photographs (prints of the photographs described above) including every child in the class. The child was asked to indicate three children whom (s)he "especially likes," and three children whom (s)he "especially dislikes." Free-choice sociometric rank was determined from the number of times a child was included in the "especially liked" group by his/her peers. The resulting sociometric ranking correlated .73 ($p < .01$) with the complete paired-comparison ranking. Both methods were in agreement as to very high-ranking and very low-ranking children. The absence of perfect agreement between the two methods is due to the fact that the free-choice method involves fewer data points per child and thus yields more tied ranks than the complete paired-comparisons task.

Correlations among Measures

In order to determine how aggression, attention structure, and sociometric data relate to one another, the children were ranked on each of the four measures (Table 3), and rank-order correlations among the measures were computed (Table 4). The object struggle data and the hazing and specific hostility data were treated as separate categories of dominance in the event that they might not have the same correlates despite the fact that both are commonly designated "aggressive behavior."

It has often been assumed that there is a single simple dominance hierarchy among members of a group such as our nursery school class. Moreover, success in object struggles, initiation of attack or threat (as in H and SH), and sociometric status have all been used as indicators of rank in this dominance hierarchy. If this were the case, we would expect significant correlations among the sociometric data, among the OS, and among H and SH data. We would then be curious as to the relationships of attention structure to "dominance." The data in Table 4 do *not* point to a simple pattern of high correlations among "dominance" behaviors. In fact, as mentioned above, OS and H and SH are only correlated .20, which is not statistically significant.

Table 3. Rank-Order Data for Behavioral and Sociometric Measures

Subjects	*Hazing and Specific Hostility*	*Object Struggles*	*Attention Structure*	*Picture Sociometric*
M1*	1	11	10	6.5
M2	2	9	11	8
M3	3	1	3	5
M4	4	6	12	13
M5	5	13	7	6.5
F1**	6	4	6	9
M6	7	3	14	12
M7	8	18	8	16
M8	9	5	1	1
M9	10	14	9	15
F2	11	8	17	14
M10	12	17	13	10
F3	13	16	5	4
F4	14	10	16	11
F5	15	7	4	2
F6	16	15	18	17
F7	17	12	15	18
M11	18	2	2	3

*M = Male
**F = Female

The correlational data suggest, instead, that attention structure (AS) and the sociometric data (PS) measure much the same thing ($r = .79$) and that hazing and specific hostility measures something different ($r_{AS,\ H\ \&\ SH} = .14$). The object-struggle data are about equally divided in relevance to the AS/PS pair ($r = .35$ and 0.43, respectively). This pattern of correlations suggests the possibility that object struggles may differ in importance depending

Table 4. Rank-Order Correlations among Behavioral and Sociometric Measures

	H & SH	*OS*	*AS*	*PS*
Hazing and specific hostility (H & SH)				
Object struggles (OS)	.20			
Attention structure (AS)	.13	.43*		
Picture sociometric (PS)	.14	.35	.79**	

*$p < .05$ (one-tailed test)
**$p < .01$ (one-tailed test)

on the situation, the individuals involved (e.g., between friends or between nonfriends) or the point in time that they occur relative to group formation. In addition, important developmental questions can be posed with respect to struggles over objects. That is, object struggles may well imply different things with 3- to 4-year-olds than with 4- to 5-year-olds, and these all may differ with their importance to school-age children. This is especially intriguing in light of the finding (e.g., Dawe, 1934) that object struggles decline as a proportion of initiated aggressive episodes over the preschool and early primary years.

Distribution of Attention, Attention Structure, and Aggression

The correlations in Table 4 reveal something about the relationships among the measures, but they are not very descriptive of the relationship between a given child's rank on a measure and his/her behavior toward other children. The authors have been especially interested in uncovering the meaning of the distribution of a child's attention to peers. On the one hand, Chance and Jolly (1970) have hypothesized that attention rather than physical dominance, access to food or females, or the ability to command a large "personal space" is the basis of primate social organization. Both Waterhouse and Waterhouse (1973) and Hold (1975) have reported behavioral correlates of visual regard by peers. Abramovitch (1976) reports that dominant children (as assessed by the outcome of object struggles) are looked at more by subordinate peers than would be expected by chance. On the other hand, some[5] have argued that attention is not the *basis* of social organization; rather it is merely a reflection of the behaviors upon which social organization depends (outcomes of various types of interactions, access to resources, and so on).

In order to determine the relationship between one's visual regard of peers and one's attention structure and "dominance" rank, the rank of each child has been related to the rank of the children to whom the attention was directed (see Figure 1). This was done in three steps. First, every look, or glance, from one child to another was given a score equal to the (attention structure or dominance) rank of the child receiving the look or glance. Second, the mean of these scores was computed for each child. If a child looked predominantly at high-ranking children, the looking scores would have a higher mean than if the child had paid more attention to low-ranking children. Third, each subject's mean

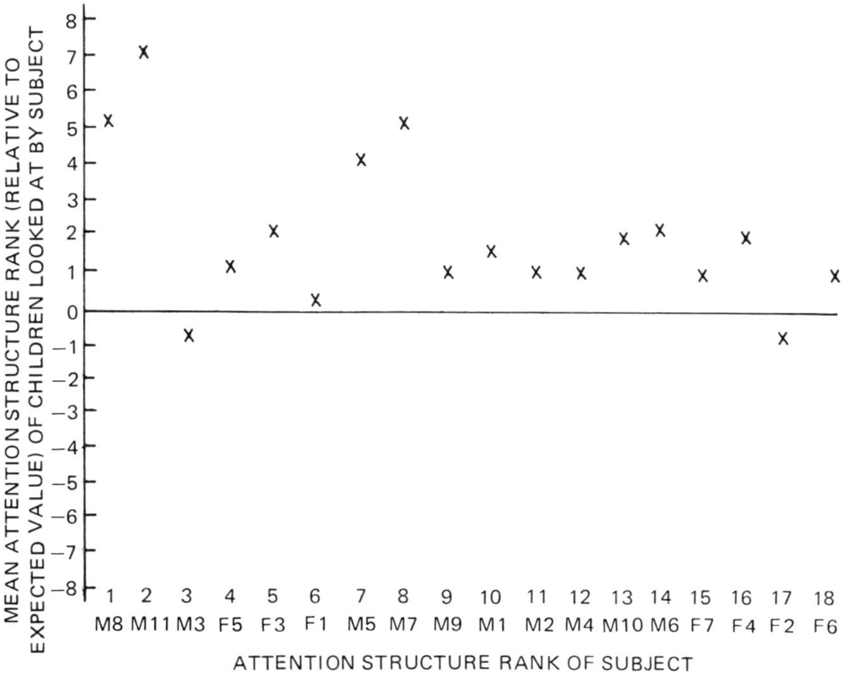

Figure 1. The relationship between attention structure rank and distribution of visual attention among peers.

looking score was compared with the mean that would be expected if the child looked equally often at each of his/her peers. If a child's attention is equally distributed among peers, that child's looking score will not differ from the expected value. If more attention is paid to high-ranking children than to others, as predicted by theories of attention structure, then mean looking scores should consistently stand above the expected value. By plotting these data in the rank order of the subjects it is possible to answer three questions: (1) Is more attention directed at high-ranking children than would be expected by chance? (2) Do children at different levels in a hierarchy distribute their attention differently? (3) Does the relationship between attention and rank depend on how we define rank?

Figure 1 illustrates the relationship between each child's rank in the attention-structure hierarchy and the distribution of his/her attention to peers. The horizontal axis is the value of looks and glances that would be expected if a child's attention were equally distributed among peers. The mean looking score is 1.87, indicating that, on the average, children do look at higher ranking peers than would be expected by chance ($t = 3.84, p < .01$). This is the kind of observation that justifies the use of the term attention *structure*. That is, it is evidence that there is something more than the sum of a child's looks and glances in the distribution of his/her attention to peers. Each child's attention seems drawn toward the children to whom his/her peers also pay the most attention; that is, there tends to be a consensus of attention.

In addition to the distribution of attention upward in the attention hierarchy, there appears to be a greater tendency for high-ranking children to look at high-ranking children than for low-ranking children to look at high-ranking children. This is reflected in a correlation of .33 ($p < .10$, one-tailed test) between a child's rank and the degree to which (s)he looks at children above the expected rank.

Similar figures illustrating the relationship between a child's ranks in the object struggle and the hazing and specific hostility hierarchies and the distribution of attention of his/her peers yielded mean looking scores of .784 and −.039, respectively. These scores indicate that, on the average, children ranked according to these criteria do not distribute their attention differently than would be expected if no hierarchy existed ($t = .107$, ns and −.007, ns, respectively). There was no indication that the distribution of an individual child's attention to his/her peers was related to her/his position in the hierarchies derived from agonistic interactions.

Discussion

The data collected contain descriptive information about the correlates of behaviors often used to determine dominance relations among individuals in small groups. Several important hypotheses about the dynamics of attention within such groups have been confirmed, but no evidence has been found that either dominance or attention can capture the entire range of social organization among preschool peers. Among the most interesting results are the high correlation between attention structure and sociometric rank, the suggestion that object struggles are perhaps in need of more detailed study, and the findings on the relationship between attention and different methods of ranking children relative to their peers.

The most noticeable "negative" result was the failure to replicate previously reported relationships between aggression-based dominance hierarchies and the deployment of attention. Both in the primate literature (Chance, 1967) and in more recent studies of human children (Abramovitch, 1976; Waterhouse and Waterhouse, 1973) there are reports that dominant animals receive large amounts of visual attention. One possible explanation for the "negative" finding could be that not enough aggressive interactions were observed to determine reliable OS and H and SH hierarchies. For example, from over 220 dominance interactions it was only possible to be certain of less than half of the 162 possible pairwise relationships (using the 2 to 1 criterion). The rest of the relationships could only be inferred after deriving the rankings in Tables 1 and 2. Clearly, the number of observations necessary to obtain reliable estimates of individual rates of aggressive interaction is far short of the number necessary to determine equally reliable dominance hierarchies from such interactions. The question arises whether the children themselves could have enough information to construct detailed models of a hierarchy based on aggressive interaction. Data presented by Omark and Edelman (1973) also suggest that preschoolers are not able to construct dominance hierarchies from their experience with peers. They report that in only 40 percent of the possible pairings of children was there agreement (using a picture sociometric-type technique) as to which of the pair was "toughest."

Group size distinguishes this study's group of preschoolers from the groups observed by Abramovitch (1976). The class numbered 19 (15 to 18 on any given day), whereas her largest group was 12. Group size is clearly a factor in the degree to which linear dominance hierarchies can occur (see Wilson, 1975). But both Hold (1975) and the present authors have been able to construct hierarchies in relatively large classes after 10 to 12 weeks of observations of interaction among children. Still, the size of the class observed is in part responsible for the fact that so many cells in the two dominance matrices are empty. In fact, when all 19 children were present, very often more than five school days would pass before a child contacted every other child in class. This was especially common among children at the bottom of the attention rank.

Waterhouse and Waterhouse (1973) have reported a study of nursery school children of similar age and in group sizes similar to that in this study, in which the highest ranked child in their dominance hierarchy did receive more attention and was more often followed by his classmates than any other child. Since a dominance hierarchy was only derived for males in the study, a clear comparison is difficult. Moreover, it is not clear how many dominance relationships were left undetermined in their data. It is interesting to note, however, that the child placed at the top of the present study's object-struggle ranking was among the top five on the attention ranking and the sociometric ranking and was ranked third on the hazing and specific hostility measure. While the similarity of the data to that of Waterhouse and Waterhouse is difficult to assess in detail, the possibility arises that either dominance and attention structure are most clearly related at the extremes of both dimensions or that one is the better model of social organization among small groups and the other more appropriate to large groups.

On the basis of comparison with Hold's (1975) data, which were derived from groups of children of much more extreme age ranges, we suspect that cross-age and same-age groups will differ significantly in many aspects of social organization. Generalizations from cross-age to same-age groups and vice versa should be avoided as carefully as generalizations from incipient groups to established groups, from preschool to school-age groups, or from same-sex to mixed-sex groups.

Another possible explanation for the low correlations between the dominance measures and attention structure of this study concerns the nature of the dominance interactions themselves. Recall that over two-thirds of these were H and SH interactions which, as already mentioned, seemed to have harm as their goal and often were initiated out of "spite" (see Wilson, 1975, p. 117). A case can be made that among nonhuman primates an animal who can assert and maintain dominance and personal space by threat and low-level aggression is fairly called the "socially competent individual." This assertion is somewhat less plausible in the nursery school. The nursery school is, after all, primarily an arena set up by adults so that their offspring can have enjoyable experiences in social interaction and at the same time experience the socializing influence of a peer group.

A child whose primary or even secondary mode of relating to peers frustrates or makes the experience of peer socialization difficult seems *less* likely to become the socially competent individual. He seems less likely to be able to use the peer groups as a resource for an advantageous developmental outcome. Here we should note the observation[6] that certain styles of aggression, especially hazing, can indicate that a child is having difficulty maintaining constructive interaction with peers. Of course, a child need not have failed entirely in the task of socializing aggression to be socially incompetent. Dominance and competence do not require the same things in children as in our beastly cousins. In a preliminary analysis of the play behaviors that were assessed along with the dominance interactions, it was found that when children are ranked on the attention measure, the top third of the hierarchy are involved significantly more often than the bottom third in interactive play and more often were noted facilitating play (i.e., keeping a game going by making suggestions about what to do next). These children of the top third also more often imitated other children.

When the rank order was based on the hazing and specific hostility hierarchy, the top and bottom thirds were *only* distinguished by activity level, involvement in "scenes" (e.g., climbing to the top of the slide then calling "Hey, look at me!") and by the category "is imitated." Interestingly, children in the *bottom* third of this hierarchy were *more* likely to be imitated ($t = -3.02$, $p < .01$, one-tailed test) than those at the top. Further, data collected by an independent observer[7] who was studying affective expressiveness and emotional development where found to distinguish children in the attention hierarchy. The rank correlations between frequency of showing positive affect, degree of involvement in ongoing activity, and the attention rank were $r_s = .47$ and .45, respectively ($p < .05$, one-tailed test). These measures were not significantly correlated to the H and SH rank order. Insofar as rank in an attention structure depends upon social competence, it is expected that more refined measures reflecting the ability to use nonverbal signals instead of outright aggression, the ability to anticipate the explosive potential of certain interactions, and the ability to modulate the aggressive behavior of one's self and one's peers will sooner be related to attention structure than will more literal interpretations of "dominance."

Conclusions

The data of this study clearly suggest that social behavior among preschool peers *is* organized. Both attention-structure and linear dominance hierarchies can be constructed from behavioral data. In view of the low frequency and predominantly disruptive nature of the aggressive dominance interactions observed, the conclusion seems justified that this type of interaction was not a major contributor to social organization in the group of children studied. The observation that success in object struggles and freedom to initiate aggression are not related to attention received from peers is in agreement with Blurton Jones (1967). It is also consistent with Kummer's observation (1957, cited in Hinde, 1974) that aggressive interactions among young (under 2 years) hamadryas baboons are poor indicators of social rank.

While we look forward to the publication of behavioral research which is underway in a number of laboratories in Europe and North America, we expect that, in addition to detailed descriptive data, a conceptualization of "social competence" appropriate to the developmental tasks of early childhood will prove necessary to a comprehensive view of the origins of human social organization.

Acknowledgments

The authors wish to express their appreciation to Drs. L. Alan Sroufe and S. K. Merrick for their contributions to this project. The special assistance provided by Dr. Shirley Moore and the nursery school staff is also very gratefully acknowledged.

Notes

1. E. A. Missakian, personal communication; F. Strayer, personal communication.
2. Both categories result from Margaret Manning's personal communication with Everett Waters.
3. Since both categories of aggressive interactions have been used by other investigators to determine dominance relationships (Abramovitch, 1976; Strayer and Strayer, 1976), a third dominance matrix was derived from a composite of the two categories. Data similar to these are reported by Strayer and Strayer (1976) as "total initiated agonism." As in Table 3, we disregarded those instances of initiated agonism in which the recipient either did not notice the attack or in which the recipient retaliated sufficiently enough to displace the attacker. This matrix yielded only 9 (of 77 possible) reversals and, interestingly, all but one of these was a cross-sex interaction. Separate male and female matrices constructed from these data yielded 100 percent linear hierarchies for both sexes. The correlations between this hierarchy and the two derived from the H and SH and the OS data were .58 and .40, respectively. In part, this pattern of correlations is due to the fact that hazing and specific hostility interactions were more frequent and thus are more heavily represented in the composite. This composite proved to be unrelated to anything but its constituent categories and need not be discussed further.

4. The original plan was to ask questions more explicitly related to social competence and dominance (e.g., "In a game of follow the leader, which of these two children would be the leader?" or "If these two children were playing together and the teacher brought in a new toy, which child would take the toy and play with it first?"). However, when pilot testing various questions in another nursery school class of 5-year-olds, we discovered that children answered these questions on the basis of which of the children they liked. Often they would explain that one child or another would be the leader "because I like him/her better." We also correlated their answers to other combinations of the pairs of children about which we asked the "like" question and found that they were truly telling whom they liked. Therefore, the standard sociometric question was selected.
5. For example, R. Hinde, personal communication.
6. M. Manning, personal communication.
7. We wish to thank Drs. L. A. Sroufe and P. Mitchell for generously sharing their unpublished data on affective expressivity for this analysis.

Chapter 22

Attention Structures in Hierarchically Organized Groups

Rona Abramovitch

One of the most important characteristics of virtually all human groups is that they are hierarchical. Some members rank higher than others. A particular hierarchy that has been of concern to psychologists and ethologists is dominance. Although dominance has been defined in different ways, its essential feature is that a less dominant member is defeated by, gives ground to, and, in general, submits to a more dominant member. This may involve actual fights, tests of strength, or less direct confrontations, but the effect is the same. Although dominance is obviously very important to the social structure of groups, it seems likely that there are other hierarchies that have considerable impact on how group members interact. Members may be ranked in terms of influence, imitation, popularity, prosocial behavior, and practically any other behavior that is crucial to group process. These other hierarchies may be related to dominance, but they are by no means equivalent to it and are not necessarily of secondary significance.

A possible difficulty with considering groups in this way is that there are an almost unlimited number of behaviors on which members could be ranked. Whereas specifying a few hierarchies might help in understanding group structure, detailing, say 25 hierarchies may provide a description of the behavior of the group but no longer deals with structure per se. In a sense, the original notion of hierarchies has lost its meaning, and all that there is is an account of how many times each member did each of 25 actions. However, we may be able to accept the fact that many hierarchies exist and still provide some structure if it can be shown that the hierarchies are closely related to each other. Ideally, it might then be possible to demonstrate that there are a small number of underlying hierarchies that account for all of the separate ones. Therefore, it would be very helpful to describe more than one hierarchy within a specific group and see if they are related.

The particular focus of this chapter is the relation of attention to other hierarchical relationships, especially dominance. It is clear that groups have attention hierarchies, with some members looked at more than others, and, as we shall see, attention is related to various group activities.

Chance (1967) and Chance and Jolly (1970) have argued that the attention structure of groups of primates is a direct indication of social relationships. They maintain that visual attention plays a predominant role in the status structure of primate groups—that the rank order of a group arises out of the persistent attention of subordinates to dominant members of the group. Rather than define dominance in terms of aggression or priority of access to goods, mates, and so forth, they have proposed that dominance structures be defined with respect to an attention structure in which the most dominant animal may be said to dominate the attention of others at most times.

According to this formulation, the only way to understand the dominance structure of a group would be to investigate the patterning of visual attention within the group. It is assumed that the dominant individual will be the primary focus of attention for a variety of reasons—that person might be most likely to determine the nature of the group's activities and the movement of the group, or might be a potential threat (i.e., likely to win in agonistic encounters), or might be a potential protector against intruders or other group members, and so on. However, whatever the reasons, the basic idea is that attention of subordinates to dominant members is the basic feature of rank-ordered behavior or dominance structures.

This chapter will examine the relation between attention and hierarchical relationships, especially dominance, primarily in groups of young children. (The nonhuman primate literature will not be reviewed.) The impetus for work in this area has come mainly from the contention by Chance that visual attention might be the basis of rank order in many primate groups.

The general idea that attention and dominance are related can be stated in several ways, from strong to weak arguments. Chance takes the strongest position when he says that attention is the underlying feature of *all* hierarchies, and specifically the dominance hierarchy. In other words, the interrelationships within the group, how they interact, and so on are based primarily and basically on attention rather than on dominance. This is not to say that some members are not stronger and more dominant than others, but that it is the attention payed them that determines the structure of interactions more than the characteristics that make them dominant in the first place. Although Chance has not gone this far, this seems to imply that any member who could call attention to himself, who could be high in terms of attention, would inevitably also be high in dominance.

A weaker and more likely position is that attention is an important feature of group interaction that depends considerably on the dominance structure. That is, the dominance hierarchy determines the attention hierarchy, which, in turn, has a variety of effects apart from those attributable directly to dominance. Thus, it might be predicted that high dominant members would be looked at more and, because of that, imitated more. The greater imitation, however, is due to the greater looking, not to their dominance behaviors per se. Similarly, attention would lead to greater influence, whereas dominance per se might not. In other words, much of the effect of dominance is indirect—it operates through the medium of first affecting attention, which, in turn, affects other behaviors.

It would seem that the most plausible position (one that also happens to be most consistent with research findings) is that attention, while related to dominance, is a separate feature that is determined by a large number of factors and that, in turn, has widespread effects. Dominant members are likely to attract more attention because less dominant members have good reasons for knowing where they are and what they are doing. But, there are other reasons for a member to attract attention. A deviant member is likely to be looked at more than a normal one; a member who engages in many prosocial behaviors will attract attention; members who are highly liked, regardless of dominance, will attract attention; and so on. Thus, attention should not be seen as dependent mainly on dominance, nor is the reverse probably true either. Rather, each hierarchy plays an important role in group interaction, and they should be substantially related—but not equivalent.

Importance of Attention

Before describing the research that has been done on the relationship between dominance and attention in children's groups, it should be pointed out that there is independent evidence of the importance of attention in children's behavior and development.

Many researchers have discussed the role played by attention as a cognitive variable. Selective attention is a major way of learning about the world—the child must learn what information is important in order to know what to attend to and vice versa. Neimark (1970) maintains that refinement of control over information input is a major developmental accomplishment. Pick, Frankel, and Hess (1975) point out that in order for individuals of all species to survive, they must exert some sort of selective control over their constant interaction with their environment. Since this environment is often social as well as physical, an important question is what sort of selectivity operates in terms of children's attention to their social environment. It is clear (see Kendon, 1967) that an individual will set up a definition of the situation he is in on the basis of information gathered about the situation. Furthermore, in order to keep it in line with the definitions of others present and to keep his own performance in line, he has to check selectively and sample the situation from time to time. Therefore, the question might arise not only to what the individual is selectively attending in social situations, but also what the process is over time. These sorts of questions have not been investigated in much detail, and certainly not developmentally.

In their review of selective attention in children, Pick *et al.* (1975) cite numerous studies that have investigated children's selective perception and memory. For the most part, this research has not focused on social situations. The one exception is their discussion of studies of observational learning, where it is pointed out that in such studies the child's memory of what he has seen a model do is assessed and what the child remembers is assumed to reflect what has been attended.

It is interesting and surprising that developmental studies of selective attention have not usually dealt with social situations even though many researchers have pointed out the importance of attention as a variable in social interactions from early infancy. For example, Robson (1967) has suggested that eye contact plays an important role in the establishment of attachment. Stern (1974) has discussed the role of eye contact and attention in the context of mother–infant interactions and games. Walters and Parke

(1965) have stressed the function of distance receptors in the development of social responsiveness. Further, Chance (1967) has pointed out that the long dependency of young mammals on their mothers or guardians requires that their attention revert there whenever there is danger.

Therefore, it seems clear that examining selective attention in social situations might be a potentially fruitful area of investigation. As noted earlier, Chance's attention structure hypothesis has generated some research on social attention in children. In the next section many of the studies that have investigated the relation between attention and various hierarchical relationships, especially dominance, in children's groups will be reviewed.

Attention and Hierarchies

Omark and Edelman (1976) investigated children's perception of dominance and other hierarchies, using various tests which involved having children rate their classmates on "toughness" (as well as smartness, niceness, and popularity). They found that by first grade children showed high agreement on where they stood in the dominance hierarchy of their class. Further, the children's accuracy of perception was best when directed at those children who ranked highest in the dominance hierarchy. The children could also reason better with hypothetical premises that concerned the children at the top rather than the bottom of the hierarchy (except when the self was involved in such reasoning). Omark and Edelman argue that these results indicate that the child's world view was oriented more toward dominance relationships at the top of the hierarchy, that children were more aware of the tougher children and better able to reason about them. In other words, as suggested by Chance, children paid more attention to those higher in the dominance hierarchy.

This is indirect evidence, however, because no independent measures of attention were obtained. The other research to be reviewed attempted to measure the attention hierarchy directly.

Waterhouse and Waterhouse (1973) observed children in a nursery school in Reading, England, and recorded three types of agonistic interactions: (1) approach–withdrawal, where one child would give up his place or flee at the approach of another child, (2) threat, and (3) struggles over objects and hitting. On the basis of these three measures, they were able to construct a dominance hierarchy for boys, but not for girls. That is, they could rank the boys so that those ranked higher consistently (most of the time) demonstrated dominance over those ranked lower. This is, of course, the criterion for a hierarchy, though it should be noted that hierarchies vary considerably in how well they satisfy this criterion. Waterhouse and Waterhouse reported that the child at the top of the hierarchy "received the highest score for being watched or being followed" (1973, p. 675). Unfortunately, it is not clear from their brief account exactly how the hierarchy was formed or what measures of attention or watching they might have used. They also mentioned only the most dominant child; therefore, there is no information on how attention is distributed throughout the hierarchy.

In an attempt to investigate the relation between attention and dominance in more detail, several groups of preschool children were observed in a study by Abramovitch (1976). The subjects for this study were 30 children 4 to 5½ years of age from three preschool classes. There were 9 children in two of the classes and 12 children in the third. In addition, one class of 3½-year-old children was observed.

Dominance was defined in terms of the number of children over whom each child won a property fight—a fight in which two children each wanted a particular toy, but only one of the children finally obtained it. Each class was divided into three groups—high, middle, and low dominance. The three groups in each of the three older classes were perfectly transitive, a child in a given group invariably won fights over children in lower ranks and invariably lost fights to children in higher ranks but might win or lose to children of the same-ranked group.

Attention or looking data were recorded using a focal individual sampling technique (Altmann,1974; Omark, Fiedler, and Marvin, 1976) in which the observer recorded every occurrence of the focal child's looking at every other child as well as the name of the recipient of the glance. In the three classes of older children, the high-ranked individuals (those who had won fights over the largest number of children) were looked at by all of the children significantly more than would be expected by chance. Middle-ranked children were looked at about as often as expected by chance and low-ranked children were looked at significantly less than expected. Further, in all three of the classes children of all ranks—high, middle, and low—looked at high-ranked children significantly more than expected and at low-ranked children significantly less than expected. In the youngest class the ranked groups were not 100 percent transitive. However, both high- and low-ranked children looked at high-ranked individuals significantly more than expected.

The attention data collected in this study clearly correspond to what would be predicted from Chance's attention-structure hypothesis. The amount of attention, both given and received, reflected status relationships with high-ranking children receiving more attention than expected from all of the children and low-ranking individuals receiving less. It should be pointed out—as will be the case for studies to be discussed subsequently—that there is a basic difference between the results of the study just described and Chance's attention-structure model. As already noted, Chance and Jolly (1970) have argued that it is the attention structure of a group that defines dominance. No evidence was found for this. Rather, the results indicate that attention structure and dominance (defined only by priority of access to objects) were related to one another, not that one could replace the other. To state it another way, there was substantial relationship between the two hierarchies, but this relationship did not even remotely approach equivalence. Clearly, a great deal of variance in attention is due to factors other than dominace, and vice versa. Thus, attention cannot be considered to define (or redefine) dominance.

These data were subsequently reanalyzed, along with dominance and attention data collected by Fred Strayer (Abramovitch and Strayer, 1978). Strayer collected detailed data on dominance relationships in a group of 17 preschoolers ranging in age from 3 to 5 years. As well as looking at the outcome of struggles over objects, he also recorded other agonistic encounters—such aggressive behaviors as hitting, kicking, biting as well as threats. Separate linear dominance hierarchies were constructed based on the outcome of attack and threat interactions and on object struggles (see Abramovitch and Strayer, 1978, Strayer and Strayer, 1976, for further details). The dominance relations in Strayer's group corresponded quite well to a linear transitivity model—if individual A dominates B and B dominates C, then A dominates C. (It is interesting that the hierarchies for attack and threat and for object struggles were not identical, although they were similar.) Furthermore, the dominance relations in this group were also fairly rigid; i.e., the usually dominant member of a dyad tended to win most if not all agonistic encounters over the usually subordinate member. It was also possible to transform the previously mentioned ranked thirds into hierarchies that were fairly linear and rigid. Overall, the linearity in all five groups ranged from 81 percent to 100 percent and the rigidity from 91 percent to 100 percent.

Once the dominance relations were ordered in linear form, it was possible to look at correlations between dominance rank and amount of attention received. Strayer had also collected attention data although glances were recorded only in the absence of any other kind of interaction (also using focal individual sampling). The correlations between attention and dominance rank for all groups were significant, ranging from .48 to .96.

The data were also analyzed in terms of episodic and dyadic wins: individuals were ranked on the basis of the total number of agonistic encounters they had won (episodic wins) and on the basis of the number of dyads in which they won the majority of encounters (dyadic wins). The findings are presented in Table 1. It is clear from Table 1 that while the correlations vary in magnitude, there is a consistent relation between receiving attention and winning agonistic encounters.

The correlations are somewhat lower for Strayer's group. This may be because the attention data were collected somewhat differently—only when the children were engaged in no interactions other than looking. Thus, Strayer's attention scores come from only a small sample of all looking behavior and may not be representative. Attention given in the absence of interaction may also serve a somewhat different function from attention given in the course of interaction and Strayer's data is thus not completely comparable with the present author's. It is worth noting that in Strayer's group the correlations between attention and status are

Table 1. Rank-Order Correlations between Attention and Group Status Criteria

	Episodic Wins	*Dyadic Wins*	*Status Rankings*
Object struggles			
Abramovitch 1	.96	.94	.96
Abramovitch 2	.94	.95	.93
Abramovitch 3	.92	.90	.95
Abramovitch 4	.80	.83	.73
Strayer	.41	.51	.48
Attack and threat			
Strayer	.68	.64	.56

higher for the rankings based on attack/threat than for those based on object struggles. Clearly, the way in which dominance is defined is quite relevant to the issue of the relation between attention and dominance.

Another study by Vaughn and Waters that considered dominance and attention in a somewhat similar manner found quite different results (this volume, Chapter 21). These researchers observed a group of 19 preschool children. Dominance was assessed by looking at object struggles and also at "hazing" and "specific hostility." The hazing category (unprovoked taunting or teasing) was unique to this study and the definition of specific hostility—out-and-out fights with an intent to injure—does not correspond completely to Strayer's attack and threat category; e.g., intention movements were probably not coded, and Strayer counted only those situations in which the target child responded to the attack or threat either by counterattacking or submitting. Further, the details of how the children were ranked were somewhat different for the two studies. Thus, it is difficult to compare the studies directly with respect to anything but object struggles. There was no appreciable correlation between attention and status based on hazing and specific hostility. Although they found a somewhat higher correlation between object struggle rank and attention (.35), this is still not significant.

There are many possible reasons for the discrepancy between these results and those of Abramovitch and Strayer (slightly different measures, group size, and so on). Group size is probably an important factor. It seems plausible that children's differential attention in terms of the status structure of the group might be somewhat less consistent in larger than in smaller groups. In larger groups more of the group members will be out of sight for a given child at any time; thus the children cannot monitor one another as well as in a small group. However, it also seems reasonable to expect variations from group to group. The fact that attention and status seem to be correlated in some groups and possibly not in others may mean that this is a somewhat variable phenomenon. Since the *N* in this case is the number of groups investigated and since this number is very low (six groups in total in which five support the model and one does not), it is not possible to make any definitive statements.

It is interesting that in contrast with the slight relationship between attention and dominance, Vaughn and Waters did find a substantial correlation between attention and a hierarchical relationship based on something other than the outcome of agonistic encounters. They found a substantial and significant correlation (.79) between attention and a child's ranking on a picture sociometric test, essentially a measure of liking or friendship. Thus, in this group of children it is the most popular and not the most dominant children who are looked at most frequently. As suggested at the beginning of this chapter, several hierarchies are important in group structure and the role of attention can be understood only by considering these other hierarchies as well as dominance.

A study by Hold (1976) provides further evidence that attention may be related to various hierarchical relationships. Hold's subjects were four groups of children from three different kindergartens (85 subjects in all) in Munich, Germany. An attention structure for these groups was obtained using an instantaneous sampling technique (Hold, 1976; see also Altmann, 1974). Each child was observed once every five minutes. The chief measure of attention was whether, at that one moment, the child was being looked at by three or more children simultaneously. No distinction was made between two lookers, one or none. The individual with the highest score was defined as the highest ranking child, and so on down the scorings. Hold also ranked the children with respect to a variety of other behaviors and then correlated the rankings on these behaviors with the child's rank in the attention structure.

Briefly, she found significant positive correlations between initiating behavior (starting a game, being imitated, proposing a game) and being the "center" of attention and between being an organizer (distributing roles, inviting, teaching, and so on) and attention. In addition, Hold also found a significant negative correlation between attention and being an onlooker. She argues that this is a low-status behavior. Being an aggressor (threaten, attack, disobey, tease) was significantly correlated with attention in two of the four groups. This latter finding is not directly comparable to the findings of the studies described previously since Hold's category of aggressor is quite different from the dominance behaviors employed by Abramovitch and Strayer or by Vaughn and Waters (for example, the inclusion by Hold of disobedience). It is clear from Hold's data, however, that being looked at is related to the child's rank in various hierarchical relationships other than just dominance, even though her groups were quite large.

One variable that Hold studied was imitation (imitating and being imitated). She reported that imitating was negatively correlated with attention in two of the four groups. Being imitated was part of her category of "initiator," which was positively correlated with receiving attention in all four groups. She concluded that imitating was a low-status behavior while being imitated was related to high status.

The relation between imitation and attention was investigated in a study conducted by Abramovitch and Grusec (1978). Five groups of children (10 to 12 children per group) ranging from 4 to 11 years of age were observed for instances of immediate imitation of peers in a free-play setting. Attention was also recorded, once again using a focal individual sampling technique. Children were ranked on the basis of how much attention they received, how often they imitated others, and how often they were imitated by others. Further, the teachers in two of the older classes were asked to rank all of the children in their class with respect to dominance.

The rank-order correlations between being imitated, imitating, attention, and dominance for the four classes in which sufficient data were collected are presented in Table 2. (The mean age of the children in groups 1 and 2 was 49 and 52 months, respectively. In groups 3 and 4 the mean age was 79 and 77 months, respectively.) It is clear that teacher's ratings of dominance and

Table 2. Correlations between Being Imitated, Imitating, Attention, and Dominance

	Imitating	*Attention*	*Dominance*
Group 1			
Being imitated	.75	.88	
Imitating		.69	
Attention			
Group 2			
Being imitated	.94	.64	
Imitating		.65	
Attention			
Group 3			
Being imitated	.27	.72	.64
Imitating		−.27	−.20
Attention			.92
Group 4			
Being imitated	.19	.56	.74
Imitating		.63	.55
Attention			.73

attention are highly correlated, lending some further support to Chance's model. Being imitated and imitating are also positively correlated with attention in most of the groups. This is only partly consistent with Hold who found a significant negative correlation between attention and imitating in two of her groups. However, it should be noted that both group size and age of children varied across these studies.

It is of course not surprising that the more a child is looked at, the more likely that child is to be imitated. It seems plausible that these (probably dominant) children were seen as sources of information about, and also engaged in, useful and interesting behavior. The fact that for three of the four groups children who were most attended to were also those who were most likely to imitate might be due to children who receive attention being generally more active in many spheres of behavior. Also, Abramovitch and Grusec argue that this accords well with the notion that imitating others may be an important social influence technique: if these children receive attention because they are influential or dominant, they are probably able to exert some control over the behavior of others, and one way in which they may do this is by increasing their attraction to their peers by imitating them.

Once again, many more groups will have to be observed before the relation between attention and imitation becomes obvious. It seems quite clear, though, that being looked at and being imitated are strongly related. More detailed research is necessary to determine whether this relation is a function of dominance, influence, or other similar factors or a more simple matter of there being a higher probability of a child imitating someone to whom he is attending a great deal. Again, we find among young children that the structure of attention is related to other social structures in the group.

Attention and Activity

The sum of the evidence seems to indicate that attention is related to a variety of hierarchical relationships. It is interesting to note that in all of the groups that have been studied some sort of attention structure has been evident. Children's attention is not deployed randomly—some children receive more attention from their peers than others. And this differential receipt of attention is often related to other factors—dominance, popularity, imitation, and so forth. Thus, it is quite clear that social attention is structured and selective; children do not simply look at others but rather at particular others who have a particular place in the structure of the group. The details of this might vary from group to group, but there was no instance in which attention was not selectively structured in some manner.

It does not seem adequate, however, simply to say that the amount of attention a child receives will be related to that child's place in one of several social hierarchies. In addition to the amount of attention an individual receives, there might be differences in terms of patterning—duration, patterns of responses, and the like (see Abramovitch and Strayer, 1978). One potentially interesting area of research is the relation between attention and the individual's specific behavior. For example, are dominant individuals looked at because they engage in attention–attracting behaviors, or are they looked at independently of the activities in which they engage? Chance and Jolly (1970) have maintained that for nonhuman primates dominant individuals might be monitored independently of their activities. "Thus the dominant animal may be said to dominate the attention of others at most, if not at all times, and usually without taking any specific actions to achieve this" (Chance and Jolly, 1970, p. 172–173). They have also suggested, however, that frequency of visual attention might depend on various other factors such as how urgently information is needed, the kinds of responses required, and the amount of movement in the group.

As a first step in answering some of these questions, five groups of preschoolers (9 to 15 children per group) were observed in order to determine to what activities children attend. When a child was sampled for attention, the observer noted not only at whom the child had looked (the target child), but also the activities in which the target child was engaged at the time of the look. Further, via an instantaneous sampling technique, each child was observed at preset intervals to determine the amount of time spent in various activities. Thus it was possible to see if certain types of activities were attended to more often than might have been expected on the basis of how often they usually occurred in the absence of attention. The child's activity, either as attention target or subject of instantaneous sampling, was coded on a checklist containing the following behaviors: vocalize, no vocalizations; sit, stand, walk, run, jump, kneel, lie, hold an object, play with an object, no involvement with an object; alone, in a group, interacting with others, no interaction; involved in an agonistic encounter.

Although many of these behaviors clearly can and do co-occur, each was looked at separately. In all five of the classes the proportion of time children spent talking or otherwise vocalizing was less than the proportion of time children were looked at when vocalizing. This was also true of interacting with others and being involved in agonistic encounters. In four classes children in groups were looked at more than would be expected on the basis of how often they usually were in groups instead of being alone. The pattern that emerges is that engaging in some sort of social interaction is attention attracting. In only two of the five groups was movement or object play attended to more than would be expected on the basis of how often children typically engaged in these activities. Thus, it would seem that looking at others is, a social act—the fact that another child is moving or playing with toys does not particularly attract attention, while vocalizing, interacting, and so forth does. Although these results can only be viewed as preliminary—many more activities have to be observed and more sophisticated analyses done—they do point to the fact that the attention structures discussed are primarily social and it is of course not surprising that they are related to other aspects of social structure.

Conclusions

In the research reviewed here members of children's groups were rank ordered on a variety of dimensions. It was found that ranks on one dimension tended to be correlated with ranks on others. In particular, attention was related to dominance, but also to popularity and imitation. These findings raise several questions concerning structures in children's groups and the role that attention plays in such groups.

First, attention appears to be an important aspect of group structure. There are consistent patterns of attention, and these are related to various other activities and structures within the group.

Second, although attention and dominance are clearly related, they are by no means equivalent. The correlations are high in some groups, not so high in others, but the general pattern gives no indication that dominance and attention are in some way interchangeable. Thus, there is little evidence to support Chance's notion that attention is the basis of dominance. While dominant group members are generally attended to more than subordinate members, this relation is far from perfect; other variables, such as popularity, are also related to attention. It seems unlikely that dominance affects the group's behavior primarily by affecting the amount of attention received by each member or that attention is the major factor producing dominance. Each is an important correlate of the other, but each is distinct and plays a different role in the group.

Third, some consideration to what is meant by a hierarchy is necessary. Sometimes the term is used to refer to any rank ordering. Yet in a dominance hierarchy, the one most often discussed, it is not only that higher ranked individuals exhibit or receive more of a particular behavior, it is also that a higher ranked individual consistently demonstrates superiority over a lower ranked one in various interpersonal encounters. This is shown by the dominant member displacing the subordinate from a position, having priority in choice of objects or mates, winning fights, and so on. It is also possible to rank order members in terms of height, but this would not be considered a hierarchy since height has no consistent effect on most interpersonal encounters. From this point of view, it might be appropriate to talk about dominance hierarchies and various other kinds of rank orderings or structures.

Dimensions other than dominance (defined as winning agonistic encounters) may be hierarchical in the sense described. For example, more popular members may consistently be chosen above less popular ones, have more influence on the group, and even have their choice of location, objects, and positions although they may not necessarily be able to win a fight on their own. Similarly, individuals ranked high in attention may consistently be looked at by lower ranking individuals, imitated by them, and even given way to by them regardless of their relative ranks in popularity and traditional dominance. The more general point is that hierarchy is typically used almost synonymously with dominance, which is usually measured by the outcomes of agonistic encounters alone. The emphasis on agonistic encounters is obviously limiting, and perhaps more emphasis should be placed on other hierarchies existing in the group.

It seems obvious that the full range of behavior on which group members can be rank ordered needs to be investigated in order to understand fully group structure, and in particular the operation of attention. While this work is certainly slow and painstaking, it would be very useful to have observational data on many different hierarchies or rank orders within a given group. It might then be possible to specify the interrelations among the various behaviors and be possible to discover some structure underlying the many different behaviors.

Chapter 23

Gender Differences in Agonistic Behavior and Dominance Relations of Synanon Communally Reared Children

Elizabeth A. Missakian

Ethological studies of gender differences in aggression have yielded contradictory findings. McGrew (1972) reported that male–male interactions among preschool children were typically agonistic or quasi-agonistic, mixed-sex interactions were agonistic, and female–female interactions were typically nonagonistic. Girls did not participate in a sufficient number of fights to enable construction of a dominance hierarchy. In an early study Blurton Jones (1967) found that preschool girls bit more frequently than boys. In a later paper (Blurton Jones, 1972b) no gender difference was found in overall frequency of aggression or frequency of specific categories of aggression (attack versus tug, for example). These contradictory findings led Blurton Jones to suggest that his nursery situation may have attracted mothers of gentler, less aggressive boys, although no evidence for this was presented. Cronin (1976) and Travis (1976) reported that in older groups of children males were more aggressive than females. Eibl-Eibelfeldt (1974b) studied aggression of !Kung Bushmen and found that both boys and girls were aggressive. It has been suggested (Crook, 1973) that high estrogen levels, contingent on the low-fat diet of

This research project was supported by National Institute of Mental Health Grant No. 1 RO1-24874.

Bushmen, may be responsible for the reduced contrast in gender differences in aggression in that culture.

Strayer and Strayer (1975) reported some girls at the top of the dominance hierarchy in one of their study groups of preschool children. Strayer (1976) reported mixed-sex dominance hierarchies among children in a laizzez-faire preschool situation (one in which there appeared to be minimal interference by adults of free play and aggression).

Most cross-cultural studies of child psychology report gender differences in aggressive behavior. McCandless, Balsbough, and Bennett (1958) compared American and Hawaiian nursery schools and found that boys were more aggressive than girls during free play. Freedman (1974) compared 5- to 7-year-old children in the United States, Bali, Hong Kong, Japan, Australia, Ceylon, India, and Kenya. Boys were more aggressive than girls. Freedman suggested that boys may be born with a lower threshold for aggressive responsiveness. White and Pope (1973) studied children 3 to 10 years of age in seven cultures and reported that although gender differences were universal, they were not as great as prior American/Western studies had suggested. Direct physical contact between children was so rarely observed that statistical analyses of sex differences were impossible. However, in all cultures studied boys engaged in more mock fighting (rough-and-tumble play) and verbal insults than girls. Omark, Omark, and Edelman (1973) studied children 4 to 10 years of age in the United States, Switzerland, and Ethiopia. In these three cultures boys were more aggressive than girls. The dominance hierarchies reported by the children placed boys at the top end and girls at the bottom end, with considerable overlap in the middle. The male portion of the hierarchy was reportedly more stable than the female portion.

Psychological/experimental studies of gender differences in agonistic behavior have indicated that males are more aggressive than females (Feshbach, 1970; Maccoby and Jacklin, 1974; Oetzel, 1966). A comprehensive review of the literature, however, revealed significant inconsistencies in the evidence regarding gender differences in aggression. Some studies demonstrated clearcut gender differences in aggressive behavior of preschool children. Green (1933a, 1933b) reported that boys have a higher quarrelsomeness index (number of quarrels divided by the number of opportunities to play with a given child) than girls. Male–male pairs quarreled more frequently than either mixed-sex or female–female pairs. Jersild and Markey (1935) found that boys were more aggressive than girls. Sears, Whiting, Nowlis, and Sears (1953) observed children 3.4 to 5.5 years of age for a total of four hours per child. Their results indicated that boys were more aggressive than girls. Pedersen and Bell (1970) studied short-term nursery school groups (5 to 6 weeks in duration) containing children 2½ years of age. Each child was observed for six days. The authors found that males were more aggressive than females. In studies of English children Brindley, Clarke, Hutt, *et al.* (1973) reported that boys in nursery schools were more aggressive than girls. These findings were confirmed by Smith and Green (1975), who studied 15 nurseries, day care situations, and play groups in England. They recorded 239 instances of aggression among children 2½ to 5 years of age and found that in 13 of 15 groups studied boys were involved in more aggression than girls.

A second set of studies have revealed only conditional gender differences. In some cases only certain categories of aggressive behaviors have revealed gender differences. Hattwick (1937) analyzed teacher ratings of 283 boys and 296 girls and reported that boys were more physically aggressive while girls were more verbally aggressive. These findings were later replicated by Bandura and Huston (1961) and McIntyre (1972). In other cases only certain age classes have revealed gender differences in aggression. Dawe (1934) analyzed 200 quarrels of 25- to 60-month-old children. With the exception of children 30 to 41 months of age, boys were more aggressive than girls. Walters, Pearce, and Dahms (1957) reported that aggression increased with age from 2 to 4 years. They observed children for 40 minutes per child. Only in the case of 2-year-olds were boys found to be more aggressive than girls. Agonistic behavior in all other age groups revealed no gender differences.

Finally, some studies have demonstrated no significant gender differences in aggression. Muste and Sharpe (1947) observed 2- to 5-year-old children in experimental play situations involving two children at a time. They reported unreliable gender differences in aggression. Peer interaction scores (hostile, friendly, associative) also revealed no gender differences (McCandless and Marshall, 1957). Sears, Maccoby, and Levin (1957) interviewed 379 mothers of American preschool children. Their data revealed no significant gender differences in aggression; yet their study is often cited as evidence that boys are more aggressive than girls. Sears (1965) found considerable overlap between boys and girls in the frequency and intensity of aggression. Some girls were more aggressive than some boys.

A review of the research on child aggression reveals several shortcomings. First, data are based primarily on short-term observations, often of unstable social groupings. Pairing together unfamiliar children or creating experiemental groups of unfamiliar children asks a very different research question than observing already established social groups. Children in long-term social groupings may show less aggression than strangers who may need to establish power relationships. In any case, it seems of little value to pair or group unfamiliar children, observe them for a short period of time, and then declare any knowledge of child social behavior/organization. Anedcotal reports are often of more value than such pseudoexperimental/scientific studies. Second, the findings themselves are inconsistent with respect to gender differences. Some studies find consistent gender differences, some report only age-dependent or behavior-dependent differences, and some report no significant gender differences in aggression. Inconsistency may be due to the various observational techniques used. Many studies used time sampling techniques, which may lead to erroneous, unrepresentative results. Finally, the cultural contexts of children observed to date are similar with respect to gender appropriate behavior; therefore, the underlying conditions for gender differences are virtually unknown.

The conflicting evidence from ethological and psychological research on gender differences in aggression leads to questions regarding the nature of a proposed "sex-linked differential readiness to respond aggressively" (Maccoby and Jacklin, 1974). Since there is no evidence that helps delineate the relative role of learning, it is not at all clear that this readiness to respond is biologically based. A more productive approach to this question may be to determine the degree to which gender differences and gender-specific behavior may be modified or manipulated by changing environments. In this sense, an environment imposing minimal sex-role stereotyping of aggressive behavior would be valuable to study in order to learn the degree to which aggression may be modified.

Long-term cross-cultural studies in peer socialization processes may reveal further differences in the social development of children. This is especially true in the case of cultures which do not share typical American/Western stereotypes of gender-specific behavior. The population of children raised in a communal situation in Synanon offers a unique opportunity for study of gender differences in aggression. Communal peer rearing with minimal adult-imposed gender-role stereotyping could provide valuable information relating to the determination of gender-specific behavioral dimorphisms.

At first glance, the group of Synanon-reared children bear a close resemblance to kibbutz-reared children. There are, however, a number of important differences between these two groups of children: (1) Synanon children live in a 24-hour-a-day school/living situation and do not have regular or daily contact with their biological parents; (2) training, curriculum, and general Synanon School environments are reportedly more highly developed than most kibbutz situations; (3) there is more continuity of single caretakers in the care of kibbutz-reared children. In Synanon, consistency is built into the child-rearing philosophy and not in individual caretakers. There is rapid turnover of personnel in the Synanon situation. It is most difficult to compare the Synanon child-rearing situation to the kibbutz because of the variability between kibbutzim in terms of both philosophy and practice of child rearing.

Preliminary observations of Synanon children, ranging in age from 6 to 30 months, carried out for a period of 17 months, indicated that social behavior in peer-reared children was not rudimentary in form and could be observed systematically.

In the more definitive study reported in this chapter, the following questions were asked:

1. Is there a difference in the frequency of aggression of boys and girls?
2. Is this difference age-dependent?
3. Is there a difference in the dominance hierarchies of boys and girls (i.e., are they overlapping hierarchies)?

The overall purpose of the study was to examine gender differences in children reared in a communal setting in the United States.

Method

Setting

Beginning with approximately 20 individuals in 1958, Synanon Foundation had grown by 1977 into a community housing over 1,300 men, women, and children. (For a detailed description of the history and current status of Synanon, see Endore, 1968; Simon, 1974; Yablonsky, 1965.) Synanon at the close of 1978 had facilities in Santa Monica, Tulare County, San Francisco, and Marin County, California. International Synanon houses are located in Germany, Indonesia, and Malaysia.

The Synanon School, established in July 1966, is a full-time 24-hour children's community where children live in peer groups from birth to 16 years of age. The school is part of the larger community of Synanon residents. When the observations were started, approximately 240 children lived in the school, with their ages ranging from newborn to 16 years. The two peer groups studied included children from 6 to 20 months (the infant program) and 20 to 45 months (the 2 to 4 program), living in the Lower School in Marin County, California. The Lower School at the time of the study consisted of one main building (487.5 m^2 floor space). Within the building were two workshops, two sleeping rooms, a kitchen/dining room, two toilet/bathing rooms, and one adult office area. The workshops were available for free play; each was 9.14 m x 15.24 m. These rooms contained areas and equipment for dramatic play, block play, reading, puzzles, art, and large-muscle activities (ramps, platforms, swings, ladders, ropes). The outdoor space was divided into three open yards totaling 1,579 m^2. They contained toys, waterbed, boxes, arts-and-crafts porch, portable swimming pool, trampolines, sandboxes, bikes and other rolling stock, jungle gym, platforms, and covered porches. During free play children had access to all equipment and play areas in the yards. In addition to the space described, children regularly (daily) were taken on field trips both within and outside of the Synanon community. The capacity of the Lower School was 29 children, and the staff–child ratio was 1 adult per 4 children.

Subjects

During the period of study the population of the Marshall Lower School fluctuated between 20 and 24 children, and children moved into and out of the program. The total number of children observed was 34, ranging in age from 7 to 45 months. Of this total, 19 were boys and 15 were girls. The mean age of children in the infant program was 10.57 months, and the mean age of children in the 2 to 4 program was 31.74 months.

Observation Schedule and Recording Methods

The daily activity schedule of the children generally consisted of periods of free indoor and outdoor play, workshops, field trips, and rest in addition to routine care, such as feeding, bathing, and toileting. Observations were made during both indoor and outdoor free-play situations. The amount of time spent in free play varied with the age of the child and the season of the year, but it averaged 3½ to 4 hours per day. Records were maintained of an adult's presence in the play space. If an adult entered and interacted with or disrupted the children's activity for more than several minutes, observations were discontinued. The observers sat stationary in the play areas in full view of the children. They moved only when the children left the general play area. No services were offered to the children nor was there any conversation with them. Only in cases of extreme aggression or danger did the observers interfere in their activities. Thus, the children seldom turned to the observers for assistance. Some children invariably offered greetings upon entrance, while others merely glanced in their direction.

Observations were direct narratives using pen and paper with appropriate abbreviated forms for speed of notation. The entire group of children present was scanned continuously, and an incident sampling procedure was used (see Smith and Green, 1975). Observations were recorded on weekdays, sampling from 0630 to 1700 hours. A total of 860 hours of observations were collected during the period from January 14 to December 31, 1974.

Procedure

Records of aggressive and submissive behaviors were maintained. Dyadic interactions in which an aggressive behavior by one child was followed by a submissive behavior in the other were recorded as "fights" and considered reliable indices of dominance. Dominance was reflected by the outcome of the interaction (i.e., submission following aggression) and not necessarily by the winner of an object in a fight involving possession of property. In other words, the critical factor in determining whether an interaction was a dominance encounter was the presence of a submissive response. The criteria employed in this study are virtually identical to that used in studies of nonhuman primates (Missakian, 1972; Sade, 1967).

Results

Table 1 lists the aggressive and submissive behaviors assembled during the year of observation. Not all individual children or age

Table 1. Behavior Categories of Aggression and Submission

	Aggression	
Physical contact		
beat	lean on	roll on
bite	mouth	rub
bump	object hit	scratch
butt	pat	shake
carry	pinch	sit on
crawl on	poke	slap
crowd	pull	spank
drag	pull at	squeeze
drop on	pull away	step on
dump	pull down	tackle
fall on	pull hair	take
grab	pull on	tap
hug	push	throw at
jump on	push down	trip
kick	push hit	tug
kick down	push off	walk on
lay on	repeat hit	
No physical contact		
bat	object bat	stare
chase	open mouth	stretch face
demand	raise arm	verbal insult
displace	roar	verbal threat
grunt	rush	wide eyes
lunge	short chase	
	Submission	
avert head	get off	walk away
beat head	release	blink
crawl away	jump away	grimace
cringe	kick feet	pout face
drop	let go	short cry
fall	pant cry	scream cry
fall prone	pull back	whine cry
flee	run away	gasp cry
flinch	step back	long cry
freeze	stomp	

groups of children exhibited all of the behaviors listed in Table 1. Double-observer reliability tests were performed on a total of 105 interactions and Spearman rho correlations on agreement were calculated. Agreement between observers regarding identification of aggressor and subordinate was +.952. Agreement of involvement of property in the interaction was +.819. Each of the observers ranked the children in terms of their dominance position. For infants 6 to 20 months of age the correlation for observer ranking was +.915; for children 20 to 45 months the correlation was +.986.

Dominance matrices based on observed interactions were prepared for the infant and 2 to 4 programs. A dominance matrix (see Tables 2 and 3) is a graph plotting frequency of aggression and submission for all possible pairs of children. Matrix linearity, the degree or percentage of interactions in a predictable direction, was determined. For example, with three individuals the relationship may be the following: A dominant to B, B dominant to C, and A dominant to C. In this example, the relationship between A, B, and C is 100 percent linear. Knowledge of the relationship between A and B enables prediction of the relationship between A and C. There are no reversals in the relationship between individuals. In the following case a triangular relationship is described: A dominant to B, B dominant to C, and C dominant to A. Knowledge of the relationship between A and B is insufficient to predict the relationship between A and C.

Knowledge of the degree of linearity in a matrix for a group of individuals enables prediction of outcomes for given dyads in the matrix. For example, a matrix with 90 percent linearity indicates that in 90 percent of cases the outcomes of an agonistic encounter between any two individuals may be correctly predicted.

Linearity was determined by the following formula:

linearity = number of interactions to right of diagonal/total number of interactions

Results

A total of 7,625 individual agonistic interactions were recorded during the period of study. Of this total, 2,247 occurred between children 6 to 20 months of age and 4,658 between children 20 to 45 months of age. The remaining 720 interactions were "mixed" (i.e., they involved children from two different age groups).

Table 2. Dominance Matrix of Aggression/Submission for Children in the Infant Program (6 to 20 Months)

Aggressor	*Subordinate*																
	M1	M2	F1	M3	M4	F2	M5	F3	M6	F4	M7	F5	M8	F6	M9	M10	Total
M1		64	34	31	8	45	13	1	4	4							204
M2	17		24	12	7	14	1	4	2	10							91
F1	3	2		5	80	218	59	80	59	75	6	13		1			601
M3	1		6		3	12	9	4	2	3							40
M4		1	30	2		32	41	37	35	55	4	6					243
F2			14	5	21		26	49	44	37	1	4		3			204
M5	1		5	1	8	12		23	30	15	1	2					98
F3					4	2	7		67	55	23	31	26	12		1	228
M6	1		2			4	5	37		46	32	41	34	19	1	2	224
F4	1		4		2	7	7	13	36		12	15	11	7	2	4	121
M7								1	1	1		24	12	39	4	9	91
F5							1	2		3	16		9	15		3	49
M8								2	1	5	9	8		11		5	41
F6											6	2	1			1	10
M9														1		1	2
M10																	
Total	24	67	119	56	133	346	169	253	281	309	110	146	93	108	7	26	2,247

Table 3. Dominance Matrix of Aggression/Submission for Children in the 2 to 4 Program (20 to 45 Months)

Aggressor	*Subordinate*																							
	M1	F1	F2	F3	M2	F4	F5	F6	M3	F7	M4	M5	M6	M7	F8	M8	F9	F10	M9	M10	M11	M12	F11	Total
M1		25	19	46	7	10	30	6	7	20	34	1	1	20	7	16		1		1		1		252
F1	4		12	11	1	9	16	5	3	18	7		4	5	8	6		2						111
F2				9	13	5	45	6	10	17	27	3	3	16	27	7	5	25	1	3	1	8	1	232
F3	15	6				8	25	6		17	34		6	21	21	5							1	165
M2							12	9	24	9	71	84		47	10	18	29	28	3	17	4	41	17	422
F4	2	4	1	4			10	10		11	11			20	12	4								89
F5	3	1	9	3	12	1		16	11	25	42	6	4	80	39	20	10	28	4	4		1	3	322
F6		2	12		1	2	4		2	44	17	3	1	3	33	9	2	12				1	7	155
M3		2	3		23		7	2		3	104	82		57	75	44	24	8	7	31	9	54		535
F7		1	2		2	1	13	7	1		52	3	5	35	39	18	11	15	6	1		9	4	226
M4	1		2	1			1	4	2	6		34	37	203	272	194	74	3	2	41	8	61	48	994
M5					1		1		2		7			14	22	19	27	6	2	24	3	39	21	188
M6								1			8			47	23	12								91
M7	2					1	4			1	28	9	5		61	51	21	1	2	6	1	10	7	210
F8				1			2			1	21	4		32		80	2	2		11		18	13	187
M8	1				3	2	3	2	3	6	3	13		8	33		53			34	2	23	23	212
F9			1							3	1	1		1	2	11		1		12	2	17	43	95
F10			1				1			2	1	1			1		2		5			2	1	17
M9							1																	1
M10															2	3	6				3	41	9	64
M11											2				1	5	7			3		3	3	24
M12										1	1	1		2	3	1	14	4	3	16			11	57
F11									1		1			2			3				1	1		9
Total	28	41	62	75	63	39	175	74	66	184	471	245	66	613	691	523	290	136	35	205	34	330	212	4,658

Dominance matrices based on observed interactions within programs were prepared and appear in Tables 2 and 3. The dominance hierarchies for both of the groups of children ranging in age from 6 to 20 months and 20 to 45 months were linear. Linearity for the matrix of the infant program was 85.7 percent and 89.5 percent for the 2 to 4 program. In other words, 85.7 percent of the outcomes of agonistic interactions in the infant program were in a predictable direction while 14.3 percent represented reversals (i.e., lower ranking children dominant to high ranking). The reversals indicated that these children live in a highly predictable environment or social structure in terms of their power relations.

Dominance matrices and resulting hierarchies are based on *whom* a child fights—not on *how much* the child fights. When children are grouped into categories of high, mid, and low rank, the results reveal that similar-ranked children fight more than children far apart in the hierarchy and that there is less fighting between children at the top of the hierarchy as compared to low-ranking and mid-ranking children.

The dominance matrices for both the infant program and the 2 to 4 program were correlated with age; the older children in each group were higher ranking than younger children ($r = +.96$ and $r = +.85$, respectively). These hierarchies were also stable over time. Only one child rose more than two rank positions over the course of the year's observations. Children tended to occupy a fixed position or move only one rank in the hierarchy. The most common cause for rise in the hierarchy was when one or more children left the program. Under these conditions all lower ranking children simply rose in rank.

Table 4 presents the total frequency of aggression, number of months observation per child, mean monthly aggression frequency, standard deviation, monthly aggression frequency rank, and dominance rank for 34 children. The range of mean monthly aggression for boys was .50 in 86.4 instances. The range for girls was 1.7 in 60.5 interactions. The mean monthly aggression frequency for boys was 31.0 ($SD = 28.2$) and 19.4 ($SD = 16.6$) for girls. In general, boys seemed to be more aggressive than girls.

Data were grouped into four age classes: 7 to 12 months, 13 to 24 months, 25 to 36 months, and 37 to 45 months. Some children appeared in two age groupings since they aged during the course of the year's observations. The four age classes were determined by the school staff on the basis of groupings for workshops and outdoor activities and were not chosen arbitrarily by the author.

Table 4. Comparison of Frequencies of Aggression for Boys and Girls

Name	*Months Observed*	*Aggression Frequency*	*Mean Aggr.*	*SD*	*Rank Aggr. Frequency*	*Dominance Rank*
Boys						
Ar	11.5	994	86.4	49.7	1	11
Jd	11.5	210	18.3	13.8	21	14
Ka	2.5	91	36.4	21.8	8	17
At	2.5	40	16.0	10.8	23	24
Jo	3.5	243	69.4	57.5	3	25
Ma	11.5	416	36.2	27.2	9	16
Da	6	252	42.0	21.5	6	1
An	4.5	98	21.8	14.4	13	27
Be	3	91	30.3	28.2	10	12
Is	11.5	224	19.5	21.8	18	29
Mx	8	91	11.4	4.6	26	31
Ch	7	64	9.1	5.4	28	21
Ao	7	422	60.3	29.2	5	5
Sm	7	57	8.1	12.3	30	23
Dm	7	535	76.4	31.9	2	9
Ja	7	188	26.9	16.5	12	13
Jr	5	41	8.2	5.5	29	33
Za	2	24	12.0	13.4	25	22
Br	2	1	.5	3.5	34	20
Girls						
Al	6	10	1.7	1.7	33	34
Em	7	17	2.4	0.9	32	19
Dl	11.5	187	16.3	11.4	22	15
Aa	11.5	696	60.5	53.7	4	18
Le	11.5	232	20.2	9.9	14	3
Mo	6	111	18.5	9.5	19.5	2
Na	11.5	226	19.7	17.1	17	10
Re	4.5	89	19.8	16.1	15.5	6
St	4.5	165	36.7	22.2	7	4
Si	11.5	213	18.5	21.3	19.5	26
Lh	11.5	121	10.5	4.5	27	20
Er	11.5	155	13.5	9.5	24	8
La	11.5	322	28.0	11.9	11	7
Jf	11.5	228	19.8	13.6	15.5	28
Ca	9	49	5.4	2.9	31	32

Table 5. Monthly Aggression Frequencies for Four Age Groups

	Boys			Girls		
Age	N	$\overline{X}$	SD	N	$\overline{X}$	SD
7–12 mos.	6	9.45	7.96	5	6.15	4.02
13–24 mos.	12	47.91	22.04	12	27.61	15.33
25–36 mos.	12	40.99	17.85	11	22.08	11.13
37–45 mos.	7	34.79	23.22	9	16.82	5.76

Table 5 presents means and standard deviations of aggression frequency for boys and girls in the four age classes.

As can be seen in Table 6, age was a significant factor ($F = 10.65$; $df = 3$; $p \leqslant .01$). Aggression for children 7 to 12 months of age was low, peaked at 13 to 24 months, and declined from 25 to 45 months of age. Gender was also a significant factor ($F = 14.61$; $df = 1$; $p \leqslant .01$). There were no gender differences in aggression for children 7 to 12 months of age. In the remaining three age classes, however, there were significant gender differences—boys were more aggressive than girls. The interaction of age and gender was not statistically significant ($F = 12$; $df = 3$; ns).

The 34 children were ranked on their mean monthly aggression frequency (see Table 4). A Mann-Whitney *U*-test revealed no significant difference between boys and girls on this ranking, although a trend was indicated ($z = 1.34$; $p \leqslant .09$). The children were assigned a dominance rank on the basis of the matrix. The ranks were classed as high, middle, and low, and boys and girls were assigned to each rank (see Table 7). The mean dominance rank of boys was 18.6 and the mean dominance rank of girls was 16.1 Girls fell into high, middle, and low ranks of the dominance hierarchy, as did the boys.

Interactions were grouped by the gender of the aggressor–

Table 6. Summary Table: Analysis of Variance for Gender × Age

Source	*SS*	*df*	*MS*	F
Gender	3,848.87	1	3,848.87	14.61*
Age	8,418.40	3	2,806.13	10.65*
Age × Gender	94.64	3	31.55	.12
Within cell	17,388.63	66	263.46	

*$p \leq .01$

Table 7. Number of Boys and Girls in High, Middle, and Low Ranks of the Dominance Hierarchy

Rank	Boys	Girls
High (ranks 1–11)	4	7
Middle (ranks 12–22)	8	3
Low (ranks 23–34)	7	5

subordinate pair: male–female, male–male, female–male, and female–female. Table 8 contains the observed and expected frequencies of aggression per gender category. The observed and expected scores were adjusted according to the specific number of pairs for each gender category. There was no significant difference in the frequency of aggression involving two females, two males, female and male, and male and female pairs ($X^2 = 1.99$; $df = 3$; ns). However, it can be seen that both female–female and male–male aggression was higher than expected by chance. Female–male aggression was lower than would be expected by chance. Male–female aggression was at chance level.

Discussion

The basic findings of this study, with respect to gender differences, are the following:

1. In overall frequency of aggression boys seem to be more aggressive than girls.
2. Ranges of aggression frequency for boys and girls overlap, and some girls are more aggressive than some boys.
3. There is no gender difference in rank aggression frequency.

Table 8. Observed Aggression between Various Gender-Pair Combinations

Aggressor	Subordinate	Observed Frequency	Expected Frequency	No. of Gender Pairs	Percent of Gender Pairs
female	female	1,600	1,425	210	18.7
female	male	1,458	1,935	285	25.4
male	male	2,551	2,323	342	30.5
male	female	2,009	1,935	285	25.4

4. Girls and boys appear in high, middle, and low dominance positions.
5. There is no significant difference in male–male, female–female, male–female, and female–male aggression.

Generally, the results showed more reduced gender differences for this group of Synanon-reared children than would be expected from most of the literature reviewed. Within-gender variability in aggression frequency was found. The fact that some girls are more aggressive than some boys has been reported frequently in the literature (see Maccoby and Jacklin, 1974). The data of this study are based on observations of children who are younger than other investigators have examined. In this sense, the data may be said to supplement rather than contradict previous research findings.

The present study's findings are not consistent with the majority of reports of male–female aggression. The primary reasons for this discrepancy appeared to be (1) the younger age of the Synanon-reared children (they are younger than home-reared children studied), (2) the high frequency of aggression among Synanon-reared girls and its subsequent effect on social structure, and (3) the use of behavioral measures of dominance rather than commonly used "report" procedures.

Aggression between two females has been reportedly lower than either male–male or mixed-gender aggression. Although the findings were not statistically significant, there was a clear trend for both male–male and female–female aggression to be higher than would be expected by chance. In other words, like-gender aggression was high, and the specific gender of the pair did not matter. Again, this finding appeared to be related to the high frequency of aggression among Synanon-reared girls. Prior research would have predicted that male–male aggression would be more frequent than male–female, which would be more frequent than female–female, which would be more frequent than female–male aggression.

The fact that Synanon boys and girls appeared in high, middle, and low dominance positions also supplemented previous reports that either (1) dominance is not a useful concept in describing the power relations of girls or (2) boys and girls have entirely separate hierarchies, hierarchies that have no overlapping. In the latter case the female hierarchy is reportedly lower in rank and less stable than the male hierarchy. Instead, a single dominance hierarchy encompassing boys and girls was found. Girls appeared in all rank categories including the high ranks.

That the study group involved peer-reared children was

perhaps the single most important reason for the discrepancy between the study's findings and those of prior investigators. Synanon communally reared children represent a radical departure from home rearing, and this difference may be responsible for variations in development, complexity, and frequency of social behavior. Because they live together 24 hours a day, these children have both the opportunity and the necessity to establish predictable social bonds and behaviors. This is especially true in the area of power relations. It is important, in stable long-term groupings, to establish predictable nonrandom patterns of aggression and to control intragroup aggression. The dominance hierarchy reflected in the observations is such a pattern of social organization. Bernstein (1974) notes that "if dominance exists as an organizing principle, then a knowledge of prior interactions should allow one to predict the outcome of future interactions" (p. 290). (A more detailed discussion of the dominance hierarchy is presented in a separate paper. See Missakian, 1976.)

The Synanon child-rearing situation is one involving minimal gender-role stereotyping. There is a deliberate attempt not to disrupt children during periods of free play—a time when much aggression ordinarily occurs. If two children are fighting, an adult interferes primarily if aggression is severe (i.e., biting, hair pulling). Since aggression is not negatively reinforced, one sees a high frequency of fighting (one instance per 6.77 minutes of observation). There is also an attempt not to reinforce aggression differentially according to gender. In other words, male aggression is not positively reinforced and female aggression not negatively reinforced. This appears to increase individual variability in aggression for both boys and girls.

Chapter 24

The Social Ecology of Toughness

Richard Parker
Donald R. Omark

Being an exploratory extrapolation, this chapter starts with previous research on children's views of their social world and adds a twist to that train of thought by introducing another variable, the sex composition of groups. Innocuous as this may sound, it becomes a potentially rewarding tack within an ethological framework and/or a mild-mannered attack on some ethological formulations, depending on one's disposition. The following text is an effort toward a more precise elaboration of these simple points.

To avoid confusion, an initial review of the research background is in order. This line of investigation began with an examination of how peers in the age range from nursery school through third grade (approximately 4 to 8 years of age) perceive themselves in relation to others, and how they perceive interrelationships between these others (Edelman and Omark, 1973; Omark and Edelman, 1976). The peer groups studied were fairly typical examples of Western-style classrooms, populated by slightly typical middle-class and upper-middle-class bright students. As part of these studies, children were asked to construct rank-order hierarchies by having each child rate every child in the class according to "Who is the toughest?" "Toughness" was considered the child equivalent to the ethologists' "dominance." The toughness hierarchies were then subjectively indicated classroom dominance hierarchies.

The approach, it should be noted, differs from that taken by others interested in human ethology. The studies in Blurton Jones' book (1972a) and McGrew (1972) all chose to approach the problem of developing an ethological method of studying children by elaborating a system of observable behavior classifications. In contrast, Omark and Edelman's main concern was to operationalize an analytical concept.

These latter authors expected that children would follow generalizations from primate studies, such as Kummer's (1971, p. 151) formulation, "The primate male, in general, is more aggressive and more dominant than the female." Indeed, in analyzing the toughness rank orders several discoveries were made. Agreement between all the raters in grade-school classes seemed sizable enough to demonstrate the existence of a "toughness" dominance hierarchy. Finer inspection of the hierarchies revealed significant sex differences. Boys tended to be at the top of the hierarchy, while girls constituted the bottom, with considerable overlap in between, as can be found in many primate troops. Also, in considering where raters placed themselves in comparison to where everyone else in the class placed them, boys overrated their personal position more frequently than did girls. This "overrating" was held to indicate the emotional salience of toughness for a sense of self. The boys' greater concern with toughness and the girls' significantly less, but generally still high, response reflected the relations in most primate species where males are the dominant animals.

The biological rootedness of these differences was taken as a working hypothesis. Extensions of the United States study to cross-cultural populations in Switzerland and Ethopia confirmed many of the original findings (Omark, Omark, and Edelman, 1975). A problem in all three cultural settings, however, was that Western-style classrooms may have produced the situations necessary for the children to interact in similar ways. In this regard the work of Draper (1975; this volume, Chapter 25) is important, as it left the working hypothesis intact. Even though the !Kung Bushmen children she observed were not in a Western-style classroom, they still exhibited many of the primatelike patterns of "free-time" interaction shown by the more Westernized groups during recess or free-play activities.

All of the above studies contained both boys and girls in the same environment, as is typical in primate groups.[1] Since human cultural groups do produce other social situations, the authors deemed it wise to examine differing arrangements in social ecology. The particular variant examined here is when boys and girls are placed in classrooms without the other sex present.[2] The basic question asked of this situation was, Do the sex differences in toughness found in mixed-sex classes also exist between same-sex classes of boys and girls?

There are two logical possibilities in comparing mixed sex and same sex classes, either the difference between boys and girls in same sex classes would be similar to the sex differences already described, or else they would not. To find the former implies fairly strongly that indeed, in a fundamental way, boys are boys and girls and girls. The differences between them are both pervasive and constant. One more instance is added in the documentation of sex differences. The import of the latter outcome is more ambiguous, but at least it would increase the range of demonstrated variability for which explanations must be sought.

Method

The first problem was finding same-sex classrooms; four were eventually located in the Chicago metropolitan area, two each for boys and girls. Of these four schools, two were not chosen because of small enrollments, multilevel classes (unlike earlier samples), and inconvenient locations. This left two Roman Catholic elementary schools with one class each in grades one through eight. Grades one, two, and three were selected for study in order to match the earlier studies.

This sample of necessity forced recognition of three methodological problems: (1) the comparability of the two same-sex schools, (2) the comparability of the same-sex schools to the private, nondenominational school studied earlier, and (3) lack of any within-grade variation. Taken together, these problems may be summarized as the nonrandomness of the sample.

Against all of this a little defensiveness should be understandable. This study might, in an ultimate sense, be categorized as "salvage anthropology." Whereas many Chicago public schools once had separate playgrounds for boys and girls, none now have such arrangements. Single-sex classrooms are also rapidly disappearing, and some information about them is preferable to none. As always, one seeks to remain cognizant of pitfalls in the path of inquiry without being intimidated from all but the shortest and safest reconnaissance. While reservations about the sampling techniques used here are appropriate, no one seemed equipped to provide better alternatives, such as conducting for three years an educational program complete with teachers, classrooms, curricula, and students assigned in suitable arrays. Neither can one confidently assume that short-term experiments with isolated individuals or small groups will show what happens in naturally occurring classrooms.

A review of the three specific issues provides a more detailed picture of the leaps of faith involved in our argument. The same-sex schools are located on separate floors of the same building, although each has a distinct administration. The students in both schools have similar family backgrounds, which may be summarized as white and middle income. Other than similarities in general geographic, ethnic, socioeconomic, and religious backgrounds, many siblings can be found in the two schools. Overall, the students in the same-sex schools seem quite comparable.

How comparable these students are to those studied by Omark and Edelman is a more difficult question to answer. Both are nonpublic schools and charge tuition. Neither school draws its student from a sharply defined neighborhood area as public schools traditionally do, but most of the students live in the general vicinity of the university in the older study or near the Catholic school in the newer one. Some out-of-school contact with a few peers can be expected in both cases. What influence the difference in religious orientation of students or the institutions may have is difficult to assess. Ideally, research would have been done in both the same-sex schools and in a nearby mixed-sex Catholic school, thus "controlling" for religious orientation. Since the authors lacked the resources for that effort, the study proceeded assuming that the difference in religious orientation was not a powerful influence on the behaviors under study. The control-group design remains appealing, and it is heartily recommended to anyone interested in settling this particular question.

Last of the three, only one same-sex class at each grade level was studied. Hence, variability between same-age/same-sex classes could not be determined. The classes are taken as representatives of their particular social structures for comparative purposes in this chapter. (The authors will gladly share their data with anyone who has the opportunity to extend the range of variation by investigating similar classes.)

Exactly the same procedures used by Omark and Edelman (1975, 1976) were used in administering the toughness "tab" hierarchy test to the same-sex classes. The task for each child was to rank order every child in the class according to "Who is the toughest?" Analysis of the responses proceeds by breaking the rank orders from each class into a matrix of all possible pairs of children. Each pair was capable of agreeing about their relative positions, or disagreeing; i.e., both individuals in the dyad may say they are tougher than the other, or both may say they are less tough then the other. The former disagreement is called overrating; the latter, underrating. Of the several statistics that can be formed from this information, two are examined here. First, the ratio of those pairs that agree relative to all possible pairs is "the percent of dyadic agreement." A class agreement score of 50 percent indicates random responses. Second, the ratio of the number of pairs that disagree by overrating to all possible pairs will be called "the percent of dyadic overrating."[3]

Hypotheses

The familiar reasoning for the adaptive value of primate dominance hierarchies claims that hierarchical organization decreases within-group conflict and promotes group cohesiveness. To the extent that classrooms of children follow a similar hierarchical organization, it was reasoned that all-girl classes would stand equally as much in need of organization as would all-boy classes. Therefore, it was hypothesized that both boy and girl same-sex classes would demonstrate an amount of agreement comparable to that demonstrated by boys in mixed-sex classes. Girls' classes were expected to show a higher level of overrating than did girls in mixed-sex classes, since the same-sex girls would be actively participating in a toughness hierarchy rather than observing one, as seemed to be the pattern in mixed-sex classes. It was also hypothesized that the same-sex classes would agree and overrate about as much as had boys in mixed-sex classes. So, there are two sets of hypotheses concerning toughness—one for the comparison between the same-sex classes and a second set for comparing the same-sex classes to mixed-sex classes.

Results

To measure these hypotheses against the evidence collected, let us first look at one statistic, the percent of dyadic agreement. As a measure of consensus between raters, it was the prime indicator of the existence of a hierarchy in Omark and Edelman's data. Part of

Table 1. Percentage of Dyadic Agreement

	Sex	
School	*Boy*	*Girl*
Mixed	66	58
Same	39	45

their findings are reproduced in Table 1. In mixed-sex classes boys agreed more about their relative positions than did girls ($p \leq .036$), and both sexes agreed substantially more than would be expected by chance (Omark, 1972). Comparable information from the same-sex classes present immediate contrasts. No significant difference exists between boys' and girls' scores in the same-sex classes, although the latter is higher in absolute terms.[4] The mixed-sex classes demonstrated a difference about agreement; the same-sex classes did not.

The same-sex classes are strikingly different from the mixed-sex ones in another regard. The same-sex class scores indicated below random agreement, the same-sex boys being significantly lower than the mixed-sex boys' scores ($p < .01$) and likewise for the girls ($p < .05$), using t-tests for difference of proportions. In summary, Edelman and Omark's classes demonstrated hierarchies and sex differences within them; the same-sex classes not only indicated no sex differences, but no hierarchies as well.

A reason for this lack of agreement can be seen in the other statistic, the percent of overrating, which is presented in Table 2. Omark and Edelman found that boys in mixed-sex classes overrate significantly more than girls ($p \leq .006$), and they inferred that toughness is a more important part of the boys' social world than of the girls. Once again, data from the same-sex classes showed no significant difference between boys and girls, even though girls rank higher in absolute terms. The equality of the sexes regarding overrating in the same-sex schools is also supported by qualitative information. Omark and Edelman reported that boys worked ex-

Table 2. Percentage of Disagreement by Overrating

	Sex	
School	*Boy*	*Girl*
Mixed	29	20
Same	47	51

citedly on the rank-ordering task while girls complied with the experimenter's requests. In contrast, in the same-sex classes the eagerness of the girls matched that of the boys. Looking at the levels of overrating between the two types of schools, it can be seen that the difference in amount of overrating is significant for both boys ($p < .01$) and girls ($p < .001$). When half of the pairs in any class overrate, it is no wonder that agreement falls below random levels. Consequently, neither boys nor girls in the same-sex classes showed much agreement about a hierarchy.

When one considers the mixed- and same-sex classrooms as different conditions under which the same measures were taken, it would seem that in same-sex classrooms it is as important for girls as for boys to be tough. Both assert their personal toughness far more than boys in mixed-sex classes, who were the previous record holders. Apparently, for boys in the same-sex classes toughness remains a valued dimension of social life; for girls it becomes one; yet for both it is amorphous and ego-centered. It was originally hypothesized that girls in the absence of boys would organize their classrooms much as did boys in mixed-sex classes. Instead, boys and girls in the absence of the opposite sex were found to evidence a condition of egocentric equality regarding dominance.

Discussion

The above set of results was certainly different than had been expected. Had Edelman and Omark found results similar to those found in the single-sex classrooms, they would have had little to report. Let us accept both the mixed-sex and same-sex findings. Principles should then be advanced that can begin to make some sense of both. Let us start with the conception of dominance. Rowell (1972) noted that the term "dominance hierarchy" is something of a misnomer; perhaps a better term would be "submission" or "subordinance hierarchy," since without submission "dominance encounters" would be nothing but a series of fights resulting in no social organization. Besides this should be placed Chance's (1967) idea that a dominance hierarchy is an attention structure. Attention is directed upward in hierarchies (see Abramovitch, this volume, Chapter 22). Emerging dominant animals may gain and maintain their positions by displays (attention-getting behavior), some of which may be acts of physical aggression.

Drawing together the ideas of submission, attention, and display, we add a fourth element, the audience. It is suggested that the presence of an opposite-sex audience is important in producing the sex differences found in mixed-sex classes. Opposite-sex others call out display behavior, facilitating the formation of a hierarchy. This organization may be dependent on a cognitive contrast of similar and different (see Omark on the Umwelt, this volume, Chapter 13). In class hierarchies solidarity is built upon acquiescence to a lower than top position. This may be most attractive when action is public, i.e., constantly under scrutiny of perceived others, such as "the boys" or "the girls."

Beyond an interest in mechanisms, the same-sex data have notable theoretical import because they increase the documented variability of behavior. The same-sex data indicate that relatively low amounts of agreement or high levels of overrating concerning toughness are not invariant characteristics of a gender, but they vary considerably depending on social ecology, for one thing. This contribution to human ethology is somewhat similar to previous developments in the study of primates. The first field studies of baboons, for example, were thought to document baboon behavior. Later studies in different areas showed that the initial reports were, in fact, descriptions of baboon behavior in particular habitat conditions (Crook, 1970b).

To handle conceptually such variability, several researchers have considered adoption of the role concept. Both Hinde (1974) and Crook (1970b) have reviewed its use in ethological studies; Hinde basically fearing that the fuzziness of the sociological concept will impede theoretical development, Crook hoping that role ideas will enable theory in ethology to catch up with the field's empirical progress. Crook suggests that roles be considered as pigeon holes into which various animals may be fitted. One can, for example, posit a leadership role and observe a new animal assume it when its incumbent dies or disappears. This assumes a constancy of the structure of pigeon holes while attending to the flexibility of individuals to fit them. Our own work implies this should be taken a step further. The very structure of pigeon holes may vary with social ecology. Thus, one may be able to observe and specify what kinds of roles or role structures exist under different ecological and social conditions.

The most concrete implication of this study springs from the overrating data. Overrating is considered as an indication of the emotional salience of toughness, and it was found that in the same-sex classes boys and girls were equal in asserting their personal toughness. This difference between mixed-sex and same-sex classes is intriguing when compared to the literature on sex differences concerning achievement motivation and aspiration. Several studies have shown that male and female subjects respond differently to achievement situations, e.g., Atkinson (1958) and French and Lesser (1964). Studies by Horner (1970, 1972) have posited a "fear of success motive" to explain such findings. Winchel, Fenner, and Shaver (1974) examined the relationship of "fear of success" to coeducation and found that the most powerful predictor of lack of fear of success among high school girls was attending an all-girl elementary school. Less rigorous information also indicates that graduates of women's colleges are disproportionately represented among successful women in business and the professions (Carnegie Commission on Higher Education, 1973).

In light of such findings, it may be speculated that the self-assertiveness tapped in relation to toughness, an early developed and salient dimension of social life, is part of an egocentricity incorporated into a developing self, a motivational egotism essential to competitiveness. If one wished to "produce" more competitive females, creating same-sex elementary classrooms should contribute to that end.[5] Ironically, coeducation may in this way increase sexual inequality. There is no evidence that same-sex classes decrease the importance of toughness to boys. Thus, methods of social engineering can be suggested which boost concern with toughness, and thereby competitiveness, but not comparable ways to lessen its salience.

If children's cognitive abilities are directly affected by their social experiences, research needs to be directed at the ecological, including social, factors that interact with cognitive development. Single-sex groups—all male—are also examined by Savin-Williams (this volume, Chapter 12). He found that even in adolescence (12 to 14 years of age) 40 percent of the boys overrated themselves by one or more positions in relatively small groups (six in size). Despite numerous years of mixed-sex classroom experience before coming to camp, these adolescent boys in single-sex groups display a great deal of egocentricity. This suggests that different-age individuals function, cognitively and behaviorally, in similar ways when in such groups. That being in single-sex groups may have ramifications at other phylogenetic levels is indicated by examinations and comparisons with the above children's data among mixed-sex and single-sex rhesus monkey populations (Goy and Omark, in preparation). The development of group formation and its relationship to conceptions of the self and others is much in need of examination from an ethological perspective.

What the roles are in single-sex groups needs further exploration. Children's groups or experimental monkey groups are obviously constrained populations that have to continue within their present social groups because of parental pressures or cage bars. But such groups also occur more naturally—gangs, army units, work groups, male bonds of primates, and so forth. In at least some of these units membership is voluntary. Some such units are also beginning to include females. Contrasts could be made between older individuals in mixed-sex and same-sex groups to see how cohesive they are and what differences in roles have developed. To survive, same-sex groups should not be expected to be leaderless (despite the fact that young children appear to be very egocentric). At the same time, if same-sex groups do exhibit more egocentricity, then there should be more pressure on the leaders that emerge and more internal bickering about positions. The results remain to be seen. The leaders that emerge might be expected to be very successful when and if they join with a mixed-sex group, given their experiences in trying to control a single-sex group. We suggest that ethologists begin to explore the adults of our species to see what really may be happening behaviorally and cognitively.

Conclusion

A relationship has been demonstrated between social ecology and individual activity that increases the documented range of variability in human behavior. This contributes a tone of modesty to formulations in human ethology, which have sometimes been given to phylogenetic sweep and genetically preprogrammed fixity. In that respect, this study lends its support to something of an "environmentalist" understanding of sex differences in toughness and their relation to human potential. Indeed, this deepest and broadest implication seems essentially harmonious with the classic statement of Rousseau, arguing in his *Discourse on the Origins of Inequality*, that social situations powerfully affect human action. At the same time the concrete referents of the variability examined here seem to increase the pervasiveness of ethological dimensions of social life. Again to cite the classic statement, Rousseau identified the roots of mankind's evil as the tendencies to dominate, to be admired, and to show off (Palmer, 1971). We have shown that such behavior may be increased dramatically.

Having reached such nice, if tenuous, conclusions, we are forced to throw ourselves back on the thorns of method to emphasize the potential contribution of additional work in this vein. There are, naturally, competing hypotheses available to explain away what has been indicated in this study. A solid ethnoscience investigation is needed to verify if toughness and dominance at higher grade levels can be linked as closely as presented here. Questions remain about cognitive influences in the rank-ordering task used to collect judgments of toughness and about now both the influence and the task may change developmentally.

As well, the problem has not been considered about the relationship of class size and children's conceptual abilities when dealing with larger or smaller groups. Other things being equal, one might expect agreement to decrease with increases in class size. Similarly, overrating, as a characteristic of children when dyads do not agree about relative position (Omark and Edelman, 1976), might be expected to increase. Abramovitch (this volume, Chapter 22), indicates that group size can produce such conceptual difficulties among preschool-age children. Comparing how boys rated boys and girls rated girls in mixed-sex classes to same-sex classes has meant comparing the ranking produced by varying numbers of children. A class-by-class inspection of the same-sex boys' data showed a relationship between class size and the relative magnitude of agreement and overrating. However, this relationship did not hold for the girls' classes (Parker, 1976).

Since the speculations developed in this chapter are based primarily upon the strangeness of the girls' same-sex classes, we may still hold to our interpretations. In addition, the earlier research on mixed-sex classes demonstrated that the boys were as accurate in perceiving the girls' portion of the hierarchy as were the girls in perceiving themselves. This finding also held for the girls perceiving the boys portion. This suggests that children in the first-to-third-grade age range can conceptually handle children's groups of about 25 to 30 with little difficulty. The problem with the results reported here would then appear not to be one of the size of the group but in its composition. This idea is developed in more detail in Chapter 13, "The Umwelt and Cognitive Development." Further research on the relationship between the self and group structure is obviously needed, and the presence or absence of the opposite sex should be counted among the conditions affecting both.

Notes

1. In primate groups of many species the juvenile males and females at some age period frequently form separate play groups during part of each day's activities.
2. Other possible groupings include raising a child in isolation, manipulating small groups for short periods of time, or structuring classes composed of all leaders. Out of some 40 classrooms so far studied for toughness, 2 were apparently inadvertent examples of the latter type. Agreement as to relative placement of the self vis-à-vis other peers was lower than expected, fighting was high and continuous across the school year, classes were noisy, and even experienced teachers had difficulty handling these classes (data in preparation).
3. The overrating statistic used in this paper differs from that originally employed by Edelman and Omark. For a discussion of this difference and methods of conversion from one to the other see Parker (1976).
4. In comparing percentage point differences between the mixed and same sex data, remember that the number of cases represented in the same sex row is quite small compared to those represented by the mixed sex digits. Not too dissimilar percentage point differences may indicate quite different variations in the two samples.
5. This is not to say that competitiveness is "good."

Chapter 25

The Interaction of Behavior Variables in the Development of Dominance Relations

Patricia Draper

Dominance interactions among individuals undoubtedly have both a biological basis and evolutionary significance. Literature on these topics has been copiously cited elsewhere in this book and will not be repeated here. Sex differences in behaviors which separately and together culminate in the ability of an individual to win agonistic encounters are also documented both for human and nonhuman primates. Less attention has been paid, however, to the process by which sex differences in the propensity to behave in a given way interact with socialization experiences of children to produce the familiar usual pattern of male dominance. This chapter proposes to outline a series of probable mechanisms whereby behaviors that are either innate or easily learned become the basis for two reasonably distinct modes of action and interaction by human males and females. Many of the behaviors have analogues, and presumably homologues, in nonhuman primate behavior.

For the purposes of this chapter the two sexual modes of action will be treated as dichotomies to emphasize certain aspects of differentiation which are pivotal for understanding the implication of sex differences for dominance interactions. In fact, with respect to most behavioral variables the sexes evince overlapping distribution, with a sizable proportion of girls and boys showing comparable frequencies of occurrence of the behavior in question. Also true, but often overlooked in such discussions, is the fact that innumerable behavioral, physiological, and psychological variables have been and could be isolated that show *no* regular pattern of sex differentiation but much individual variation. The notion to be developed here is that there are target areas of behavior which are not only sex differentiated but also differentiated in ways which portend interestingly for inequality in certain types of interaction situations.

When dealing with humans and the issue of the biological basis of any or all behavior, one must consider whether cultural conditioning may not account for some part of the behavior or behavior complex which is under scrutiny. In the analysis which follows the subject will be examined at an elemental level so that the question of cultural influences can be sidestepped. In a later section the topic of sex-differentiated behavior in the context of a particular culture (!King Bushman) will be considered.

There are in the individual a bundle of sex-specific proclivities to behave. These proclivities are furthered by various forces, each of which in isolation carries no directional value but which, in the context of other forces contribute to the characteristic path of the proclivities.

Characteristic Behavior of Infants and Toddlers

In the case of the female there are several behaviors which emerge relatively soon after birth. These, if interpreted in a certain light, are harbingers of things to come. Girl infants (1) smile more frequently than do boys of the same age (Korner, 1974), (2) seem to show preference (as measured by longer attention span) for gazing at human faces rather than other objects and designs (Lewis, Kagan, and Kalafat, 1966), and (3) seem to prefer to listen to unusual and nonrepetitive sounds rather than monotonous, sonorous sounds (Kagan and Lewis, 1965). In all likelihood, these represent responses in the infant which are initially unconditioned but which, given the usual situation of an infant being surrounded by at least one attentive parent, rapidly become conditioned responses. That is, willy-nilly, the infant learns that to give a smile, or a stare, or to show the attitude of listening (separately or all together) brings the caretaker closer and holds her there for longer. Caretaker response as well can become conditioned by different infant signals. Mothers, regardless of their views on the desirability of treating girls and boys equally, may find that what works for a girl with respect to quieting, or predicting, or intervening will be less likely to work with a boy (Moss, 1967; Lewis and Rosenblum, 1974). In addition, as early as 1½ years of age girls show marked linguistic superiority over boys in terms of vocabulary, length of utterances, and understanding of grammatical rules (Maccoby, 1966; but see Maccoby and Jacklin, 1974). By these means an Every Girl has been set in motion. The four elements just mentioned are a bundle of proclivities which when released by birth upon an unsuspecting human audience will start a characteristic set of experiences for the girl.

By the time girls and boys can crawl and walk another aspect of sex-differentiated behavior appears. Girls explore a space that is more restricted than that utilized by boys. In this sense they are less active—that is, they cover less ground in the same amount of time. A number of studies show that girl toddlers in experimental and naturalistic settings remain closer to the major caretaker or the home base than does the average boy of the same age. While staying closer to the caretaker in spatial terms, girls show a heightened interest in the caretaker as evidenced by their greater frequency of looking back at the caretaker and physically returning to touch that adult before moving out again (Goldberg and Lewis, 1969).

Boy toddlers do these things markedly less. It appears that they are not uninterested but less interested in the caretaker as an orientation point. One forms the impression that boys enjoy their own bodily activity and the play with and manipulation of inanimate objects. They do not disregard the caretaker, but in comparison with girls they are freer of influence from that quarter. On the average they are farther away from a caretaker and when the opportunity provides, often out of sight. In this way, and apparently at their own instigation, boys inhabit a socialization environment that is, in comparison with the convivial niche preferred by girls, freer of adult supervision and ultimately adult control. It appears that continued closeness and eye contact with the familiar caretaker is less rewarding to boys, and there is not the same gravitational pull on them back to a place of more intimate socialization. A boy puts himself on a longer leash at an earlier age. Remembering his lesser social orientation and lesser command of language, we can best understand Every Boy as around the corner, out of sight, pulling on his elongated tether and wearing earphones in which the volume has been turned down.[1]

Every Mother stands outside the evolving of these patterns, but she learns, consciously or unconsciously, what will work. Daughters (in the 2- to 5-year age range) can be controlled relatively easily because they are more receptive to social cues (verbal and nonverbal) and because they are spatially closer to the mother. The socialization of a girl (or the schedule of her rewards and punishments) will be more consistent, contingent, and hence more efficient. The attribute of compliance in the female behavioral pattern can in this way be seen as the end point in a snowballing trajectory.

These daughters know what is expected of them by the adults in charge, although, of course, they are often disobedient. Nevertheless, they are astute observers and develop a quality of carefulness in their deceits that follows from their peculiar type of young female intelligence. Boys, on the other hand, ply their outlawry in an innocent vacuum. When apprehended, they have a look of astonishment, not guilt.

Peer Interaction

The literature on children and young primates shows parallel types of sex differentiation in other behaviors. With increasing age juveniles of both sexes move into greater peer interaction, though males exceed females in frequency and duration of peer contact and in the distance they put between themselves and their caretakers. Males outdistance their female age-mates in aggressive encounters and rough-and-tumble play. Males consistently group themselves with larger numbers of their peers than girls do. Girls prefer smaller groups where the interaction is usually quieter and of a less competitive and physically challenging nature (Omark, Omark, and Edelman, 1975).

The male preference for aggressive display and encounter is surely biologically based. Whether the human male's drive to compete—to strive to excel another in various skills and performances—is innate or easily learned is uncertain, but all of the evidence indicates that for boys it is more easily developed or acquired than for girls. An ethological explanation would class male striving as merely one of a number of epigamic behaviors. (I.e., display behaviors which have been selected for in the male because they call the attention of others, especially females, to him and enable females to judge vigor and potential worth as a mate. Such display also signals to other males his competitive ability.)

The social dominance which accrues differentially to human males rather than to females seems an outgrowth of the types of innate predispositions and easily learned behaviors described earlier. At young ages adults are clearly dominant to both girls and boys. According to the analysis developed here, however, boys' attention is directed away from adult signals and thus only a fraction of the signals is received by the boys. As a result, even in their interactions with adults, boys receive less training in submission than do girls.

As they mature, boys are drawn to peer interactions where aggressive display and, in most societies, competitive striving are common. While studies repeatedly show a developmental trend in preference for same-sex playmates by both girls and boys, these same studies, and others too, reveal that boys have greater preference for interaction with same-sex individuals than do girls. This finding is not surprising in view of sex differences in reaction to agonistic encounters. Typically, when assaulted by a peer, a young male returns threat display in kind, and a bout of rough-and-tumble play ensues. However, with increasing age, girls (and juvenile female primates) return a threatening initiation from an age-mate with another type of signal—not necessarily with a signal of submission but with a signal of disengagement—one that says, "I'm not playing your silly games!" Unfortunately for the female, she thereby loses another larger battle.

Much of what a male, age 6 to 20 years, learns in the context of peer interaction is where he stands in comparison with others and with respect to various skills. By his willingness to threaten, to attempt to dominate, to challenge another, he gains practice in dominance—whether or not he is himself dominant in a particular social situation. This know-how, fueled, of course, by ancient urges, is what accounts in large measure for the social dominance of males in most heterosexual settings. It also accounts for the widely reported finding of comparatively well-developed dominance hierarchies in male groups.

Groups of girls have dominance hierarchies which are less well developed and less stable (see Tiger and Shepher, 1975, for interesting findings on hierarchies in women's groups). In part, this would be accounted for by the fact that females prefer to operate in smaller numbers, perhaps in numbers below some threshold for which some hierarchical organization would be favored. Furthermore, if girls are less aggressive and less competitive, the mechanism for establishing a rank ordering among individuals with respect to some skill is clearly not operative or only weakly so.

It is the present author's contention that underlying biologically based predispositions in interaction with constants of socialization will produce more or less universally observable styles of action and interaction by males and females. Individuals can, and do, diverge from this pattern, either because of constitutional factors or because the socialization environment in which they were reared was constructed in an unusual manner. The discussion thus far has been aimed at a probable path for a particular behavior. Indeed a fascinating subject, but one outside the scope of this chapter, is what modern parents of non-sex stereotyping persuasion could do to minimize the risk of oversocializing their daughters (Bardwick, 1971; Freedman and Omark, 1973; Parker and Omark, this volume, Chapter 24).

Behavior Patterns of !Kung Bushmen Children

Data on the behavior of children among the hunting and gathering !Kung Bushmen[2] of Botswana and South West Africa bear on the kinds of interactions that can occur between cultural and hereditary factors leading to sex-role differentiation[3] (Draper, 1975a). The !Kung (at least those who are still foraging and living nomadically) are an excellent "test case" because they do not put differential socialization pressures on girls and boys.

In most other traditional societies girls and boys from an early age are treated differently, chiefly in anticipation of the work and skills they will need to master as adults. In these technologically simple societies subsistence work is the time-consuming preoccupation of most adults, and, whenever possible, children are enlisted to help in the actual food-producing chores, e.g., managing stock and weeding gardens. In some cases children are excluded from primary food production itself, as when the location of adult work is too far away for children to travel or where the work is dangerous or in some other way unsuitable. In these cases children are asked to remain at home doing other kinds of work which free the parents (chiefly the mother) for other chores. Tending small siblings and cousins and light household work, such as cleaning, washing, drawing water, and the like are typical of this work. Around the world such work falls unequally on girls, who have been shown to experience obedience and responsibility pressure earlier and more severely than boys (Barry, Bacon, and Child, 1957). The usual explanation is that such domestic chores are ultimately the work of women; therefore, when circumstances permit, girls and not boys will be called upon to help. As indicated earlier, and as will be illustrated more empirically below, this is only one possible interpretation of the regularities observed cross-culturally in the sex-role socialization of children.

The data on the behavior of !Kung children of nomadic groups constitute an apparent paradox in the context of the environmentalist argument that sex-role differences are a result of cultural patterning. In the hunting and gathering setting (which, as noted, does not put obvious differential socialization pressures on the sexes) there are, nevertheless, some consistent differences in the behavior of girls and boys which are not exploited or intensified by sex-specific socialization practices. The particular dimensions of contrast in the behaviors of the foraging children are such that, without too much effort, one can see how girls in particular are "preadapted" to the character requirements of the female role stereotype.

Before proceeding further, some general ethnographic information will be provided about the !Kung as well as an explanation of the circumstances in that society which free children from sex-role typing early in life. The Kalahari !Kung are well described in the anthropological literature (Lee and DeVore, 1976; Marshall, 1976). Like other hunter-gatherer populations which were once more populous and wider in territorial extent, the !Kung are much reduced in numbers and today there are only a few thousand people, most of whom have given up their traditional hunting and gathering and live sedentary lives, often in association with the dominant, Bantu-speaking pastoral peoples of the area. There is only a small minority of !Kung-speakers who still live by nomadic hunting and gathering. Of this group, the author personally knew about 120 individuals, who, in various band groupings, were living on the international border between Botswana and South West Africa in 1969.

There are several aspects of !Kung economy and ecology which must be understood and which relate to the assertion that differential cultural pressure does not account for the observed sex differences in the behavior of the nomadic children. The practice of assigning useful tasks to children is essentially nonexistent among the foraging !Kung. Also, older bush children of either sex are not made responsible for tending younger children. Therefore, neither the practice of divergent and sex-specific task assignment nor the unequal assignment of child tending to girls rather than boys can contribute to different experiences for boys and girls of this society.

The nomadic !Kung are a remarkably "leisured" society. Men and women work on the average only about three days per week in quest of food (Lee, 1968). This is due in part to the extremely low human population density in the area of the Kalahari where they roam (one person per ten square miles). Thus there is minimal pressure on available resources. In addition, given their mobility and their complex knowledge of edible plant and animal foods, the !Kung find an extremely rich food supply in the Kalahari. Since food is sufficiently abundant and so readily found, there is no economic need for bringing children into the labor force. In a series of systematic and randomly scheduled behavior observations[4] on !Kung children it was found that girls and boys ranging from 2 to 14 years of age did, on the average, one minute of work per hour of observation. Increasing age had no effect on the amount of work these children were observed to do.

Not only is there no economic necessity for bringing children into the labor force, but there are reasons which the !Kung themselves give for not encouraging children to participate voluntarily in adult work. Both men and women in the course of hunting and gathering, respectively, travel out from camp over a distance of miles. These treks take them over waterless territory and through scorching daytime temperatures. The !Kung realize that if older children went with their parents to hunt or gather during many months of the year, water would have to be brought from camp for them to drink. The children themselves would grow weary and want to be carried. Carrying water and/or carrying the children themselves would substantially reduce the efficiency of the adults. Therefore, and not surprisingly, the !Kung typically discourage children from going on these treks and the children are quite happy to stay in camp.[5]

At the home camp there is no lack of supervision of the children whose parents are working on a given day. Adults alternate their days of food collection with one or more days of rest in camp. There is a tradition among the !Kung to watch over each other's children, and this procedure requires no special arrangements and creates no indebtedness among parents. As a result, neither girls nor boys in this society have a delegated role as child nurse. This is due to the interaction of the economic factors described above with other demographic factors. For example, the average birth spacing between siblings is approximately four years (Howell, 1976). This fact, together with the !Kung practice of nursing into the child's third year, means that the mother–child bond remains intense and strong. So accustomed are these mothers to being able to nurse their young regularly during the day that the mothers routinely carry 2- and 3-year-olds with them on day-long gathering trips. Given the ubiquitous presence of other adults in camp, the !Kung rely on adults, not older children, to do the regular and reliable supervision of the children who are no longer nursed or back-carried. The older children show a great deal of interest in toddlers and infants, but they are not charged with the responsibility of caring for them for periods of any duration. The author found that girls and boys 2 to 6 years of age showed an average number of child-caring acts per hour of observation of zero and .5, respectively; whereas girls and boys 7 to 14 years of age rated an average of 1 and .42 child-caring acts per hour of observation.[6]

On the basis of the previous discussion, one can assert that children of the nomadic !Kung do not work, and the different types of skills, attitudes, and experiences which girls and boys might acquire in such work is not a factor in their sex-role socialization.

The bulk of the behavioral observations done on 36 nomadic !Kung children reveals, however, areas in which the behavior of the 15 girls and 21 boys is differentiated along familiar lines. The sex differences are small, but in the various measures the differences are in the same direction alluded to in the first half of the chapter. (1) Among the !Kung too girls were spatially closer to adults or home base (in this case closer to the center of the camp), whereas boys on the average were farther away. (2) Girls showed greater preference for the close society of adults and less for peer involvement, while boys generally included fewer adults in their close proximity and sought peers instead. (3) Girls had more frequent physical contact with another person than did boys (which is taken to be an indirect measure of their less physically active recreational style). (4) In addition, girls sustained more verbal and nonverbal directions and interferences from adults than did boys (Draper, 1972, 1975a, 1975b, 1976).

The spot observations showed boys to be beyond eye and ear contact with the adults in camp over twice as frequently as girls. Analysis of the same set of observations revealed that girls on the average were found in social interaction with 2.5 adults whereas boys were with 1.9 adults. Girls were found in groups composed of children only about 25 percent of the time. The corresponding figure for boys was 34 percent. The average frequency of being in physical contact with another person was 70 percent for girls aged up to 5 years. Boys of the same age range showed the much lower figure of 42 percent. Older children showed a decrease in physical contact with others, but girls remained higher than boys at all ages.

Several behaviors which are characteristic of the !Kung girls have implications for the extent to which they come under the influence of adults. The girls are closer at hand from an adult point of view; they have a greater preference for adult society, and they have less interest in their peers in comparison with boys. Looking at girls' social interactions shows that they interact more with adults, particularly women, but also with men. This is no doubt related both to their spatial proximity to supervising adults and to their apparent preference for adult company. It is also possible that adults call on them more frequently because of the two factors cited above. Such a pattern of being interrupted and redirected frequently by others may well reinforce a pattern of staying close to adults.

In another study Whiting and Edwards (1973) report similar findings and suggest that it may be related to higher female compliance. During the course of the observations this researcher perceived that !Kung adults did not deliberately select female targets from among the children; the adults simply interacted more frequently with those children who were closest.

The findings for !Kung children of a more restricted spatial range of females is, of course, not novel. Findings of this type are one of the most stable in behavior research which attends to sex differences. Primatologists find a comparable sex difference (see Harlow, 1965; Harlow and Harlow, 1962; Harlow, Harlow, and Hansen, 1963; Jay, 1963; Jensen and Bobbitt, 1965; Jensen, Bobbitt, and Gordon, 1967a, 1967b, 1968; Poirier, 1972). The notion that females are more sensitive to social cues and to the needs of others may have its origins in the restricted mobility and greater orientation to adults of females. If girls do follow such a pattern, this would set the stage for a more consistent socialization environment in which positive rewards could be used to a greater extent than with boys who are, on the average, farther away and less aware of what an adult may want and whose behavior, therefore, would be less easy to shape consistently.[7]

Sex differences in rough-and-tumble play with peers are reliably reported with male participation in such activities much greater than female participation (Blurton Jones, 1967; Blurton Jones and Konner, 1973; Freedman and Omark, 1973). In a sampling of ten-minute observations on !Kung children,[8] girls exceeded boys in rough-housing in the 4- to 7-year age group, but for the 8-

to 14-year age group the more usual picture of males higher in rough-housing than females is seen (Draper and Cashdan, 1974). Another behavioral study of Bushman children from the same population found !Kung girls to be higher than a sample of London girls matched for age. Both these findings regarding high levels of rough-and-tumble play in girls may be attributable to the fact that !Kung children nearly always play in heterosexual play groups rather than in same-sex groups. This is because the usual living group for Bushmen is about 35 to 40 people—a group too small to provide many children of the same sex who are also close in age. It is possible that in this society young girls develop a rougher style of interaction due to their interaction with boys, as is suggested by Blurton Jones and Konner (1973).

A contrasting line of interpretation about the effects of mixed-sex groups is offered by Parker and Omark (this volume, Chapter 24). They suggest that the variable of audience (in particular whether the audience contains opposite-sex others) may affect display behaviors and increase sex differentiation in the variables of overrating and hierarchy formation. It seems that an important task for researchers in this area of ethology is to sort out which behavioral variables are sensitive to this aspect of variation in the social ecology of behavior.

Conclusions

The data on the !Kung children are of interest to the issues raised in this volume for several reasons. Sex differences that occur in the behavior of children have been widely reported elsewhere for human young in a variety of cultural settings and have been reported in very similar form for nonhuman primate young. Among the !Kung children the sex differences remain largely unexploited by differential socialization practices for girls and boys. The economic reasons for this rather unusual circumstance have been discussed. To this author the best explanation for the findings about sex differences in behavior is that biological events in intrauterine life deliver different levels of sex hormones to the brain of male and female fetuses. As a result, certain parts of the brain are preprogrammed in such a way that with respect to certain classes of stimuli, males and females are likely to respond differently (Levine, 1966; McEwen, 1976).

The present discussion makes at least four assumptions. (1) Only a relatively small proportion of the total behaviors of males and females are noticeably different. (2) Of this minority, a few are crucial for their later role in differentiating the sexes in social dominance. (3) One should not be misled when behavioral variables of the minority type show consistent but not "statistically" significant sex differentiation. It is the combination of sex-differentiated elements which contribute to the familiar sex differences in style. In other words, small differences concatenate so that there are strong reliable sex differences in a multivariate field. These are easily obscured by univariate statistical testing. (4) The concept of evolved ease of learning is essential for understanding the postnatal unfolding of sex-differentiated behaviors. For example, girls undoubtedly can be encouraged to engage in rough-and-tumble play, and it is quite likely that !Kung girls have a rougher style because they are thrown more with male playmates. However, if these girls had mostly girls to play with, they would likely choose different pastimes.

As suggested earlier, males and females given the same social ecological milieu in which to develop will seek out different stylistic niches. Individual behaviors of males and females are often not impressively different. What is impressive is the different organization of these behaviors and for different ends. The final causes of the phenomena discussed are to be sought in the domain of the theory of behavioral evolution, in particular with reference to evolutionary forces affecting sex differences in reproductive strategy.

Notes

1. What is argued here has to do with variables of sensitivity in the child to the caretaker during the caretaker presence and the resulting implications of the variation in this sensitivity for the socialization experience of the child. There is a substantial literature on the behavior of children when they are separated from mothers, and these studies show no clear effect of sex. Interesting differences appear to characterize the types of reactions of girls and boys when the mother is present but out of the child's reach (Goldberg and Lewis, 1969; but see Jacklin, Maccoby, and Dick, 1973). For a good recent review of separation and attachment literature, see Weinraub and Lewis, 1977.
2. The exclamation point represents a click phoneme in the !Kung language. Click sounds are characteristic of this and other Bushman (Khoisan) languages.
3. Fieldwork for this research was supported by NIMH Grant No. MH-136111 to Irven DeVore and Richard B. Lee.

4. See Draper, 1972 and 1975a for a more complete discussion of the observational methods employed in gathering these data. Several observation techniques were used in collecting these data on child behavior. Spot observations were taken on each child on repeated occasions and at different times of the day. Each time the order in which the children of a particular group were to be observed was randomized. Noted was such information as the child's location in space, the names of others in the child's vicinity and in his immediate presence, the whereabouts of the child's parents, and whether or not the child was in physical contact with another person. Along with the spot observations, a series of systematic, randomized "elapsed-time" observations was taken on the same children. (For a discussion of "focal child" observations see Altmann, 1974; Omark, Fiedler, and Marvin, 1976). These time observations extended from ten minutes to one hour in length. In all cases only the subject child was the focus of the observations. This technique recorded commands issued by and received by the child as well as the age and sex characteristics of the individual with whom the child was interacting. Careful notations were also made on the location and location changes during the observation, personnel changes during the observation, along with a running description of the activity, if any, which absorbed the child's interest.
5. Children occasionally do go on gathering trips, but this most often happens when a relatively brief outing is planned and/or when water will be found along the way and temperatures are cool. Children were observed on gathering trips, and they do little if any serious gathering.
6. Child caretaking behaviors include such items as feeds, amuses, wipes face, dresses, comforts, and helps a younger child. Each episode of holding or carrying a child was counted once. It was originally intended to score child caretaking by a subject child in number of minutes per hour of observation. However, this behavior was so fleeting that the other variable of "episodes of child care" was devised as a more meaningful way of representing the frequency of this behavior.
7. See Frieze, Parsons, Johnson, *et al.*, 1978, for a parallel discussion of how sex differences in irritability may interact with socialization.
8. In these ten-minute observations there were 40 observations on 12 males and 29 observations on 7 females.

Part VI

A Final Overview

Chapter 26

Current Problems in the Study of Human Dominance

F. F. Strayer

The research presented in the preceding chapters provides impressive evidence for the potential value of social dominance as an analytic concept in the comparative study of human social behavior. These same studies reveal, however, a surprising diversity both in the use of social dominance as a descriptive term and in the specification of necessary and sufficient criteria for its evaluation. Such a variety of conceptual and methodological approaches is not uncommon during the early phases of exploratory research in any new field. In fact, such diversity has certain important functions during this formative period. For example, it provides a broader empirical view of the phenomena under study; it permits a preliminary evaluation of the potential scope and possible limits of the basic concepts; and, perhaps most importantly, it helps to assure that initial results are not merely artifacts of an excessively narrow theoretical focus or overly technical methodological procedures. From this perspective, the present diversity of analytic approaches helps demonstrate the robustness of dominance as an organizational aspect of human social adaptation within a variety of stable social settings as well as in diverse cultural and linguistic contexts.

However, having demonstrated the potential value of social dominance as an analytic concept in human ethology, we now must begin a more critical reexamination of both conceptual and methodological problems associated with its evaluation. The healthy plurality of views that characterized the discovery and documentation phases of exploratory research must give way to a more integrated and systematic analytic approach, if we wish to avoid the conceptual confusion which all too often emerges from an unnecessary lack of precision in the social sciences.

Problems in the Description of Social Dominance

A major problem in virtually all studies of social behavior involves adequately specifying the precise referents for our more abstract descriptive terms. This problem is most important for concepts like social dominance which have had an extended, but often controversial, history of application in social research. At various times dominance has been used to refer to forms of social behavior, classes of social interaction, differences between individuals, qualities of social relationships, and structural properties of social groups. With such a diversified list of possible referents it is often difficult, or nearly impossible, to determine which of these meanings is most appropriate within the context of a given dominance study. Such conceptual ambiguity has often been as troublesome for researchers in the field as it has been for their audiences. Three quite different approaches to establishing the behavioral referent of social dominance are evident in current ethological research.

The first approach views the analysis of dominance as primarily a problem of definition and measurement. The implicit research strategy is to develop an acceptable operational definition of social dominance and, subsequently, to specify the different analytic procedures that permit specific inferences from this basic measure. In primate research induced competition provides a good example of this strategy. The researcher operationally defines social dominance as priority of access to a selected incentive and then evaluates both social relations and group structure with an apparently valid, but somewhat arbitrary, objective measure. Similarly, the sociometric approach to human dominance uses the child's verbal report as a basic measure of his relationship with other group members. The agreement between two individuals concerning their relative "toughness" is operationally defined as an established dyadic relationship. Subsequent analysis of all established dyads permits the derivation of a social structure which reflects relative status among group members. Since niether of these methods includes a behavioral analysis of social interaction, at a descriptive level they provide less direct information concerning the nature of dominance relations and social structure within the natural group setting.

The second approach to dominance assessment places less emphasis upon the derivation of a single operational definition; instead it attempts to identify a set of behavioral patterns which can provide an empirical and ecologically valid basis for the evaluation of individual differences in social status. This research strategy involves demonstrating that certain behavioral patterns in the social repertoire share important characteristics, and thus can be regrouped into more comprehensive, descriptive categories. Although in ethology, such higher order descriptive categories are usually determined according to morphological, causal, and/or functional similarities among behavioral patterns, dominance activities are generally regrouped only according to their immediate social effect.

Even with the relatively restricted criterion of immediate social effect, ethologists differ substantially in the final set of behaviors which they use as a basis for the description of social dominance. The major problem here is the specification of which social effects are necessary and sufficient for the identification of specific activities which must be included in the category of dominance gestures. For example, Chance (1967) suggests that activities which attract social attention are often expressions of social dominance. Attracting attention seems to be the most important function of what have been called "dominance displays" (e.g., locomotor postures, the elevation of a monkey's tail, or the position a child's chin). However, it seems less clear that all types of attention-getting activities necessarily express social dominance. In contrast, Rowell (1974) places primary emphasis upon social submission as the necessary effect of dominance gestures. Such a criterion substantially reduces the number of behavioral patterns that are included in a category of individual dominance gestures.

Although these two examples provide an indication of the current problem in delineating the descriptive basis for social dominance assessments, they also pose important empirical questions about the organization of social behavior within stable groups. Do individuals who successfully elicit submissive reactions from conspecifics also engage in more attention-getting display activity? Similar questions are implicit in suggestions that all forms of social influence, or social control, represent expressions of social dominance. If the behavioral basis for the assessment of social dominance is to include any form of social influence, we must demonstrate that our various subcategories of behavior which are grouped according to their immediate social effect (i.e., submission, attention, change of another's behavior, etc.), all provide essentially convergent information about social dominance in different groups and in different observational settings. Our present inability to select the most appropriate index among these various possible subcategories clearly reveals the limits of current ethological research on dominance behaviors.

Finally, the third approach to the assessment of social dominance avoids the development of general descriptive categories of individual behaviors; instead the attempt is to regroup specific forms of social exchange as the primary observational data necessary for the derivation of dominance relations and group status. Rowell's (1974) claim that dominance cannot be discussed without simultaneous reference to social submission provides the best example of this final approach. Although a specific behavioral activity may have a high probability of leading to submission, only when it actually produces a submissive reaction from the social partner are we justified in concluding that it has functioned as a dominance signal. From this perspective, the identification of dominance activities requires the description of how patterns of behavior are coordinated between two individuals as they participate in social interaction. Descriptive listings of individual action patterns as examples of dominance gestures provide only a summary of behaviors which *may potentially function* to express dominance. To determine whether such behaviors have actually been used as dominance signals, we must always examine how such acts have functioned in the specific sequence of social exchange. One of the more important advantages of this final descriptive approach is that it forces the observer to pay more attention to how behavior patterns are organized within a social sequence. Such emphasis upon the temporal and behavioral context of social actions may eventually reveal important causal and functional differences among activities which are difficult to distinguish at a morphological level (e.g., wrestling versus rough-and-tumble play, cooperative role play versus social control, etc.).

Although each of the three approaches to the description of social dominance provides objective information, there remain important qualitative differences in the types of data which each employs to assess dyadic dominance relations and group status structures. Such differences pose a number of basic problems which may be critical for advancements in our understanding of social dominance. Such problems cannot be resolved simply by redefining our basic descriptive terminology. Without the essential empirical information, problems of definition remain ultimately matters of personal opinion. The elaboration of a more comprehensive, and systematic descriptive basis for the observational evaluation of social dominance must continue to be one of the more important research objectives in human ethology.

Problems in the Analysis of Dominance Status

A review of recent dominance research shows that two quite different analytic procedures are often employed for the derivation of an apparently similar group status ranking. The first approach emphasizes individual differences in the organization of personal response styles. Hinde (1974) provides one of the clearest formulations of this analytic perspective when he argues that dominance, like aggressiveness, is best understood as an intervening variable that summarizes a complex set of relations between certain antecedent conditions experienced by the individual (i.e., independent variables) and characteristic patterns of response to such stimulation (i.e., dependent variables). Individuals who respond to such conditions with confident and assertive reactions are assumed to have a higher social status than others who respond with more timid and self-effacing behaviors. From this perspective, social dominance is conceptualized as a dimension of personality. Differences in dominance status reflect basically different styles of social participation which result from differences in individual adaptation in a variety of social situations.

The second analytic approach to social status places greater emphasis upon the dynamics of social interaction within the stable group. For example, Bernstein (this volume, Chapter 3) argues that social dominance refers to a social relationship between two individuals and should be understood as an emergent property of dyadic social exchange. Even though the analysis of dominance relations requires noting differential responsiveness of two individuals during social interaction, dominance should not be conceptually reduced to a calculus of individual differences in general behavioral traits. Furthermore, since the group status hierarchy summarizes a network of such observed dyadic relationships, it cannot be equated with simple rank orderings of group members according to individual differences in some predetermined behavioral characteristic.

Individual rankings on aggressiveness or submissiveness can be more or less correlated with status rank in the group dominance hierarchy, but the correspondence between dominance status and ranking according to frequency of social activity will always depend upon the particular dynamics which characterize the social group. Tables 1 and 2 provide an interesting hypothetical illustration of this problem. When the majority of dominance struggles occur between high-status individuals (as in Table 1), it is evident that these same group members will have higher rates of both aggressive and submissive acts. In such a case rank orderings of group members according to their total frequency of either activity will correlate positively with higher position in the status hierarchy ($r = 1.00$ for aggressive rank and dominance status; $r = 1.00$ for submissive rank and dominance status). However, in a second group, where lower status individuals engage more often in social conflict, we might find exactly the opposite correlations between behavioral rankings and social status. An example of the latter is shown in Table 2. Although at a structural level this second status hierarchy is identical to our earlier example from Table 1, it is clear that the distribution of social conflict has shifted within the matrix. More importantly, the correlations between dominance status and rankings of individuals according to frequency of behaviors are now negative ($r = -1.00$ for aggressive rank and dominance status; $r = -1.00$ for submissive rank and dominance status).

Table 1. Illustration of Group Dominance Structure Where Dominant Individuals Participate in the Majority of Conflict

Initiators		*Targets*								
Status	*Name*	*A*	*B*	*C*	*D*	*E*	*F*	*G*	*H*	*Total Aggression*
1	A	XX	24	17	7	1	1	1	1	52
2	B	15	XX	5	7	5	1	1	1	35
3	C	10	—	XX	7	3	4	1	1	26
4	D	—	—	—	XX	4	4	1	1	10
5	E	—	—	—	—	XX	2	2	1	5
6	F	—	—	—	—	—	XX	2	1	3
7	G	—	—	—	—	—	—	XX	1	1
8	H	—	—	—	—	—	—	—	XX	0
Total Submission		25	24	22	21	13	12	8	7	132

Table 2. Illustration of Group Dominance Structure Where Subordinate Individuals Participate in the Majority of Conflict

Initiators		*Targets*								
Status	*Name*	*A*	*B*	*C*	*D*	*E*	*F*	*G*	*H*	*Total Aggression*
1	A	XX	1	1	1	1	1	1	1	7
2	B	—	XX	2	2	1	1	1	1	8
3	C	—	—	XX	2	4	4	1	1	12
4	D	—	—	—	XX	4	3	5	1	13
5	E	—	—	—	—	XX	7	7	7	21
6	F	—	—	—	—	—	XX	5	17	22
7	G	—	—	—	—	—	—	XX	24	24
8	H	—	—	—	—	—	10	15	XX	25
Total Submission		0	1	3	5	10	26	35	52	132

Although these two hypothetical examples reveal some of the important differences between an individual and social approach to the derivation of a group status ranking, in actual studies it is often quite difficult to determine which analytic procedure has been used. Such confusion results from the fact that with either procedure primary emphasis is usually given to describing a single rank ordering of group members. Even researchers who employ the social network analysis of dominance relations often fail to describe adequately the hierarchical organization of a group dominance structure.

The recent development of rigidity and linearity statistics for observed dominance relations represents an initial step toward a more adequate descriptive treatment of group social structure. Nevertheless, these two measures still fail to capture much of the important information that distinguishes different dominance hierarchies. For example, the status structures represented in Table 3 and 4 are equivalent in terms of the rigidity and linearity of dominance relations. In Table 3, though, the social structure is a complete hierarchical network where each member of the group engages in dominance interaction with every other individual. In contrast, each group member is dominated only by the individual immediately above him in the social status hierarchy summarized in Table 4. This latter type of social organization represents a potentially different form of hierarchical organization since the amount of connectedness between individuals is substantially

Table 3. Illustration of Highly Connected Linear Dominance Structure

Initiators		*Targets*								
Status	*Name*	*A*	*B*	*C*	*D*	*E*	*F*	*G*	*H*	*Total Aggression*
1	A	XX	2	2	2	2	2	2	2	14
2	B	—	XX	2	2	2	2	2	2	12
3	C	—	—	XX	2	2	2	2	2	1
4	D	—	—	—	XX	2	2	2	2	8
5	E	—	—	—	—	XX	2	2	2	6
6	F	—	—	—	—	—	XX	2	2	4
7	G	—	—	—	—	—	—	XX	2	2
8	H	—	—	—	—	—	—	—	XX	0
Total Submission		0	2	4	6	8	10	12	14	56

less than in Table 3. Once again, these two examples illustrate extremes from a continuum of possible status structures. However, greater attention to qualitative differences such as the amount of connectedness in social status structures may provide important measures in future studies which attempt to monitor either developmental or comparative differences in the organization of naturally occurring social conflict among group members.

Table 4. Illustration of Minimally Connected Linear Dominance Structure

Initiators		*Targets*								
Status	*Name*	*A*	*B*	*C*	*D*	*E*	*F*	*G*	*H*	*Total Aggression*
1	A	XX	14	—	—	—	—	—	—	14
2	B	—	XX	12	—	—	—	—	—	12
3	C	—	—	XX	10	—	—	—	—	10
4	D	—	—	—	XX	8	—	—	—	8
5	E	—	—	—	—	XX	6	—	—	6
6	F	—	—	—	—	—	XX	4	—	4
7	G	—	—	—	—	—	—	XX	2	2
8	H	—	—	—	—	—	—	—	XX	0
Total Submission		0	14	12	10	8	6	4	2	56

Limits in the Use of Dominance As an Explanatory Concept

The preceding discussion of current descriptive and analytic problems in the evaluation of social dominance indicates many of the important limits in the immediate use of dominance relations and group status for the explanation of other social phenomena. Once these conceptual and methodological problems have been resolved, it seems likely that social dominance will have an important explanatory role in ethological analyses of both individual social adaptation and the patterning of social communication within stable human groups.

At the level of group processes important questions remain to be answered concerning the social function of group status structures. From an ethological perspective, we expect that the emergence of the social dominance hierarchy facilitates the reduction of overt aggressive conflict within a social unit. Theoretically, once the social structure of the group has stabilized, individual group members should adopt different interaction strategies, and thus avoid severe forms of social conflict. Essentially, an ethological view of dominance relations predicts a gradual transition from more overt forms of aggressive contact to ritualized expressions of threat during the stabilization of dyadic dominance. Such predictions seem to contrast with expectations derived from a social conditioning view of aggressive activity. This perspective would predict an increase in aggressive behaviors for individuals who were more successful in dominating their peers. The differences between these two theoretical perspectives can be resolved directly in an observational study of the process of dominance stabilization during the period of group formation.

Additional questions remain to be explored concerning the coordination of social dominance with other social activities which occur within stable groups. Although preliminary findings suggest that dominance covaries with individual differences in receipt of social attention, more systematic studies are required which analyze both attentional relationships and attention structure. In this sense, the current procedures for the analysis of dominance that distinguish rates of individual social conflict, dyadic dominance relations, and group status hierarchies provides an important descriptive model that can be profitably applied to other forms of social behavior. A rank ordering of group members according to frequency of received attention seems less interesting as a measure of social structure than a network analysis of dyadic attentional relations that are defined by both the frequency and the direction of visual regard within a stable group. Certainly a similar multileveled analytic approach should be applied to other classes of social activity, such as affiliative gestures, forms of social control, imitation, cooperation, and altruistic behaviors. When we are able to specify the descriptive basis for determining the organization of these latter social activities, we shall be in a position to pose much more interesting questions about how differences in the organization of social dominance may help explain the organization and coordination of other social behavior within the natural group. Answers to these questions will establish both the value and the limits of social dominance as an explanatory concept in human ethology.

References

Abramovitch, R. 1976. The relation of attention and proximity to dominance in preschool children. In *The social structure of attention*, eds., M. R. A. Chance and R. R. Larsen. London: Wiley.

Abramovitch, R., and Grusec, J. E. 1978. Peer imitation in a natural setting. *Child Dev.* 49: 60–65.

Abramovitch, R., and Strayer, F. 1978. Preschool social organization: Spacing and attentional behaviors. In *Aggression, dominance and individual spacing: Advances in the study of communication and affect*, vol. 4, eds., J. L. Krames, P. Pliner, and T. Alloway. New York: Plenum.

Ainsworth, M. D. S., and Bell, S. M. V. 1970. Attachment, exploration, and separation: Illustrated by the behavior of one-year-olds in a strange situation. *Child Dev.* 41: 49–67.

Aldis, O. 1975. *Play fighting*. New York: Academic Press.

Aldrich-Blake, F. P. G., and Chivers, D. J. 1973. On the genesis of a group of siamang. *Am. J. Phys. Anthropol.* 38: 631–36.

Alexander, B. K., and Bowers, J. M. 1969. Social organization of a troop of Japanese monkeys in a two-acre enclosure. *Folia Primatol.* 10: 230–242.

Alexander, R. D. 1961. Aggressiveness, territoriality, and sexual behavior in field crickets *(Orthoptera: gryllidae)*. *Behaviour* 17: 130–223.

Alexander, R. D. 1971. The search for an evolutionary philosophy of man. *Proc. Roy. Soc. Victoria* 84: 99–120.

Alexander, R. D. 1974. The evolution of social behavior. *Ann. Rev. Ecol. Systematics* 5: 325–383.

Allee, W. C. 1977. Concerning the origin of sociality in animals. In *Vertebrate social organization*, ed., E. M. Banks. Stroudsberg, Penn.: Dowden, Hutchison, and Ross.

Altman, I., and Haythorn, W. W. 1967. The ecology of isolated groups. *Behav. Sci.* 12: 169–182.

Altmann, J. 1974. Observational study of behavior: Sampling methods. *Behaviour* 49: 227–267.

Altmann, S. A. 1962. A field study of the sociobiology of rhesus monkeys (*Macaca mulatta*). *Ann. N.Y. Acad. Sci.* 102: 338–435.

Altmann, S. A. 1967. The structure of primate social communication. In *Social communication among primates*, ed., S. A. Altmann. Chicago: University of Chicago Press.

Altmann, S. A., and Altmann, J. 1970. *Baboon ecology: African field research*. Chicago: University of Chicago Press.

Anastasi, A. 1958. *Differential psychology*. New York: Macmillan.

Anderson, H. H. 1940. An examination of the concepts of domination and integration in relation to dominance and ascendance. *Psychol. Rev.* 47: 21–37.

Anderson, H. H., and Brandt, H. F. 1939. Study of motivation involving self-announced goals of fifth-grade children and the concept of level of aspiration. *J. Soc. Psychol.* 10: 209–232.

Anderson, S., and Messick, S. 1974. Social competency in young children. *Dev. Psychol.* 10: 282–293.

Andrew, R. J. 1963. The displays of the primates. In *Genetic and evolutionary biology of the primates*, ed., J. Buettner-Janusch. New York: Academic Press.

Apsler, R. 1975. Effects of embarrassment on behavior toward others. *J. Pers. Soc. Psychol.* 32: 145–153.

Arnheim, R. 1965. *Art and visual perception*. Berkeley: University of California Press.

Aronfreed, J. 1968. *Conduct and conscience*. New York: Academic Press.

Atkinson, J. W. 1958. *Motives in fantasy, action, and society*. New York: Van Nostrand, Reinhold.

Bales, R. F. 1953. The equilibrium process in small groups. In *Working papers in the theory of action*, eds., T. Parsons, R. Bales, and E. Shils. New York: Free Press.

Bandura, A., and Huston, A. C. 1961. Identification as a process of incidental learning. *J. Abnorm. Soc. Psychol.* 63: 311–318.

Barchas, P. R., and Barchas, J. D. 1975. Physiological sociology: Endocrine correlates of status behaviors. In *New psychiatric frontiers*, eds., D. A. Hamburg and K. Brodie. New York: Basic Books.

Bardwick, J. 1971. *The psychology of women*. New York: Harper & Row.

Barkow, J. H. 1970. Hausa and Maguzawa: Processes of group differentiation in a rural area of North Central State, Nigeria. Ph.D. dissertation, University of Chicago.

Barkow, J. H. 1973a. Muslims and Maguzawa in North Central State, Nigeria: An ethnographic comparison. *Can. J. Africa Studies* 7: 59–76.

Barkow, J. H. 1973b. Darwinian psychological anthropology: A bio-social approach. *Curr. Anthropol.* 14: 373–388.

Barkow, J. H. 1974a. Evaluation of character and social control among the Hausa. *Ethos* 2: 1–14.

Barkow, J. H. 1974b. Human ethology and intra-individual systems. Paper read at the Human Ethology Meeting, London.

Barkow, J. H. 1975a. Strategies for self-esteem and prestige in Maradi (Niger Republic). In *World Congress: Psychological anthropology*, ed., T. R. Williams. The Hague: Mouton.

Barkow, J. H. 1975b. Social prestige and culture: A biosocial interpretation. *Contemp. Anthropol.* 16: 553–572.

Barkow, J. H. 1976. Attention structures and evolution of human psychological characteristics. In *The social structure of attention*, eds., M. R. A. Chance and R. R. Larsen. London: Wiley.

Barkow, J. H. 1978a. Building on the basis of A. I. Hallowell: Social norms, the self, and sociobiology. *Curr. Anthropol.* 19: 99–118.

Barkow, J. H. 1978b. Sociobiology: Is this the new theory of human nature? *Am. Anthropol.* 80: 5–20.

Barry, H. A., Bacon, M. K., and Child, I. L. 1957. A cross-cultural survey of some sex differences in socialization. *J. Abnorm. Soc. Psychol.* 55: 327–332.

Bartmann, T. 1963. Der Einfluss von Zeitdruck auf die Leistung und das Denkverhaulten bei Volksschuleren. *Psychol. Forsch.* 27: 1–61.

Bass, B. M. 1949. An analysis of the leaderless groups discussion. *J. Appl. Psychol.* 33: 527–533.

Bateman, A. J. 1948. Intra-sexual selection in *Drosophila*. *Heredity* 2: 349–368.

Bates, E. 1976. *Language and context*. New York: Academic Press.

Bauer, H. R. 1976. Behavioral changes about the time of reunion in parties of chimpanzees in the Gombe National Park. In *Contemporary primatology*, ed., S. Kondo. New York: Karger.

Bauer, H. R. 1977a. Ethological aspects of Gombe chimpanzee aggregations with implications for hominization. Ph.D. dissertation, Stanford University. Ann Arbor, Mich.: University Microfilms.

Bauer, H. R. 1977b. Chimpanzee bipedal locomotion in the Gombe National Park, East Africa. *Primates* 18: 913–921.

Bauer, H. R. 1979. Agonistic and grooming behavior in the reunion context of Gombe Stream chimpanzees. In *Perspectives on human evolution*, vol. 4, eds., D. A. Hamburg, E. R. McGown, and J. Goodall. Menlo Park, Ca.: Addison Wesley.

Bauer, H. R. In press. Notes on chimpanzee communication and cognition.

Bauer, H. R., and Philip, M. In preparation. Chimpanzee individual recognition.

Beach, F. A. 1976. Sexual attractivity, proceptivity, and receptivity in female mammals. *Horm. Behav.* 7: 105–136.

Bell, D. V. J. 1975. *Power, influence and authority*. New York: Oxford University Press.

Bell, S. M. 1970. The development of the concept of object as related to infant–mother attachment. *Child Dev.* 41: 291–311.

Berger, J., Haynes, M., and Stettner, L. J. 1979. Social organization in a female-led captive Guinea baboon troop. Paper presented at the American Psychological Association Convention, New York.

Berkson, G., and Schusterman, R. J. 1964. Reciprocal food sharing of gibbons. *Primates* 5: 1–10.

Bernstein, I. S. 1964. The integration of rhesus monkeys introduced to a group. *Folia Primatol.* 2: 50–63.

Bernstein, I. S. 1970a. Primate status hierarchies. In *Primate behavior: Developments in field and laboratory research,* vol. 1, ed , L. A. Rosenblum. New York: Academic Press.

Bernstein, I. S. 1970b. Some behavioral elements of the Cercopithecoidea. In *Old World monkeys: Evolution, systematics, and behavior,* eds., J. Napier and P. Napier. New York: Academic Press.

Bernstein, I. S. 1974. Principles of primate group organization. In *Perspectives in primate biology,* ed., A. B. Chiarelli. New York: Plenum.

Bernstein, I. S., and Draper, W. A. 1974. The behavior of juvenile rhesus monkeys in groups. *Anim. Behav.* 12: 92–96.

Bernstein, I. S., and Gordon, T. P. 1974. The function of aggression in primate societies. *Am. Sci.* 62: 304–311.

Bernstein, I. S., and Smith, E. O. 1978. Ecological influences on social organization. *Curr. Anthropol.* 19: 381–382.

Bettelheim, B. 1943. Individual and mass behavior in extreme situations. *J. Abnorm. Soc. Psychol.* 36: 417–452.

Bettelheim, B. 1959. Joey, a mechanical boy. *Sci. Am.,* March, 116–130.

Bigelow, R. 1973. The evolution of aggression, cooperation and self-control. In *Nebraska symposium on motivation,* vol. 21. eds., J. R. Cole and D. D. Jensen, Lincoln: University of Nebraska.

Bischof, N. 1975. Comparative ethology of incest avoidance. In *Biosocial anthropology,* ed., R. Fox. New York: Halsted.

Block, J. H. 1976. Issues, problems, and pitfalls in assessing sex differences: A critical review of *The Psychology of Sex Differences. Merrill-Palmer Q.* 22: 283–308.

Blurton Jones, N. G. 1967. An ethological study of some aspects of social behavior in children in nursery school. In *Primate ethology,* ed., D. Morris. London: Weidenfeld and Nicholson.

Blurton Jones, N. G. 1968. Observations and experiments on causation of threat displays of the great tit *(Parus major). Anim. Behav. Monogr.* 1: 75–158.

Blurton Jones, N. G. 1972a. Characteristics of ethological studies of human behaviour. In *Ethological studies of child behavor,* ed., N. G. Blurton Jones. London: Cambridge University Press.

Blurton Jones, N. G. 1972b. Categories of child-child interaction. In *Ethological studies of child behavior,* ed., N. G. Blurton Jones. London: Cambridge University Press.

Blurton Jones, N. G. 1972c. Comparative aspects of mother–child contact.

In *Ethological studies of child behavior*, ed., N. G. Blurton Jones. London: Cambridge University Press.

Blurton Jones, N. G. 1972d. Non-verbal communication in children. In *Non-verbal communication*, ed., R. A. Hinde. London: Cambridge University Press.

Blurton Jones, N. G. 1975. Ethology, anthropology and childhood. In *Biosocial anthropology*, ed., R. Fox. New York: Halsted.

Blurton Jones, N. G., and Konner, M. J. 1973. Sex differences in the behavior of Bushman and London two-to-five-year-olds. In *Comparative ecology and the behavior of primates*, eds., R. P. Michael and J. H. Crook. New York: Academic Press.

Bock, R. D. 1973. Word and image: Sources of the verbal and spatial factors in mental test scores. *Psychometrika* 38: 437–457.

Bock, R. D., and Kolokowski, D. P. 1973. Further evidence of sex-linked major-gene influence in human spatial visualizing ability. *Am. J. Hum. Genet.* 25: 1–14.

Boehm, C. 1978. Rational preselection from hamadryas to *Homo sapiens:* The place of decisions in adaptive process. *Am. Anthropol.* 80: 265–297.

Bowlby, J. A. 1969. *Attachment and loss*, vol. 1. *Attachment*. New York: Basic Books.

Bramblett, C. A. 1976a. *Patterns of primate behavior*. Palo Alto, Calif.: Mayfield.

Bramblett, C. A. 1976b. Ethology and primates: Some new directions for the 1970's. *Am. Anthropol.* 78: 593–607.

Brannigan, C. R., and Humphries, D. A. 1972. Human non-verbal behavior, a means of communication. In *Ethological studies of child behavior*, ed., N. G. Blurton Jones. London: Cambridge University Press.

Brenner, M., and Omark, D. R. 1979. The effects of sex, structure and social interaction on preschoolers' play behaviors in a naturalistic setting. *Instruct. Sci.* 8: 91–105.

Bridges, K. M. B. 1932. Emotional development in early infancy. *Child Dev.* 3: 324–341.

Brindley, C., Clarke, P., Hutt, C., *et al.* 1973. Sex differences in the activities and social interactions of nursery children. In *Comparative ecology and the behavior of primates*, eds., R. P. Michael and J. H. Crook. New York: Academic Press.

Brown, R. W. 1963. A determinant of the relationship between ridgitiy and authoritarianism. *J. Abnorm. Soc. Psychol.* 48: 469–476.

Bruner, J. S., Jolly, A., and Sylva, K. 1976. *Play: Its role in development and evolution*. New York: Penguin Books.

Bryant, P. E. 1974. *Perception and understanding in children*. London: Methuen.

Bryant, P. E., and Kopytynska, H. 1976. Spontaneous measurement by young children. *Nature* 260: 773.

Burghardt. G. M. 1977. Ontogeny of communication. In *How animals communicate,* ed., T. A. Sebeok. Bloomington: Indiana University Press.

Burlinghame, D., and Robertson, J. 1967. *Nursery school for the blind.* (Film) London: Hampstead Child Therapy Clinic.

Bygott, J. D. 1972. Cannibalism among wild chimpanzees. *Nature* 238: 410–411.

Bygott, J. D. 1979. Agonistic behavior, dominance, and social structure in wild chimpanzees of the Gombe National Park. In *Perspectives on human evolution,* vol. 4, eds., D. A. Hamburg, E. R. McGown, and J. Goodall. Menlo Park, Ca.: Addison Wesley.

Byrne, D., Clore, G. L., and Worchel, P. 1966. Effect of economic similarity–dissimilarity on interpersonal attraction. *J. Pers. Soc. Psychol.* 4: 220–224.

Callan, H. 1970. *Ethology and society: Towards an anthropological view.* Oxford, Eng.: Clarendon.

Campbell, J. K. 1966. Honour and the devil. In *Honour and shame: The values of Mediterranean society,* ed., J. G. Peristiany. Chicago: University of Chicago Press.

Camras. L. 1975. The role of children's facial expressions in a conflict situation. Ph.D. dissertation, University of Chicago.

Camras, L. 1976. The role of children's facial expressions in a conflict situation. Paper read at the meeting of the Animal Behavior Society, Boulder, Colorado.

Camras, L. 1977. Facial expressions used by children in a conflict situation. *Child Dev.* 48: 1431–1435.

Carnegie Commission on Higher Education. 1973. *Opportunities for women in higher education.* New York: McGraw-Hill.

Carpenter, C. R. 1958. The concept of territory. In *Behavior and evolution,* eds., A. Roe and G. G. Simpson. New Haven: Yale University Press.

Carpenter, C. R. 1964. *Naturalistic behavior of nonhuman primates.* University Park: Pennsylvania State University Press.

Caspari, E. 1972. Sexual selection in human evolution. In *Sexual selection and the descent of man,* ed., B. Campbell. Chicago: Aldine.

Chagnon, N., and Asch, T. 1976. *The ax fight* (Film). Pennsylvania Cinema Register.

Chagnon, N., and Irons, W. 1979. *Evolutionary biology and human social behavior: An anthropological perspective.* North Scituate, Mass.: Duxbury.

Chance, M. R. A. 1961. The nature and special features of the instinctive social bond of primates. In *Social life of early man,* ed., S. L. Washburn. Chicago: Aldine.

Chance, M. R. A. 1963. The social bond of primates. *Primates* 4: 1–22.

Chance, M. R. A. 1967. Attention structures as the basis of primate rank orders. *Man* 2: 503–518.

Chance, M. R. A. 1976. Attention structures as the basis of primate rank orders. In *The social structure of attention*, eds., M. R. A. Chance and R. R. Larsen. London: Wiley.

Chance, M. R. A., and Jolly, C. J. 1970. *Social groups of monkeys, apes, and men*. London: Jonathan Cape.

Chance, M. R. A., and Larsen, R. R., eds. 1976. *The social structure of attention*. London: Wiley.

Chapple, E. D. 1976. Ethology without biology. *Am. Anthropol.* 78: 590–593.

Chevalier-Skolnikov, S. 1973. Facial expression of emotion in nonhuman primates. In *Darwin and facial expression*, ed., P. Ekman. New York: Academic Press.

Child, I. L., and Whiting, J. W. M. 1949. Determinants of level of aspiration: Evidence from everyday life. *J. Abnorm. Soc. Psychol.* 44: 303–314.

Chittenden, G. E. 1942. An experimental study in measuring and modifying assertive behavior in young children. *Monogr. Soc. Res. Child Dev.* 7: iii–87.

Christopher, S. B. 1972. Social validation of an objective measure of dominance in captive monkeys. *Behav. Res. Methods Instrument.* 4: 19–20.

Clark, G., and Birch, H. G. 1945. Hormonal modifications of social behavior. 1. The effect of sex-hormone administration on the social status of a male-castrate chimpanzee. *Psychosom. Med.* 7: 321–329.

Clutton-Brock, T. H., and Harvey, P. H. 1976. Evolutionary rules and priamte societies. In *Growing points in ethology*, eds., P. P. G. Bateson and R. A. Hinde. Cambridge, Eng.: Cambridge University Press.

Cohen, R. 1972. Altruism; Human, cultural or what? *J. Soc. Issues* 3: 39–57.

Collias, N. E. 1944. Aggressive behavior among vertebrate animals. *Physiolog. Zool.* 17: 83–123.

Collins, B. E., and Raven, B. E. 1969. Group structure: Attraction, coalitions, communication, and power. In *Handbook of social psychology*, vol. 4, eds., G. Lindzey and E. Aronson. Reading, Mass.: Addison-Wesley.

Conger, J. J. 1977. *Adolescence and youth: Psychological development in a changing world*, 2nd ed. New York: Harper and Row.

Cook, K. 1975. *Sharing by orangutans in a zoo*. Unpublished manuscript.

Coombs, R. H. 1969. Social participation, self-concept and interpersonal valuation. *Sociometry* 32: 273–286.

Coopersmith, S. 1960. Self-esteem and achievement need as determinants of selective recall and repetition. *J. Abnorm. Soc. Psychol.* 60: 310–317.

Coopersmith. S. 1967. *The antecedents of self-esteem*. San Francisco: Freeman.

Coopersmith, S. 1968. Studies in self-esteem. *Sci. Am.* 218: 96–106.

Crawford, M. P. 1940. The relation between social dominance and the menstrual cycle in female chimpanzees. *J. Comp. Psychol.* 30: 483–513.

Cronin, C. L. 1976. Playground behavior as a function of rank in the dominance hierarchy. Paper read at the meeting of the Animal Behavior Society, Boulder, Colorado.

Crook, J. H. 1970a. Social organization and the environment: Aspects of contemporary social ethology. *Anim. Behav.* 18: 197–209.

Crook, J. H. 1970b. The socio-ecology of primates. In *Social behaviour in birds and mammals*, ed., J. H. Crook. New York: Academic Press.

Crook, J. H. 1973. Darwinism and the sexual politics of primates. *Academia Nazionale Dei Lincei*, Rome, Italy.

Crook, J. H., and Goss-Custard, J. 1972. Social ethology. *Ann. Rev. Psychol.* 23: 277–312.

D'Andrade, R. G. 1966. Sex differences and cultural institutions. In *The development of sex differences*, ed., E. E. Maccoby. Palo Alto: Stanford University Press.

D'Aquili, E. 1972. *The biopsychological determinants of culture.* Reading, Mass.: Addison-Wesley.

Darwin, C. 1859. *The origin of species.* Reprinted—New York: Collier, 1962.

Darwin, C. 1872. *The expression of the emotions in man and animals.* Reprinted—New York: Philosophical Library, 1955; Chicago: University of Chicago Press, 1965.

Davenport, R. K. 1978. Some behavioral disturbances of great apes in captivity. In *Perspectives on human evolution*, vol. 5, eds., D. A. Hamburg and J. Goodall. New York: Holt, Rinehart and Winston.

Davenport, R. K., and Rogers, C. M. 1970. Differential rearing of the chimpanzee: A project survey. In *The chimpanzee*, vol. 3, ed., G. H. Bourne. New York: Karger.

Dawe, H. C. 1934. An analysis of two hundred quarrels of preschool children. *Child Dev.* 5: 139–157.

Dawkins, R. 1976. *The selfish gene.* New York: Oxford University Press.

Devons, E. and Gluckman, M. 1964. Introduction. In *Closed systems and open minds: The limits of naivete in social anthropology*, ed., M. Gluckman. Chicago: Aldine.

DeVore, I. 1965. *Primate behavior: Field studies of monkeys and apes.* New York: Holt, Rinehart and Winston.

DeVore, I., and Hall, K. R. L. 1965. Baboon ecology. In *Primate behavior: Field studies of monkeys and apes*, ed., I. DeVore. New York: Holt, Rinehart and Winston.

DeVore, I., and Washburn, S. L. 1963. Baboon ecology and human evolution. In *African ecology and human evolution*, eds., F. C. Howell and F. Bourliere. Viking Fund Publications in Anthropology, 36.

De Vos, G. A. 1968. Achievement and innovation in culture and personality. In *The study of personality*, eds., E. Norbeck, D. Price-Williams, and W. M. McCord. New York: Holt, Rinehart and Winston.

Diebschlag, E. 1940. Psycholische Beobachtungen über die Rangordnung bei der Haustaube. *Z. Tierpsychol.* 4: 173–188.

Dion, K. K. 1972. Physical attractiveness and evaluations of children's aggressions. *J. Pers. Soc. Psychol.* 24: 207–213.

Dion, K. K., Berscheid, E., and Walster, E. 1972. What is beautiful is good. *J. Pers. Soc. Psychol.* 24: 285–290.

Dolhinow, P. J. 1972. The North Indian langur. In *Primate patterns,* ed., P. J. Dolhinow. New York: Holt, Rinehart and Winston.

Dolhinow, P. J., and Bishop, N. 1972. The development of motor skills and social relationships among primates through play. In *Primate patterns,* ed., P. J. Dolhinow. New York: Holt, Rinehart and Winston.

Draper, P. 1972. !Kung Bushman childhood. Ph.D. dissertation, Harvard University.

Draper, P. 1975a. Cultural pressure on sex differences. *Am. Ethnolog.* 2: 602–616.

Draper, P. 1975b. !Kung women: Contrasts in sex egalitarianism in the foraging and sedentary contexts. In *Toward an anthropology of women,* ed., R. Reiter. New York: Monthly Review Press.

Draper, P. 1976. Social and economic constraints on child life among the !Kung. In *Kalahari hunter gatherers,* eds., R. B. Lee and I. DeVore. Cambridge: Harvard University Press.

Draper, P., and Cashdan, E. 1974. The impact of sedentism of !Kung socialization. Paper read at the meeting of the American Anthropology Association, Mexico City.

Dweck, C. S., and Reppucci, N. D. 1973. Learned helplessness and reinforcement responsibility in children. *J. Pers. Soc. Psychol.* 25: 109–116.

Eaton, G. G., and Resko, J. A. 1974. Plasma testosterone and male dominance in a Japanese macaque *(Macaca fuscata)* troop compared with repeated measures of testosterone in laboratory males. *Horm. Behav.* 5: 251–259.

Edelman, M. S. 1973. Peer-group formation in young children: Perception. Ph.D. dissertation, University of Chicago.

Edelman, M. S., and Omark, D. R. 1973. Dominance hierarchies in young children. *Soc. Sci. Info.* 12: 103–110.

Edelman, M. S., and Omark, D. R. 1976. An ethological approach to the study of cognitive development. In *Symposium on early growth,* ed., C. Smock, Athens, Ga.: Mathemagnetic Activities Program.

Eibl-Eibesfeldt, I. 1970. *Ethology: The biology of behavior.* New York: Holt, Rinehart and Winston.

Eibl-Eibesfeldt, I. 1972. *Love and hate: The natural history of behavior patterns.* New York: Holt, Rinehart and Winston.

Eibl-Eibesfeldt, I. 1974a. The myth of the aggression-free hunter and gatherer. In *Primate aggression, territoriality, and xenophobia,* ed., R. L. Holloway. New York: Academic Press.

Eibl-Eibesfeldt, I. 1974b. Aggression of the !Kung Bushmen. In *Aggression.* Baltimore: Williams & Wilkins.

Ekman, P. In press. About brows: Emotional and conversational signals. In *Human ethology,* eds., J. Aschoff, M. Cranach, W. Lepennies, and I. Eibl-Eibesfeldt. London: Cambridge University Press.

Ekman, P., and Friesen, W. 1975. *Unmasking the face.* Englewood Cliffs, N.J.: Prentice-Hall.

Ekman, P., and Friesen, W. 1978. *The facial action coding system.* Palo Alto, Calif.: Consulting Psychology Press.

Ellefson, J. O. 1968. Territorial behavior in the common white-handed gibbon, *Hylobates lar Linn.* In *Primates: Studies in adaptation and variability,* ed., P. C. Jay. New York: Holt, Rinehart and Winston.

Ember, C. R. 1977. Cross-cultural cognitive studies. *Ann. Rev. Anthropol.* 6: 33–56.

Emlen, S. T., and Oring, L. W. 1977. Ecology, sexual selection, and the evolution of mating systems. *Science* 197: 215–223.

Endore, G. 1968. *Synanon.* Garden City, N.Y.: Doubleday.

Epstein, S., and Taylor, S. 1967. Instigation to aggression as a function of degree of defeat and perceived aggressive intent of the opponent. *J. Pers.* 35: 264–289.

Esser, A. H. 1968. Dominance hierarchy and clinical course of psychiatrically hospitalized boys. *Child Dev.* 39: 147–157.

Etkin, W. 1964. *Social behavior and organization among vertebrates.* Chicago: University of Chicago Press.

Evans-Pritchard, E. E. 1965. *The position of women in primitive societies and other essays.* New York: Free Press.

Exline, R. A. 1971. Visual interaction: The glance of power and preference. In *Nebraska symposium on motivation,* vol. 19, ed., R. Jones. Lincoln: University of Nebraska Press.

Farina, A., Fischer, E. H., Sherman, S., *et al.* 1977. Physical attractiveness and mental illness. *J. Abnorm. Psychol.* 86: 510–517.

Ferguson, L. R. 1970. Dependency motivation in socialization. In *Early experiences and the processes of socialization,* eds., R. A. Hoppe, G. A. Milton, and E. C. Simmel. New York: Academic Press.

Feshbach, S. 1970. Aggression. In *Carmichel's manual of child psychology,* vol. 2, ed., P. H. Mussen. New York: Wiley.

Festinger, L. 1950. Informal social communication. *Psychol. Rev.* 57: 271–282.

Festinger, L. 1957. *The theory of cognitive dissonance.* New York: Harper & Row.

Finney, B. R. 1977. Voyaging canoes and the settlement of Polynesia. *Science* 196: 1277–1285.

Flavell, J. H. 1963. *The developmental psychology of Jean Piaget.* New York: Van Nostrand Reinhold.

Form, W. H., and Nosow, S. 1958. *Community in disaster.* New York: Harper & Row.

Free, J. B. 1961. The social organization of the bumblebee colony. Paper read to the Central Association of Bee-keepers. Fleet, England: North Hants Printing.

Freedman, D. G. 1974. *Human infancy: An evolutionary perspective.* Hillsdale, N.J.: Erlbaum.

Freedman, D. G. 1975. *Comment* on J. H. Barkow, Prestige and culture: A biosocial interpretation. *Curr. Anthropol.* 16: 564–565.

Freedman, D. G. 1976. Infancy, biology and culture. In *Developmental psychobiology,* ed., L. P. Lipsitt. Hillsdale, N.J.: Erlbaum.

Freedman, D. G. 1979. *Human sociobiology: A holistic approach.* New York: Free Press.

Freedman, D. G., and Omark, D. R. 1973. Ethology, genetics, and education. In *Cultural relevance and educational issues,* eds., F. A. J. Ianni and E. Storey. Boston: Little Brown.

Freeman, D. 1977. On sociobiology and anthropology. Paper read at the meeting of the Australian Society for the Study of Animal Behavior, Canberra, Australia.

French, E. G., and Lesser, G. S. 1964. Some characteristics of the achievement motive in women. *J. Abnorm. Soc. Psychol.* 68: 119–128.

Freud, A. 1966. *The ego and mechanisms of defense.* New York: International Universities Press.

Frieze, I. H., Parsons, J. E., Johnson, P. B., *et al.* 1978. *Women and sex roles: A social psychological perspective.* New York: Norton.

Frisch, J. 1968. Individual behavior and intertroop variability in Japanese monkeys. In *Primates: Studies in adaptation and variability,* ed., P. C. Jay. New York: Holt, Rinehart and Winston.

Furuya, Y. 1960. An example of fission of a natural troop of Japanese monkeys at Gagyusan. *Primates* 2: 149–179.

Gallagher, F. R., and Brouha, L. 1943. A simple method of testing the physical fitness of boys. *Res. Q.* 14: 23–31.

Gallup, G. G. 1975. Towards an operational definition of self-awareness. In *Socioecology and psychology of primates,* ed., R. H. Tuttle. The Hague: Mouton.

Gardner, R. A., and Gardner, B. T. 1971. Two-way communication with an infant chimpanzee. In *Behavior of nonhuman primates,* vol. 4, eds., A. Schrier and F. Stollnitz. New York: Academic Press.

Gartlan, J. S. 1968. Structure and function in primate society. *Folia primatol.* 8: 89–120.

Gartner, K., Reznik-Schuller, H., and Reznik, G. 1973. The influence of overcrowding on spermatogenesis, size of Leydig cell nuclei (histometrical investigation) and the adrenal corticosterone contents in mice. *Acta Endocrinol.* 74: 783–791.

Garvey, C. 1974. Some properties of social play. *Merrill-Palmer Q.* 20: 163–180.

Gates, A. I. 1937. *A list of spelling difficulties in 3876 words*. New York: Columbia University.

Geen, R., and Berkowitz, L. 1967. Some conditions facilitating the occurrence of aggression after observation of violence. *J. Pers.* 35: 666–676.

Geertz, C. 1965. The transition to humanity. In *Horizons of anthropology*, ed., S. Tax. London: Allen and Unwin.

Gellert, E. 1961. Stability and fluctuation in the power relationships of young children. *J. Abnorm. Soc. Psychol.* 62: 8–15.

Gewirtz, H. B. 1959. Generalization of children's preferences as a function of reinforcement and task similarity. *J. Abnorm. Soc. Psychol.* 58: 111.

Ginsburg, B. E. 1966. All mice are not created equal: Recent findings on genes and behavior. *Soc. Sci. Rev.* 40: 121–134.

Ginsburg, B. E. 1976. Evolution of communication patterns in animals. In *Communicative behavior and evolution*, eds., M. E. Hahn and E. C. Simmel. New York: Academic Press.

Ginsburg, B. E., and Allee, W. C. 1942. Some effects of conditioning on social dominance and coordination in inbred strains of mice. *Physiol. Zool.* 15: 485–506.

Ginsburg, H. J., and Pollman, V. 1975. The functional relationship between spatial density and mode of aggression in male elementary school children. Paper read at the meeting of the Society for Research in Child Development, Denver.

Goldberg, S., and Lewis, M. 1969. Play behavior in the year-old infant: Early sex differences. *Child Dev.* 40: 21–31.

Goldstein, A. 1964. *Biostatistics: An introductory text*. New York: Macmillan.

Goodall, J. 1965. Chimpanzees of the Gombe Stream Reserve. In *Primate behavior: Field studies of monkeys and apes*, ed., I. DeVore. New York: Holt, Rinehart and Winston.

Goodall, J. 1975. The behavior of the chimpanzee. In *Hominization und Verhalten*, ed., G. Kurth and I. Eibl-Eibesfeldt. Stuttgart: Gustav Fisher Verlag.

Goodall, J. 1977. Infant killing and cannibalism in free living chimpanzees. *Folia Primatol.* 28: 259–282.

Graham. C. E. 1970. Reproductive physiology of the chimpanzee. In *The chimpanzee*, vol. 3, ed., G. H. Bourne. New York: Karger.

Grant. E. C. 1969. Human facial expression. *Man* 4: 525–536.

Green, E. H. 1933a. Group play and quarreling among preschool children. *Child Dev.* 4: 302–307.

Green, E. H. 1933b. Friendships and quarrels among preschool children. *Child Dev.* 4: 237–252.

Greenberg, M. S., and Shapiro, S. P. 1971. Indebtedness: An adverse aspect of asking and receiving help. *Sociometry* 34: 290–301.

Greene, H. A. 1954. *The new Iowa spelling scale*. Iowa City: Bureau of Educational Research and Service, State University of Iowa.

Griffin, D. R. 1976. *The question of animal awareness*. New York: Rockefeller University Press.

Griffin, D. R. 1977. Expanding horizons in animal communication behavior. In *How animals communicate*, ed., T. A. Sebeok. Bloomington: Indiana University Press.

Gronlund, N. E. 1959. *Sociometry in the classroom*. New York: Harper & Row.

Guthrie, R. D. 1976. *Body hot spots*. New York: Van Nostrand Reinhold.

Gutmann, D. 1973. Men, women, and the parental imperative. *Commentary*, December, 59–64.

Hahn, E. 1970. Performance in sports as a criterion of social approach in school classes: A sociometric investigation. In *Contemporary psychology of sports*, eds., G. S. Kenyon and T. M. Grogg. Chicago: Athletic Institute.

Hahn, M. E., and Simmel, E. C. 1976. *Communicative behavior and evolution*. New York: Academic Press.

Hall, K. R. L. 1965. Behavior and ecology of the wild patas monkey, *Erythrocebus patas*, in Uganda. *J. Zool.* 148: 15–87.

Hall, K. R. L. 1968. Social learning in monkeys. In *Primates: Studies in adaptation and variability*, ed., P. C. Jay. New York: Holt, Rinehart and Winston.

Hall, K. R. L., and DeVore, I. 1965. Baboon social behavior. In *Primate behavior: Field studies of monkeys and apes*, ed., I. DeVore. New York: Holt, Rinehart and Winston.

Hallowell, A. I. 1955. *Culture and experience*. Philadelphia: University of Pennsylvania Press.

Hallowell, A. I. 1960. Self, society, and culture. In *The evolution of man*, ed., S. Tax. Chicago: University of Chicago Press.

Hamburg, D. A. 1971. Crowding, stranger contact and aggressive behavior. In *Society, stress and disease*, ed., L. Levi. London: Oxford University Press.

Hamburg, D. A., McGown, E. R., and Goodall, J., eds. 1979. *Behavior of great apes: Perspectives on human evolution*, vol. 4. Menlo Park, Ca.: Addison Wesley.

Hamilton, W. D. 1971. Selection of selfish and altruistic behavior in some extreme models. In *Man and beast: Comparative social behavior*, eds., J. F. Eisenberg and W. S. Dillon. Washington, D.C.: Smithsonian Institution Press.

Hanson, B. 1978. Hannah: A study of the first use of words. Ph.D. dissertation, University of Illinois, Urbana.

Harlow, H. F. 1965. Sexual behavior in the rhesus monkey. In *Sex and behavior*, ed., F. A. Beach. New York: Wiley.

Harlow, H. F., and Harlow, M. K. 1962. The effect of rearing conditions on behavior. *Bull. Menninger Clin.* 26: 213.

Harlow, H. F., Harlow, M. K., and Hansen, E. W. 1963. The maternal affectional system of rhesus monkeys. In *Maternal care in mammals*, ed., H. L. Rheingold. New York: Wiley.

Harré, R. 1974. The conditions for a social psychology of childhood. In *The integration of a child into a social world*, ed., M. P. M. Richards. Cambridge, Eng.: Cambridge University Press.

Harris, D. B. 1963. *Children's drawings as measures of intellectual maturity: A revision and extension of the Goodenough Draw-A-Man test*. New York: Harcourt Brace Jovanovich.

Hartup, W. W. 1970. Peer interaction and social organization. In *Carmichel's manual of child psychology*, Vol. II. ed., P. Mussen. New York: Wiley.

Hartup, W. W. 1974. Aggression in childhood: Developmental perspectives. *Am. Psychol.* 29: 336–341.

Harvard Computation Laboratory Staff. 1955. *Tables of the cumulative binomial probability distribution*. Cambridge: Harvard University Press.

Harvey, O. J. 1953. An experimental approach to the study of status relations in informal groups. *Sociometry* 18: 357–367.

Hattwick, L. A. 1937. Sex differences in behavior of nursery school children. *Child Dev.* 8: 343–355.

Hausfater, G. 1974. Dominance and reproduction in baboons (Papio cynocephalus): A quantitative analysis, Ph.D. dissertation, University of Chicago.

Hausfater, G. 1975a. Estrous females: Their effects on the social organization of the baboon group. *Proceedings from the symposia of the fifth congress of the international primatological society*. Tokyo: Japan Science Press.

Hausfater, G. 1975b. *Dominance and reproduction in baboons. Contributions to primatology*, vol. 7. Basel, Switz.: Karger.

Hausfater, G., and Watson, D. F. 1976. Social and reproductive correlates of parasite ova emissions by baboons. *Nature* 262: 688–689.

Hebb, D. O. 1971. Comment on altruism: The comparative evidence. *Psychol. Bull.* 76: 490–510.

Heckhausen, H. 1968. Achievement motive research: Current problems and some contributions towards a general theory of motivation. In *Nebraska symposium on motivation*, vol. 16, ed., W. J. Arnold. Lincoln: University of Nebraska Press.

Hediger, H. 1965. Environmental factors influencing the reproduction of zoo animals. In *Sex and behavior*, ed., F. A. Beach. New York: Wiley.

Heider, F. 1958. *The psychology of interpersonal relations*. New York: Wiley.

Hersen, M., and Barlow, D. H. 1976. *Single case experimental studies*. New York: Pergamon.

Hess, E. H. 1973. *Imprinting*. New York: Van Nostrand Reinhold.

Hilgard, E. R., and Sait, E. M. 1941. Estiamtes of past and future performances as measures of aspiration. *Am. J. Psychol.* 54: 102–108.

Hinde, R. A. 1970. *Animal behavior. A synthesis of ethology and comparative psychology*, 2nd ed. New York: McGraw-Hill.

Hinde, R. A. 1974. *Biological basis of human social behavior*. New York: McGraw-Hill.

Hinde, R. A. 1975. Interactions, relationships and social structure in non-human primates. *Proceedings from the symposia of the fifth congress of the international primatological society*. Tokyo: Japan Science Press.

Hinde, R. A., and Rowell, T. 1962. Communication by postures and facial expressions in the rhesus monkey *(Macaca mulatta)*. *Proc. Zool. Soc. London* 138: 1–21.

Hinde, R. A., and Stevenson-Hinde, J. 1976. Towards understanding relationships: Dynamic stability. In *Growing points in ethology*, eds., P. P. G. Bateson and R. A. Hinde. Cambridge, Eng.: Cambridge University Press.

Hirsch, J. 1963. Behavior genetics and individuality understood. *Science* 142: 1436–1442.

Hirsch, J. 1970. Behavior-genetic analysis and its biosocial consequences. *Semin. Psychiatry* 2: 89–105.

Hirsch, J., and Vetta, A. 1978. The misconceptions of behavior genetics. *Ricerche di Psicologia*, February.

Hockett, C. F. 1973. *Man's place in nature*. New York: Wiley.

Hodges, L. A., and Byrne, D. 1972. Verbal dogmatism as a potentiator of intolerance. *J. Pers. Soc. Psychol.* 21: 312–317.

Hoffman, L. R., and Maier, N. R. F. 1966. Social factors influencing problem solving in women. *J. Pers. Soc. Psychol.* 4: 382–390.

Hoffman, P. J., Festinger, L., and Lawrence, D. H. 1954. Tendencies toward group comparability in competitive bargaining. *Hum. Rel.* 7: 141–159.

Hold, B. 1975. Soziale Rangordnung in Kingergruppen unter besonderer berücksichtigung der rangspezifischen Verhaltensweisen. Ph.D. dissertation, Ludwig-Maximillian-Universitat, Munich.

Hold, B. 1976. Attention structures and rank specific behavior in preschool children. In *The social structure of attention*, eds., M. R. A. Chance and R. R. Larsen. London: Wiley.

Homans, G. C. 1974. *Social behavior: Its elementary forms*, 2nd ed. New York: Harcourt Brace Jovanovich.

Horner, M. S. 1970. Femininity and successful achievement: A basic inconsistency. In *Feminine personality and conflict*, eds., J. Bardwick, E. M. Douvan, M. S. Horner, and D. Gutmann. Belmont. Calif.: Brooks-Cole.

Horner, M. S. 1972a. Toward an understanding of achievement-related conflicts in women. *J. Soc. Issues* 28: 157–175.

Horner, M. S. 1972b. The motive to avoid success and changing aspirations of college women. In *Readings on the psychology of women*, ed., J. Bardwick. New York: Harper & Row.

Howell, F. C. 1967. Recent advances in human evolutionary studies. *Quart. Rev. Biol.* 42: 471–513.

Howell, N. 1976. The population of the Dobe area !Kung. In *Kalahari hunter gatherers,* eds., R. B. Lee and I. DeVore. Cambridge: Harvard University Press.

Hoyenga, K. B., and Hoyenga, K. T. 1979. *The question of sex differences.* Boston: Little, Brown.

Hrdy, S. B. 1977a. Infanticide as a primate reproductive strategy. *Am. Sci.* 65: 40–49.

Hrdy, S. B. 1977b. *The langurs of Abu.* Cambridge: Harvard University Press.

Hsu, F. L. K. 1971. Psychological homeostasis and jen: Conceptual tools for advancing psychological anthropology. *Am. Anthropol.* 73: 23–44.

Hutt, C. 1966. Exploration and play in children. *Symp. Zool. Soc. London* 18: 61–81.

Hutt, C. 1972. *Males and Females.* Middlesex, Eng.: Penguin Books.

Huxley, J. 1942. *Evolution: The modern synthesis.* London: Allen and Unwin.

Imanishi, K. 1957. Social behavior in Japanese monkey *Macaca fuscata. Psychologia: Int. J. Orient. Psychol.* 1: 47–54. (Also reproduced in *Primate social behavior,* ed., C. H. Southwick. New York: Van Nostrand Reinhold, 1963.)

Inkeles, A. 1963. Social change and social character: The role of parental mediation. In *Personality and social systems,* eds., N. J. Smelser and W. T. Smelser. New York: Wiley.

Irons, W. 1976. Emic and reproductive success. Paper read at the meeting of the American Anthropology Association, Washington, D.C.

Isen, A., Horn, N., and Rosenhan, D. H. 1973. Effects of success and failure on children's generosity. *J. Pers. Soc. Psychol.* 27: 239–247.

Itani, J. 1963. Paternal care in the wild Japanese monkey, *Macaca fuscata.* In *Primate social behavior,* ed., C. H. Southwick. New York: Van Nostrand Reinhold. (Reprinted from *J. of Primatology* 2: 61–93, 1959.)

Itani, J., and Suzuki, A. 1967. The social unit of wild chimpanzees. *Primates* 8: 355–381.

Izard, C. E. 1971. *The face of emotion.* New York: Appleton-Century-Crofts.

Jack, L. M. 1934. An experimental study of ascendant behavior in preschool children. *Child Dev.* 5: 139–157.

Jacklin, C. N., and Maccoby, E. E. 1978. Social behavior at thirty-three months in same-sex and mixed-sex dyads. *Child Dev.* 49(3): 547–556.

Jacklin, C. N., Maccoby, E. E., and Dick, A. E. 1973. Barrier behavior and toy preference: Sex differences (and their absence) in the year-old child. *Child Dev.* 44: 196–220.

Jacob, F. 1977. Evolution and tinkering. *Science* 196: 1161–1166.
Jay. P. C. 1963. Mother–infant relations in langurs. In *Maternal behavior in mammals,* ed., H. Rheingold. New York: Wiley.
Jay, P. C. 1965. The common langur of North India. In *Primate behavior: Field studies of monkeys and apes,* ed., I. DeVore. New York: Holt, Rinehart and Winston.
Jay, P. C. 1968. *Primate studies in adaptation and variability.* New York: Holt, Rinehart and Winston.
Jensen, G. D., and Bobbitt, R. A. 1965. On observational methodology and preliminary studies of mother–infant interaction in monkeys. In *Determinants of infant behavior,* 3rd ed., ed., B. M. Foss. London: Methuen.
Jensen, G. D., Bobbitt, R. A., and Gordon, B. N. 1967a. Sex differences in social interaction between infant monkeys and their mothers. *Rec. Adv. Biolog. Psychiatry* 9: 283–293.
Jensen, G. D., Bobbitt, R. A., and Gordon, B. N. 1967b. The development of mutual independence in mother–infant pigtail monkeys. In *Social communication among primates,* ed., S. A. Altmann. Chicago: University of Chicago Press.
Jensen, G. D., Bobbitt, R. A., and Gordon, B. N. 1968. Sex differences in the development of independence of infant monkeys. *Behavior* 30: 1–14.
Jersild, A. T., and Markey, F. V. 1935. Conflict between preschool children. *Child Dev.* 21.
Johannesson, I. 1967. *Effects of praise and blame.* Stockholm: Almquist and Wiksell.
Jolly, A. 1972. *The evolution of primate behavior.* New York: Macmillan.
Jones, E. E., and Nisbett, R. E. 1972. The actor and the observer. In *Attribution: Perceiving the causes of behavior,* eds., E. E. Jones, D. E. Kanouse, H. H. Kelley, *et al.* Morristown, N.J.: General Learning Press.
Jones, L. V., and Fiske, D. W. 1953. Models for testing the significance of combined results. *Psychol. Bull.* 50: 375–382.
Joslyn, W. E. 1973. Androgen-induced social dominance in infant female rhesus monkeys. *J. Child Psychol. Psychiatry* 14: 137.

Kagan, J., and Lewis, M. 1965. Studies of attention in the human infant. *Merrill-Palmer Q.* 11: 95–127.
Kalmus, H. 1969. Animal behavior and theories of games and of language. *Anim. Behav.* 17: 607–617.
Kaplan, A. 1970. Unpublished paper, University of Chicago (in Omark, 1972).
Kaufman, I. C. 1974. Discussion with R. A. Hinde. In *Ethology and psychiatry,* ed., N. F. White. Toronto: University of Toronto.
Kawabe, M. 1966. One observed case of hunting behavior among wild chimpanzees in the savanna woodland of western Tanzania. *Primates* 7: 393–396.

Kellogg, R., and O'Dell, S. 1967. *The psychology of children's art*. New York: Random House.
Kendon, A. 1967. Some functions of gaze-direction in social interaction. *Acta Psychol.* 26: 22–63.
Kimura, D. 1969. Spatial localization in left and right visual fields. *Can. J. Psychol.* 23: 445–458.
Kirk, R. E. 1968. *Experimental design: Procedures for the behavioral sciences*. Belmont, Calif.: Brooks Cole.
Klopfer, P. H. 1976. Evolution, behavior and language. In *Communicative behavior and evolution*, eds., M. E. Hahn and E. C. Simmel. New York: Academic Press.
Kohlberg, L. 1964. Development of moral character and moral ideology. In *Review of child development research*, eds., M. Hoffman and L. W. Hoffman. New York: Russell Sage Foundation.
Kohlberg, L., and Turiel, E. 1971. *Research in moral development: The cognitive-developmental approach*. New York: Holt, Rinehart and Winston.
Kohler, W. 1927. *The mentality of apes*. London: Routledge and Kegan Paul.
Korner, A. F. 1974. The effect of the infant's state, level of arousal, sex, and ontogenetic stage on the caregiver. In *The effect of the infant on its caregiver*, eds., M. Lewis and L. A. Rosenblum. New York: Wiley.
Kortlandt, A. 1962. Chimpanzees in the wild. *Sci. Am.* 206: 128–138.
Kortlandt, A., Orshoven, J., Pfeÿffers, R., and Zon, J. C. J. *Chimpanzees in the wild, Guinea 1966–1967* (Film).
Krebs, D. L. 1970. Altruism—an examination of the concept and a review of the literature. *Psychol. Bull.* 73: 258–302.
Krebs, D. L. 1971. Infrahuman altruism. *Psychol. Bull.* 76: 411–414.
Krebs, K. 1972. An observational study of dominance behavior in young children. Ph.D. dissertation, Oxford University, England.
Kronhausen, P., and Kronhausen, E. 1964. *The sexually responsive woman*. New York: Grove.
Kruuk, H. 1972. *The spotted hyena*. Chicago: University of Chicago Press.
Kühme, W. 1965. Communal food distribution and division of labour in African hunting dogs. *Nature* 205: 443–444.
Kummer, H. 1957. Soziales verhalten einer Mantelpaviangruppe. *Schweiz z. Psychol.* 33: 1–91.
Kummer, H. 1968. *Social organization of hamadryas baboons*. Chicago: University of Chicago Press.
Kummer, H. 1971. *Primate societies: Group techniques of ecological adaptation*. Chicago: Aldine.

Labov, W. 1971. The study of language in its social context. In *Advances in the sociology of language*, ed., J. A. Fishman. The Hague: Mouton.
Latané, B. 1966. Studies in social comparison. *J. Soc. Psychol.*, Suppl. 1.
Laughlin, W. S. 1968. Hunting: An integrating bio-behavioral system and

its evolutionary importance, in *Man the hunter*, eds., R. B. Lee and I. DeVore. Chicago: Aldine.
Lazarus, R. S. and Ericksen, C. W. 1952. Effects of failure stress upon skilled performance. *J. Exp. Psychol.* 43: 100.
Lebo, D. 1962. Aggressiveness and expansiveness in children. *J. Genet. Psychol.* 100: 227–240.
Lee, R. B. 1968. Hunter gatherer ecology, or how to get along on scarce resources. In *Man the hunter*, eds., R. B. Lee and I. DeVore. Chicago: Aldine.
Lee, R. B. and DeVore, I., eds., 1976. *Kalahari hunter gatherers.* Cambridge: Harvard University Press.
Leeds. R. 1963. Altruism and the norm of giving. *Merrill-Palmer Q.* 9: 229–240.
Lerner, M. J., and Lichtman, R. R. 1968. Effects of perceived norms on attitudes and altruistic behavior toward a dependent other. *J. Pers. Soc. Psychol.* 9: 226–232.
Leventhal, G. S., Weiss, T., and Long, G. 1969. Equity, reciprocity and reallocating rewards in the dyad. *J. Pers. Soc. Psychol.* 13: 300–305.
LeVine, R. A. 1966. *Dreams and deeds.* Chicago: University of Chicago Press.
LeVine, R. A. 1973. *Culture, behavior and personality: An introduction to the comparative study of psychosocial adaptation.* Chicago: Aldine.
LeVine, R. A., and Campbell, D. T. 1972. *Ethnocentrism: Theories of conflict, ethnic attitudes and group behavior.* New York: Wiley.
Levine, S. N. 1966. Sex differences in the brain. *Sci. Am.* 214: 84–90.
Lewin, K., Dembo, T., Festinger, L., and Sears, P. S. 1944. Level of aspiration. In *Personality and the behavior disorders*, vol. 1, ed., J. McVicker Hunt. New York: Ronald.
Lewis, M., Kagan, J., and Kalafat, J. 1966. Patterns of fixation in infants. *Child Dev.* 37: 331–341.
Lewis, M., and Rosenblum, L. A. 1974. *The effect of the infant on its caregiver.* New York: Wiley.
Lewontin, R. 1978. Adaptation. *Sci. Am.* 239: 212–230.
Lorenz, K. 1964. Ritualized aggression. In *The history of aggression*, eds., J. D. Carthy and F. J. Ebling. London: Academic Press.
Lorenz, K. 1965. *Evolution and modification of behavior.* Chicago: University of Chicago Press.
Lorenz, K. 1966. *On aggression.* New York: Harcourt Brace Jovanovich.
Lorenz, K. 1971. *Studies in animal and human behavior*, vol. 2. London: Methuen.
Lowenfeld, V. 1970. *Creative and mental growth.* New York: Macmillan.

Macaulay, J., and Berkowitz, L. 1970. *Altruism and helping behavior.* New York: Academic Press.
Maccoby, E. E. 1966. Sex differences in intellectual functioning. In The

development of sex differences, ed., E. E. Maccoby. Stanford, Calif.: Stanford University Press.

Maccoby, E. E., and Jacklin, C. N. 1974. *The psychology of sex differences.* Stanford, Calif.: Stanford University Press.

Machover, K. 1964. Drawing of the human figure: A method of personality investigation. In *An introduction to projective techniques,* eds., H. H. Anderson and G. L. Anderson. Englewood Cliffs, N.J.: Prentice-Hall.

Maclay, G., and Knipe, H. 1972. *The dominant man.* New York: Delacorte.

Maddi, S. R. 1972. *Personality theories: A comparative analysis.* Homewood, Ill.: Dorsey.

Maehr, M. L. 1974. Culture and achievement motivation. *Am. Psychol.* 29: 887–96.

Marks. J. B. 1957. Interests and leadership among adolescents. *J. Genet. Psychol.* 91: 163–172.

Marler, P. 1976. On animal aggression. *Am. Psychol.* 31: 239–246.

Marler, P. 1977. The evolution of communication. In *How animals communicate,* ed., T. A. Sebeok. Bloomington: Indiana University Press.

Marsh, R. M. 1970. Evolution and revolution: China. In *Essays in comparative social stratification,* eds., L. Plotnicov and A. Tuden. Pittsburgh: University of Pittsburgh Press.

Marshall, L. 1976. *The !Kung of Nyae Nyae.* Cambridge: Harvard University Press.

Marvin, R. S. 1977. An ethological-cognitive model for the attenuation of mother–child attachment behavior. In *Attachment behavior,* eds., T. Alloway, P. Pliner, and L. Krames. New York: Plenum.

Marvin, R. S., and Greenberg, M. T. n.d. Preschoolers' changing conceptions of their mothers: A social-cognitive study of mother–child attachment.

Marvin, R. S., VanDevender, T. L., Iwanaga, M. I., *et al.* 1977. Infant-caregiver attachment among the Hausa of Nigeria. In *Ecological factors in human development,* ed., H. McGurk. Amsterdam: North Holland.

Maslow, A. H. 1937. Dominance-feeling, behavior, and status. *Psychol. Rev.* 44: 404–429.

Mason, W. A. 1967. Motivational aspects of social responsiveness in young chimpanzees. In *Early behavior: Comparative and developmental approaches,* eds., H. W. Stevenson, E. H. Hess, and H. L. Rheingold. New York: Wiley.

Mason, W. A. 1973. Regulatory functions of arousal in primate psychosocial development. In *Behavioral regulators of behavior in primates,* ed., C. R. Carpenter. University Park, Pa.: Bucknell University Press.

Mason, W. A. 1976. Review of *The question of animal awareness. Science* 194: 930–931.

Mason, W. A., Green, P. C., and Posepanko, C. J. 1959. Sex differences in affective responses of rhesus monkeys. *Psychonomic Sci.* 15: 74–83.

Maudry, M., and Nekula, M. 1939. Social relations between children of the same age during the first two years of life. *J. Gen. Psychol.* 54: 193–215.

Maynard Smith, J., Ridpath, M. G. 1972. Wife sharing in the Tasmanian native hen, *Tribonyx mortierii*: A case of kin selection? *Am. Naturalist* 106: 447–452.

Mayr, E. 1970. *Populations, species, and evolution*. Cambridge: Harvard University Press.

Mayr, E. 1977. *Animal species and evolution*. Cambridge: Harvard University Press.

Mayr, E. 1978. Evolution. *Sci. Am.* 239: 47–55.

Mazur. A. 1973. A cross-species comparison of status in small established groups. *Am. Sociol. Rev.* 38: 513–530.

McCandless, B. R., Balsbough, C., and Bennett, H. L. 1958. Preschool socialization and maternal control techniques. *Am. Psychol.* 13: 320.

McCandless, B. R., and Marshall, H. R. 1957. Sex differences in social acceptance and participation of preschool children. *Child Dev.* 421–426.

McClelland, D. C. 1951. Measuring motivation in phantasy: The achievement motive. In *Groups, leadership, and men*, ed., H. Guetzkow. Pittsburgh: Carnegie Press.

McClelland, D. C. 1955. Some consequences of achievement motivation. In *Nebraska symposium on motivation*, vol. 3, ed., M. R. Jones. Lincoln: University of Nebraska Press.

McClelland, D. C. 1958a. Risk-taking in children with high and low need for achievement. In *Motives in fantasy, action and society*, ed., J. W. Atkinson. New York: Van Nostrand Reinhold.

McClelland, D. C. 1958b. Methods of measuring human motivation. In *Motives in fantasy, action and society*, ed., J. W. Atkinson. New York: Van Nostrand Reinhold.

McClelland, D. C. 1961. *The achieving society*. New York: Van Nostrand Reinhold.

McClelland, D. C., Atkinson, J. W., Clark, R. A., and Lowell, E. L. 1953. *The achievement motive*. New York: Appleton-Century-Crofts.

McClelland, D. C., and Winter, D. G. 1969. *Motivating economic achievement*. New York: Free Press.

McDonald, A. L., Heimstra, N. W., and Dankot, D. K. 1968. Social modification of agonistic behavior in fish. *Anim. Behav.* 16: 437–441.

McDougall, W. 1921. *An introduction to social psychology*. Boston: Luce.

McEwen, B. S. 1976. Interaction between hormones and nerve tissue. *Sci. Am.* 235: 48–58.

McGinnis, P. R. 1979. Sexual behavior in free living chimpanzees: Consort relationships. In *Perspectives on human evolution*, vol. 4, eds., D. A. Hamburg, E. R. McGown, and J. Goodall. Menlo Park, Ca.: Addison Wesley.

McGrew, W. C. 1969. An ethological study of agonistic behavior in preschool children. *Proceedings from the symposia of the second congress of the international primatological society.* New York: Karger.

McGrew, W. C. 1972. *An ethological study of children's behavior.* New York: Academic Press.

McGuire, M. T. 1974. The St. Kitts vervet. In *Contributions to primatology,* ed., H. Kuhn. Basel: Karger.

McGuire, T. R., and Hirsch, J. 1977. General intelligence (g) and heritability (H^2, h^2). In *The structuring of experience,* eds., I. Č. Užgiris and F. Weizmann. New York: Plenum.

McIntyre, A. 1972. Sex differences in children's aggression. *Proceedings of the 80th annual convention of the American Psychological Association* 7: 93–94.

McMillen, D. L. 1971. Transgression, self-image, and compliant behavior. *J. Pers. Soc. Psychol.* 29: 336–341.

Mech, D. 1966. *The wolves of Isle Royale.* Washington, D.C.: U.S. Government Printing Office.

Megargee, E. I. 1969. Influence of sex roles on the manifestation of leadership. *J. Appl. Psych.* 53: 377–382.

Mehrabian, A. 1968. Inference of attitude from the posture, orientation, and distance of a communicator. *J. Consult. Clin. Psychol.* 10: 26–30.

Menzel, E. W., Jr. 1971. Communication about the environment in a group of young chimpanzees. *Folia Primatol.* 15: 220–232.

Miller, G. A., Galanter, E., and Pribram, K. H. 1960. *Plans and the structure of behavior.* New York: Holt, Rinehart and Winston.

Mischel, W., Coates, B., and Raskoff, A. 1968. Effects of success and failure on self-gratification. *J. Pers. Soc. Psychol.* 10: 381–390.

Miscione, J., Marvin, R. S., O'Brien, R., and Greenberg, M. 1978. Preschool children's understanding of the words "know" and "guess." *Child Dev.* 49: 1107–1113.

Missakian, E. A. 1972. Genealogical and cross-genealogical dominance relations in a group of free-ranging rhesus monkeys *(Maccaca mulatta)* on Cayo Santiago. *Primates* 13: 169–180.

Missakian, E. A. 1976. Dominance relations in communally reared children in Synanon. Paper read at the meeting of the Animal Behavior Society, Boulder, Colorado.

Miyadi, D. 1959. On some new habits and their propagation in Japanese monkey groups. *Proc., XV Int. Cong. Zool.* 857–860.

Money, J., and Ehrhardt, A. A. 1972. *Man and woman, boy and girl.* Baltimore: Johns Hopkins University Press.

Moore, S. 1967. Correlates of peer acceptance in nursery school children. In *The young child,* eds., W. W. Hartup and N. L. Smothergill. Washington, D.C.: NAEYC.

Moore, S. 1973. A sociometric status test for young children: Manual of instructions. Minneapolis: Institute of Child Development, University of Minnesota. Mimeo.

Moore, B. S., Underwood, B., and Rosenhan, D. L. 1973. Affect and altruism. *Dev. Psychol.* 8: 99–104.

Morris, D. 1967. *The naked ape.* New York: McGraw-Hill.

Morris, J. 1966. Propensity for risk-taking as a determinant of vocational choice. *J. Pers. Soc. Psychol.* 3: 328–325.

Moss, H. A. 1967. Sex, age and state as determinants of mother–infant interaction. *Merrill-Palmer Q.* 13: 19–36.

Moynihan, M. 1955a. Some aspects of the reproductive behavior of the black-headed gull *(Larus ridibundus Linn.)* and related species. *Behav. Suppl.* 4: 1–201.

Moynihan, M. 1955b. Types of hostile displays. *Auk* 72: 247–259.

Moynihan, M. 1956. Notes on the behavior of some North American gulls. I. Aerial hostile behavior. *Behavior* 10: 126–178.

Moynihan, M. 1960. Some adaptations which help to promote gregariousness. *Proc., 12th Int. Cong. Ornithol. Cong.* 12: 523–541.

Munroe, R. L., and Munroe, R. H. 1971. Effect of environmental experience on spatial ability in an East African society. *J. Soc. Psychol.* 83: 15–22.

Mussen, P. H., Conger, J. J., and Kagan, J. 1974. *Child development and personality.* New York: Harper & Row.

Muste, M., and Sharpe, D. 1947. Some influential factors in the determination of aggressive behavior in preschool children. *Child Dev.* 18: 11–28.

Nafziger, E. W. 1968. Inter-regional economic relations in the Nigerian footwear industry. *J. African Studies* 6: 531–542.

Nagel, U., and Kummer, H. 1974. Variation in cercopithecoid aggressive behavior. In *Primate aggression, territoriality, and xenophobia,* ed., R. L. Holloway. New York: Academic Press.

Narins, P. M., and Capronica, R. R. 1976. Sexual differences in the auditory system of the tree frog. *Science* 193: 278–280.

Neimark, E. D. 1970. Model for a thinking machine: An information processing framework for the study of cognitive development. *Merrill-Palmer Q.* 16: 345–368.

Netting, R. McC. 1971. *The ecological approach in cultural anthropology.* Reading, Mass.: Addison-Wesley.

Neugarten, B. N. 1946. Social class and friendship among school children. *Am. J. Sociol.* 51: 305–313.

Nisbett, R. E., and Wilson, T. DeC. 1977. Telling more than we can know: Verbal reports on mental processes. *Psychol. Rev.* 84: 231–259.

Nishida, T. 1979. The social structure of chimpanzees in the Mahali Mountains. In *Perspectives on human evolution,* vol. 4, eds., D. A. Hamburg, E. R. McGown, and J. Goodall. Menlo Park, Ca.: Addison Wesley.

Nishida, T., and Kawanaka, K. 1972. Inter-unit-group relationships among wild chimpanzees of the Mahali Mountains. In *Kyoto University African studies,* vol. 7, ed., T. Umesao. Kyoto, Japan: Kyoto University Press.

Nissen, H. W., and Crawford, M. P. 1936. A preliminary study of food-sharing behavior in young chimpanzees. *J. Comp. Psychol.* 22: 383–419.

Nowlis, V. 1941. Companionship preference and dominance in social interactions of young chimpanzees. *Comp. Psychol. Monog.* 17: 1–57.

Oetzel, R. M. 1966. Classified summary of research in sex differences. In *The development of sex differences,* eds., E. E. Maccoby and C. B. Jacklin. Stanford: Stanford University Press.

Ollendick, T. H. 1974. Level of N achievement and persistence behavior in children. *Dev. Psychol.* 10: 457.

Omark, D. R. 1972. Peer group formation in young children: Action. Ph.D. dissertation, University of Chicago.

Omark, D. R. 1976. The mind/body problem from an ethological perspective. Paper read at the meeting of the American Anthropology Association, Washington, D.C.

Omark, D. R. In press. a. Psychological testing and bilingual education: The need for reconceptualization. In *The futures of bilingual education: Interdisciplinary perspectives,* ed., H. Trueba. Rowley, Mass.: Newbury House.

Omark, D. R. n.d. b. Genetically inferior or just different: Conceptualizations of the bilingual child. In *The bilingual exceptional child,* eds., D. R. Omark and J. G. Erickson.

Omark, D. R. In press. c. Approaches to language assessment in naturalistic and clinical settings. In *Communicative assessment of bilingual children: Issues and guidelines,* eds., J. G. Erickson and D. R. Omark. Baltimore: University Parks Press.

Omark, D. R., and Brenner, M. 1976. Institutional ecological factors affecting cognitive and social behaviors. Paper read at the meeting of the Animal Behavior Society, Boulder, Colorado.

Omark, D. R., and Edelman, M. S. 1975. A comparison of status hierarchies in young children: An ethological approach. *Soc. Sci. Info.* 14: 87–107.

Omark, D. R., and Edelman, M. S. 1976. The development of attention structures in young children. In *The social structure of attention,* eds., M. R. A. Chance and R. R. Larsen. London: Wiley.

Omark, D. R., Fiedler, M. L., and Marvin, R. S. 1976. Dominance hierarchies: Observational techniques applied to the study of children at play. *Instruct. Sci.* 5: 403–423.

Omark, D. R., and Marvin, R. S. Submitted for publication. Observational methods: Some applications and limitations.

Omark, D. R., Omark, M., and Edelman, M. S. 1973. Dominance hierarchies in young children. Paper read at the International Congress of Anthropological and Ethnological Sciences, Chicago.

Omark, D. R., Omark, M., and Edelman, M. S. 1975. Formation of dominance hierarchies in young children: Action and perception. In *Psychological anthropology,* ed., T. Williams. The Hague: Mouton.

Osgood, C. E., Miron, M. S. and Archer, W. K. 1963. *Progress report: The cross-cultural generality of effective meaning systems.* Urbana: Center for Comparative Linguistics, University of Illinois.

Osgood, C. E., Miron, M. S., and Archer, W. K. 1975. *Cross-cultural universals of affective meaning.* Urbana: University of Illinois Press.

Palmer, R. R. 1971. *The world of the French Revolution.* New York: Harper & Row.

Parker, R. 1976. Social hierarchies and same-sex peer groups. Ph.D. dissertation, University of Chicago.

Parsons, T., and Bales, R. F. 1955. *Family socialization and interaction process.* New York: Free Press.

Patchen, M. A. 1961. A conceptual framework and some empirical data regarding comparisons of social rewards. *Sociometry* 24: 136–156.

Pedersen, F. A., and Bell, R. Q. 1970. Sex differences in preschool children without histories of complications of pregnancy and delivery. *Dev. Psychol.* 3: 10–15.

Pepitone, A. 1964. *Attraction and hostility.* New York: Atherton.

Peplau, L. A. 1973. Impact of sex-role attitudes and opposite-sex relationships on women's achievement: An experimental study of dating couples. Ph.D. dissertation, Harvard University.

Peplau, L. A. 1976. Fear of success in dating couples. *Sex Roles* 2: 249–258.

Petry, P., and Omark, D. R. In preparation. Self-esteem, status relationships and social interactions in a bilingual classroom.

Pfeiffer, J. E. 1972. *The emergence of man.* New York: Harper & Row.

Piaget, J. 1952. *The child's conception of number.* London: Routledge and Kegan Paul.

Piaget, J. 1962. *The child's conception of space.* New York: Norton.

Piaget, J., and Inhelder, B. 1969. *The psychology of the child.* New York: Basic Books.

Pick, A. D., Frankel, D. G., and Hess, V. L. 1975. Children's attention: The development of selectivity. In *Review of child development research,* vol. 5, ed., E. M. Hetherington. Chicago: University of Chicago Press.

Piers, G., and Singer, M. B. 1953. *Shame and guilt: A psychoanalytic and a cultural study.* Springfield, Ill.: Thomas.

Pitcairn, T. K., and Eibl-Eibesfeldt, I. 1976. Concerning the evolution of nonverbal communication in man. In *Communicative behavior and evolution,* eds., M. E. Hahn and E. C. Simmel. New York: Academic Press.

Pitt-Rivers, J. 1966. Honour and social status. In *Honour and shame: The values of Mediterranean society,* ed., J. G. Peristiany. Chicago: University of Chicago Press.

Plotnicov, L. 1967. *Strangers to the city: Urban man in Jos, Nigeria.* Pittsburgh: University of Pittsburgh Press.

Poirier, F. E. 1972. *Primate socialization.* New York: Random House.

Premack, D. 1971. Language in chimpanzees? *Science* 172: 808–822.

Rachlin, H. 1976. *Introduction to modern behaviorism.* San Francisco: Freeman.

Redican, W. K. 1975. Facial expression in nonhuman primates. In *Primate behavior: Developments in field and laboratory research,* vol. 4, ed., L. A. Rosenblum. New York: Academic Press.

Regan, J. W. 1971. Guilt, perceived injustice, and altruistic behavior. *J. Pers. Soc. Psychol.* 18: 124–132.

Reynolds, V., and Reynolds, F. 1965. Chimpanzees of the Budongo Forest. In *Primate behavior: Field studies of monkeys and apes,* ed., I. DeVore. New York: Holt, Rinehart and Winston.

Rhine, R. J. 1975. The order of movement of yellow baboons *(Papio cynocephalus). Folia Primatol.* 23: 72–104.

Richards, S. M. 1972. Tests for behavioral characteristics in rhesus monkeys. Ph.D. dissertation, Cambridge University.

Richards, S. M. 1974. The concept of dominance and methods of assessment. *Anim. behav.* 22: 914–930.

Ripley, S. 1967. Intertroop encounters among Ceylon gray langurs. In *Social communication among primates,* ed., S. A. Altmann. Chicago: University of Chicago Press.

Robson, K. S. 1967. The role of eye contact in maternal–infant attachment. *J. Psychol. Psychiatry* 8: 13–25.

Rose, R. M., Holiday, J. W., and Bernstein, I. S. 1971. Plasma testosterone, dominance rank, and aggressive behavior in male rhesus monkeys. *Nature* 231: 366–368.

Rosen, B. C. 1962. Socialization and achievement in Brazil. *Am. Sociol. Rev.* 27: 612–624.

Rosen, B. C., and D'Andrade, R. 1959. The psychological origins of achievement motivation. *Sociometry* 22: 185–218.

Rosen, S., and Bielefeld, R. J. 1967. Help received by a needy competitor as contingent upon the deference he shows. Paper read at the meeting of the American Psychological Association, Milwaukee.

Rowell, T. E. 1966. Hierarchy in the organization of a captive baboon troop. *Anim. Behav.* 14: 430–443.

Rowell, T. E. 1972. *The social behavior of monkeys.* Harmondsworth, Eng.: Penguin Books.

Rowell, T. E. 1974. The concept of social dominance. *Behav. Biol.* 11: 131–154.

Rumbaugh, D. M., ed. 1977. *Language learning in a chimpanzee: The LANA project.* New York: Academic Press.

Ryan, D. E., and Lakie, W. L. 1965. Competitive and noncompetitive performance in relation to achievement motive and manifest anxiety. *J. Pers. Soc. Psychol.* 1: 342–345.

Sade, D. S. 1967. Determinants of dominance in a group of free-ranging rhesus monkeys. In *Social communication among primates,* ed., S. A. Altmann. Chicago: University of Chicago Press.

Sade, D. S. 1968. Inhibition of son–mother mating among free-ranging rhesus monkeys. *Sci. Psychoanal.* 12: 18–38.

Sade, D. S. 1972. A longitudinal study of social behavior of rhesus monkeys. In *The functional and evolutionary biology of the primates*, ed., R. Tuttle. Chicago: Aldine.

Sands, D. E., and Chamberlain, C. H. A. 1952. Treatment of inadequate personality in juveniles by dehydroisoandrosterone. *Br. Med. J.* 2: 66–68.

Savin-Williams, R. C. 1975. *Lord of the Flies:* An ethological study of dominance ordering in a group of human adolescents. Paper read at the meeting of the Society for Research in Child Development, Denver.

Savin-Williams, R. C. 1976. An ethological study of dominance formation and maintenance in a group of human adolescents. *Child Dev.* 47: 972–979.

Savin-Williams, R. C. 1977a. Dominance in a human adolescent group. *Anim. Behav.* 25: 400–406.

Savin-Williams, R. C. 1977b. Dominance-submission behaviors and hierarchies in young adolescents in summer camp: Predictors, effects and sex differences, Ph.D. dissertation, University of Chicago.

Savin-Williams, R. C. 1979. The developmental study of dominance during adolescence. Paper presented at the Society for Research in Child Development Convention, San Francisco.

Savin-Williams, R. C. In press. Social interactions of adolescent females in natural groups. In *Friendship and social relations in children*, eds., H. Foot, T. Chapman, and J. Smith. Sussex, Eng.: Wiley.

Savin-Williams, R. C., and Freedman, D. G. 1977. A bio-social approach to human development. In *Primate bio-social development: Biological, sociological, and ecological determinants*, eds., S. Chevalier-Skolnikoff and F. E. Poirier. New York: Garland.

Schaller, G. B. 1963. *The mountain gorilla. Ecology and behavior.* Chicago: University of Chicago Press.

Schaller, G. B. 1972. *The Serengeti lion.* Chicago: University of Chicago Press.

Schenkel, R. 1967. Submission: Its features and function in the wolf and dog. *Am. Zool.* 7: 319–329.

Schjelderup-Ebbe, T. 1913. Honsenes stemme. Bidvag til honsenes Psykologi. *Naturen* 37: 262–276.

Schjelderup-Ebbe, T. 1922. Soziale Verhaltnisse bei Vogeln. *Z. Psychol.* 90: 106–107.

Schjelderup-Ebbe, T. 1935. Social behavior of birds. In *A handbook of social psychology*, ed., C. Murchinson. Worcester, Mass.: Clark University Press.

Schneider, D. M., and Gough, K. 1961. *Matrilineal kinship.* Berkeley: University of California Press.

Scott, J. P. 1976. Genetic variation and the evolution of communication. In *Communicative behavior and evolution*, eds., M. E. Hahn and E. C. Simmel. New York: Academic Press.

Searle, J. R. 1969. *Speech acts: An essay in the philosophy of language.* Cambridge, Eng.: Cambridge University Press.

Sears, P. S. 1940. Levels of aspiration in academically successful and unsuccessful children. *J. Abnorm. Soc. Psychol.* 35: 498–536.

Sears, R. R. 1937. Initiation of the repression sequence by experienced failure. *J. Exper. Psychol.* 20: 570–580.

Sears, R. R. 1965. Development of gender role. In *Sex and behavior,* ed., F. A. Beach. New York: Wiley.

Sears, R. R., Maccoby, E. E., and Levin, H. 1957. *Patterns of child rearing.* New York: Harper & Row.

Sears, R., Whiting, J., Nowlis, V., and Sears, P. 1953. Some child-rearing antecedents of aggression and dependency in young children. *Genetic Psychol. Monographs* 47: 135–236.

Sebeok, T. A. 1977. *How animals communicate.* Bloomington: Indiana University Press.

Shafton, A. 1976. Conditions of awareness: Subjective factors in the social adaptations of man and other primates. Portland, Or.: Riverstone Press.

Shaw, R., and Bransford, J. 1977. Introduction: Psychological approaches to the problem of knowledge. In *Perceiving, acting, and knowing,* eds., R. Shaw and J. Bransford. Hillsdale, N.J.: Erlbaum.

Sherif, M. 1956. Experiments in group conflict. *Sci. Am.* 195: 54–58.

Sherif, M., Harvey, O. J., White, B. J., *et al.,* 1961. *Intergroup conflict and cooperation: The Roberts Cave experiment.* Norman, Okla.: Institute of Group Relations.

Sherif, M., and Sherif, C. W. 1953. *Groups in harmony and tension.* New York: Harper & Row.

Sherif, M., and Sherif, C. W. 1956. *An outline of social psychology.* New York: Harper & Row.

Sherif, M., and Sherif, C. W. 1964. *Reference groups.* Chicago: Regnery.

Siegel, S. 1956. *Nonparametric statistics for the behavioral sciences.* New York: McGraw-Hill.

Simon, S. I. 1974. The Synanon game. Ph.D. dissertation, Harvard University.

Simpson, G. G. 1953. *The major features of evolution.* New York: Simon and Schuster.

Simpson, G. G. 1977. An adversary view of sociobiology, book review. *Science* 195: 773–774.

Sluckin, A. 1975. Three ways of measuring dominance in preschool children. Paper read at the Third International Human Ethology Workshop, Sheffield, England.

Sluckin, A. M., and Smith, P. K. 1977. Two approaches to the concept of dominance in preschool children. *Child Dev.* 48: 917–923.

Smith, M., Harris, P. G., and Strayer, F. F. 1977. Laboratory methods for the assessment of dominance among captive squirrel monkeys. *Primates* 18: 977–984.

Smith, P. K. 1974a. Aggression in a preschool playgroup: Effects of varying physical resources. In *Determinants and origins of aggressive behavior,* eds., J. de Wit and W. W. Hartup. The Hague: Mouton.

Smith, P. K. 1974b. Ethological methods. In *New perspectives in child development*, ed., B. M. Foss. Harmondsworth, Eng.: Penguin Books.
Smith, P. K., and Green, M. 1975. Aggressive behavior in England nurseries and play groups: Sex differences and response of adults. *Child Dev*. 46: 211–214.
Smith, W. J. 1965. Message, meaning, and context in ethology. *Am. Naturalist* 99: 405–409.
Smith, W. J. 1977. *The behavior of communicating*. Cambridge: Harvard University Press.
Solzhenitsyn, A. 1974. *The Gulag Archipelago*. New York: Harper & Row.
Spuhler, J. M. 1959. *The evolution of man's capacity for culture*. Detroit: Wayne State University Press.
Stern, C. 1960. *Principles of human genetics*. San Francisco: Freeman.
Stern, C., and Sherwood, E. R. 1966. *The origin of genetics*. San Francisco: Freeman.
Stern, D. 1974. Mother and infant at play: The dyadic interaction involving facial, vocal and gaze behavior. In *The effect of the infant on its caregiver*, eds., M. Lewis and L. Rosenblum. New York: Wiley.
Strayer, F. F. 1976. Social-ecological factors during childhood: Caregivers, peers, and institutions. Paper read at the Human Ethology Workshop, Animal Behavior Society meeting, Boulder, Colorado.
Strayer, F. F. In press. Learning and imitation as a function of social status in macaque monkeys (*Macaca nemestrina*). *Anim. Behav*.
Strayer, F. F., Bovenkerk, A., and Koopman, R. F. 1975. Dominance and affiliation in captive squirrel monkeys (*Saimiri sciureus*). *J. Comp. Physiol. Psychol*. 89: 308–318.
Strayer, F. F., and Strayer, J. 1975. An ethological analysis of dominance relations among young children. Paper read at the meeting of the Society for Research in Child Development, Denver.
Strayer, F. F., and Strayer, J. 1976. An ethological analysis of social agonism and dominance relations among preschool children. *Child Dev*. 47: 980–989.
Stroebe, W., Insko, C. A., Thompson, V. D., and Layton, B. D. 1971. Effects of physical attractiveness, attitude similarity and sex on various aspects of interpersonal attraction. *J. Pers. Soc. Psychol*. 18: 79–91.
Struhsaker. T. 1967. Behavior of vervet monkeys (*Cercopithecus aethiops*). *Univ. Calif., Berkeley, Publ. Zool*. 82: 1–74.
Suchowsky, G. K., Pegrassi, L., and Bonsignori, A. 1969. The effect of steroids on aggressive behaviour in isolated male mice. In *Aggressive behavior*, eds., S. Garattini and E. B. Sigg. Excerpta Medica Foundation.
Sugiyama, Y. 1965a. Behavioral development and social structure in two troops of Hanuman langurs (*Prebytis entellus*). *Primates* 6: 213–247.
Sugiyama, Y. 1965b. On social change of Hanuman langurs in their natural condition. *Primates* 6: 381–418.
Sugiyama, Y. 1969. Social behavior of chimpanzees in the Budongo Forest, Uganda. *Primates* 10: 197–225.

Suttles, G. D. 1968. *The social order of the slum.* Chicago: University of Chicago Press.

Suttles, W. 1962. Variation in habitat and culture on the Northwest Coast. *Proc. 30th Int. Cong. Americanists.* Horn-Vienna: Ferdinand Berger.

Syme, G. J. 1974. Competitive orders as measures of social dominance. *Anim. Behav.* 22: 931–940.

Szal, J. A. 1972. Sex differences in the cooperative and competitive behaviors of nursery school children. Master's thesis, Stanford University.

Tanner, J. M. 1962. *Growth at adolescence.* Oxford, Eng.: Blackwell.

Tanner, O. 1976. *Bears and other carnivores.* New York: Time-Life Books.

Teleki, G. 1973. The omnivorous chimpanzee. *Sci. Am.* 228: 33–42.

Teleki, G. 1975. Primate subsistence patterns: Collector-predators and gatherer-hunters. *J. Hum. Evol.* 4: 125–184.

Thompson, D. M. 1967. An ethological study of dominance hierarchies in preschool children. Unpublished manuscript, University of Wisconsin (Zoology, 405).

Thompson, G. G., and Hunnicut, C. W. 1944. The effect of praise or blame on the work achievement of "introverts" and "extroverts." *J. Ed. Psychol.* 35: 257–266.

Thornton, D. A., and Arrowood, A. J. 1966. Self-evaluation, self-enhancement, and the locus of social comparison. *J. Exp. Soc. Psychol.*, Suppl. 1, 40–48.

Thrasher, F. 1927. *The gang.* Chicago: University of Chicago Press.

Tiger, L. 1969. *Men in groups.* New York: Random House.

Tiger, L., and Fox, R. 1971. *The imperial animal.* New York: Holt, Rinehart and Winston.

Tiger, L., and Shepher, J. 1974. A four-year study of three generations of female social behavior in the Israeli kibbutz movement. Paper read at the Third International Human Ethology meeting, Sheffield, England.

Tiger, L., and Shepher, J. 1975. *Women in the kibbutz.* New York: Harcourt Brace Jovanovich.

Tinbergen, N. 1951. *The study of instinct.* Oxford, Eng.: Clarendon.

Tinbergen, N. 1952. A note on the origin and evolution of threat display. *Ibis* 94: 160–162.

Tinbergen, N. 1953. *Social behavior in animals.* London: Chapman and Hall.

Tinbergen, N. 1963. On aims and methods of ethology. *Z. Tierpsychol.* 20: 410–429.

Tinbergen, N. 1972. *The animal in its world,* vol. 1 Cambridge: Harvard University Press.

Tinbergen, N. 1973. *The animal in its world,* vol. 2. London: Allen and Unwin.

Tomkins, S. S. 1963. *Affect, imagery consciousness,* vol. 2. *The negative effects.* New York: Springer.

Torrance, E. P. 1966. *Thinking creatively with words.* Princeton, N.J.: Personnel Press.

Travis, C. B. 1976. Social behavior of children in a structural setting. Paper read at the meeting of the Animal Behavior Society, Boulder, Colorado.

Trivers, R. L. 1971. The evolution of reciprocal altruism. *Q. Rev. Biol.* 46: 35–57.

Trivers, R. L. 1972. Parental investment and sexual selection. In *Sexual selection and the descent of man, 1871–1971,* ed., B. Campbell. Chicago: Aldine.

Tsumori, A. 1967. Newly acquired behaviors and social interactions in Japanese monkeys. In *Social communication among primates,* ed., S. A. Altmann. Chicago: University of Chicago Press.

Tuttle, R. H. 1970. Postural, propulsive and prehensile capabilities in the cheiridia of chimpanzees and other great apes. In *The chimpanzees,* vol. 2, ed., G. H. Bourne. New York: Karger.

Vaitl, E. n.d. A re-evaluation of the concept of dominance in primate research. University of California, Los Angeles, Mimeo.

van Hooff, J.A. R. A. M. 1967. The facial displays of the catarrhine monkeys and apes. In *Primate ethology,* ed., D. Morris. London: Weidenfeld and Nicolson.

van Hoof, J.A. R. A. M. 1972. A comparative approach to the phylogeny of laughter and smiling. In *Non-verbal communication,* ed., R. A. Hinde. Cambridge, Eng.: Cambridge University Press.

van Lawick-Goodall, J. 1967. *My friends the wild chimpanzees.* Washington, D.C.: National Geographic Society.

van Lawick-Goodall, J. 1968. The behavior of free-living chimpanzees in the Gombe Stream Reserve. *Anim. Behav. Monog.* 1: 161–311.

van Lawick-Goodall, J. 1971. *In the shadow of man.* London: Collins.

van Lawick-Goodall, J., and van Lawick, H. 1971. *Innocent killers.* Boston: Houghton Mifflin.

von Bertalanffy, L. 1964. The mind–body problem: A new view. *Psychosom. Med.* 26: 29–45.

von Fürer-Haimendorf, C. 1967. *Morals and merit.* London: Weidenfeld and Nicolson.

von Uexküll, J. 1926. *Theoretical biology.* London: Kegan Paul.

von Uexküll, J. 1957. Streifzüge durch die Umwelten von Tieren und Menschen. Berlin: Springer. Translated in *Instinctive behavior,* ed., C. H. Schiller. New York: International Universities Press, 1957.

Waddington, C. H. 1975. *The evolution of an evolutionist.* Ithaca, N.Y.: Cornell University Press.

Wallace, A. F. C. 1970. *Culture and personality,* 2nd ed. New York: Random House.

Wallace, J., and Sadalla, E. 1966. Behavioral consequences of transgression: The effects of social recognition. *J. Exp. Res. Pers.* 1: 187–194.

Walters, J., Pearce, D., and Dahms, L. 1957. Affectional and aggressive behavior of preschool children. *Child Dev.* 28: 15–26.

Walters, R. H., and Parke, R. D. 1965. The role of distance receptors in the development of social responsiveness. In *Advances in child development and behavior*, vol. 2, eds., L. P. Lipsett and C. C. Spiker. New York: Academic Press.

Warren, J. M., and Maroney, R. J. 1958. Competitive social interaction between monkeys. *J. of Soc. Psychol.* 48: 223–233.

Warren, R. A. 1978. Analysis of a spelling contest. Master's thesis, University of Chicago.

Warren, R. A. 1979. Competitive behavior and posture displays. University of Chicago, M.A. thesis.

Washburn, S. L. 1961. *Social life of early man: A symposium*. Chicago: Aldine.

Washburn, S. L., and Avis, V. 1958. Evolution of hunting. In *Behavior and evolution*, eds., A. Roe and G. G. Simpson. New Haven: Yale University Press.

Washburn, S. L., and Hamburg, D. A. 1965. The implications of primate research. In *Primate behavior: Field studies of monkeys and apes*, ed., I. DeVore. New York: Holt, Rinehart and Winston.

Washburn, S. L., and Hamburg, D. A. 1968. Aggressive behavior in Old World monkeys and apes. In *Primates: Studies in adaptation and variability*, ed., P. C. Jay. New York: Holt, Rinehart and Winston.

Washburn, S. L., and Lancaster, C. S. 1968. The evolution of hunting. In *Perspectives on human evolution*, eds., S. L. Washburn and P. C. Jay. New York: Holt, Rinehart and Winston.

Waterhouse, M. J., and Waterhouse, H. B. 1973. Primate ethology and human social behavior. In *Comparative ecology and behavior of primates*, eds., R. P. Michael and J. H. Crook. New York: Academic Press.

Weber, M. 1947. *Theory of social and economic organization*. New York: Free Press.

Wechsler, D. 1965. *Manual of the Wechsler intelligence scale for children ages 5–15*. New York: Psychological Corporation.

Weinraub, M., and Lewis, M. 1977. The determinants of children's responses to separation. *Monog. Soc. Res. Child Dev.* 43.

Weisfeld, G. E. 1973. Violations of social norms as inducers of aggression. *Int. J. Group Tensions* 2: 53–70.

Weisfeld, G. E. 1978. Determinants and behavioral correlates of dominance in high school boys. Ph.D. dissertation, University of Chicago.

Weisler, A., and McCall, R. R. 1976. Exploration and play: Resumé and redirection. *Am. Psycholog.* 31: 492–508.

Werner, H. 1961. *Comparative psychology of mental development*. New York: Science Editions.

Wheeler, L. 1966. Motivation as a determinant of upward comparison. *J. Exp. Soc. Psychol.*, Suppl. 1, 27–31.

White, R. W. 1959. Motivation reconsidered: The concept of competence. *Psychol. Rev.* 66: 297–333.
Whiting, B., and Pope, C. P. 1973. A cross-cultural analysis of sex differences in the behavior of children aged three through eleven. *J. Soc. Psychol.* 91: 171–188.
Whiting, B., and Edwards, C. P. 1973. A cross-cultural study of sex differences in the behavior of children aged three thorugh eleven. *J. Soc. Psychol.* 91: 171–188.
Whiting, B. B. and Edwards, C. P. 1977. *The effects of age, sex and modernity on the behavior of mothers and children.* Progress report to the Ford Foundation.
Whittemore, I. C. 1925. The influence of competition on performance. *J. Abnorm. Soc. Psychol.* 20: 17–33.
Wickler, W. 1972. *The sexual code.* New York: Doubleday.
Williams, G. C. 1966. *Adaptation and natural selection.* Princeton, N.J.: Princeton University Press.
Williams, L. 1967. *Man and monkey.* London: Deutsch.
Wilson, E. O. 1975. *Sociobiology: The new synthesis.* Cambridge: Harvard University Press.
Winchel, R., Fenner, D., and Shaver, P. 1974. Impact of coeducation on "fear of success" imagery. *J. Ed. Psychol.* 66: 726–730.
Winter, D. G. 1973. *The power motive.* New York: Free Press.
Witherspoon, G. 1977. *Language and art in the Navajo universe.* Ann Arbor: University of Michigan Press.
Wrangham, R. W. 1977. Feeding behaviour of chimpanzees in Gombe National Park, Tanzania. In *Primate ecology,* ed., T. H. Clutton-Brock. New York: Academic Press.
Wright, H. F. 1960. Observational child study. In *Handbook of research methods,* ed., P. H. Mussen. New York: Wiley.
Wynne-Edwards, V. C. 1962. *Animal dispersion in relation to social behavior.* New York: Hafner.

Yablonsky, L. 1965. *Synanon: The tunnel back.* New York: Penguin Books.
Yerkes, R. M. 1948. *Chimpanzees, a laboratory colony,* 4th ed. New Haven: Yale University Press.
Yolles, S. F. 1967. How different are they? *New York Times Magazine,* February 5.

Zihlman, A. L., and Hunter, W. S. 1972. A biomechanical interpretation of the pelvis of *Australopithecus. Folia Primatol.* 18: 1–19.
Zilvin, G. n.d. "Preschoolers" facial-postural status messages. Mimeo.
Zuckerman, S. 1963. Human sociology and the sub-human primates. In *Primate Social Behavior,* ed., C. H. Southwick. New York: Van Nostrand Reinhold.

Author Index

Subject Index

Page numbers followed by *t* indicate tables.